COMPUTATIONAL METHODS FOR REPRODUCTIVE AND DEVELOPMENTAL TOXICOLOGY

QSAR in Environmental and Health Sciences

Series Editor

James Devillers

CTIS-Centre de Traitement de
l'Information Scientifique
Rillieux La Pape, France

Aims & Scope

The aim of the book series is to publish cutting-edge research and the latest developments in QSAR modeling applied to environmental and health issues. Its aim is also to publish routinely used QSAR methodologies to provide newcomers to the field with a basic grounding in the correct use of these computer tools. The series is of primary interest to those whose research or professional activity is directly concerned with the development and application of SAR and QSAR models in toxicology and ecotoxicology. It is also intended to provide the graduate and postgraduate students with clear and accessible books covering the different aspects of QSARs.

Published Titles

COMPUTATIONAL METHODS FOR REPRODUCTIVE AND DEVELOPMENTAL TOXICOLOGY

EDITED BY

Donald R. Mattison

National Institutes of Health
Bethesda, Maryland, USA, and
Risk Sciences International
Ottawa, Ontario, Canada

CRC Press
Taylor & Francis Group
Boca Raton London New York

CRC Press is an imprint of the
Taylor & Francis Group, an informa business

First published in paperback 2024

First published 2016 by CRC Press
2385 NW Executive Center Drive, Suite 320, Boca Raton FL 33431

and by CRC Press
4 Park Square, Milton Park, Abingdon, Oxon, OX14 4RN

CRC Press is an imprint of Taylor & Francis Group, LLC

© 2016, 2024 Taylor & Francis Group, LLC

Publisher's Note
The publisher has gone to great lengths to ensure the quality of this reprint but points out that some imperfections in the original copies may be apparent.

ISBN: 978-1-4398-6107-3 (hbk)
ISBN: 978-1-03-292189-1 (pbk)
ISBN: 978-0-429-10659-0 (ebk)

DOI: 10.1201/b19189

**Visit the Taylor & Francis Web site at
http://www.taylorandfrancis.com**

**and the CRC Press Web site at
http://www.crcpress.com**

Contents

SECTION I Introduction

SECTION II Reproduction and Development: Biological Processes and Endpoints

SECTION III   Reproduction and Development: Biological and Computational Methods

Series Introduction

The correlation between the toxicity of molecules and their physicochemical properties can be traced back to the nineteenth century. Indeed, in a French thesis entitled *Action de l'alcool amylique sur l'organisme* (Action of Amyl Alcohol on the Body), which was presented in 1863 by A. Cros before the Faculty of Medicine at the University of Strasbourg, an empirical relationship was made between the toxicity of alcohols and their number of carbon atoms, as well as their solubility. In 1875, Dujardin-Beaumetz and Audigé were the first to stress the mathematical character of the relationship between the toxicity of alcohols and their chain length and molecular weight. In 1899, Hans Horst Meyer and Fritz Baum, at the University of Marburg, showed that narcosis or hypnotic activity was in fact linked to the affinity of substances to water and lipid sites within the organism. At the same time, at the University of Zurich, Ernest Overton came to the same conclusion providing the foundation of the lipoid theory of narcosis. The next important step was made in the 1930s by Lazarev in St. Petersburg who first demonstrated that different physiological and toxicological effects of molecules were correlated with their oil–water partition coefficient through formal mathematical equations in the form $\log C = a \log P_{oil/water} + b$. Thus, the quantitative structure–activity relationship (QSAR) discipline was born. Its foundations were definitively fixed in the early 1960s by the seminal works contributed by C. Hansch and T. Fujita. Since that period, the discipline has gained tremendous interest, and now the QSAR models represent key tools in the development of drugs as well as in the hazard assessment of chemicals. The new Registration, Evaluation, Authorization, and Restriction of Chemicals (REACH) legislation on substances, which recommends the use of QSARs and other alternative approaches instead of laboratory tests on vertebrate species, clearly reveals that this discipline is now well established and is an accepted practice in regulatory systems.

In 1993, the journal *SAR and QSAR in Environmental Research* was launched by Gordon and Breach to focus on all the important works published in the field and to provide an international forum for the rapid publication of structure–activity relationship (SAR) and QSAR models in (eco)toxicology, agrochemistry, and pharmacology. Today, the journal, which is now owned by Taylor & Francis and publishes three times more issues per year, continues to promote research in the QSAR field by favoring the publication of new molecular descriptors, statistical techniques, and original SAR and QSAR models. This field continues to grow very rapidly, and many subject areas that require larger developments are unsuitable for publication in a journal due to space limitation.

This prompted us to develop a series of books entitled *QSAR in Environmental and Health Sciences* to act in synergy with the journal. I am extremely grateful to Colin Bulpitt and Fiona Macdonald for their enthusiasm and invaluable help in making the project become a reality.

This fifth book in the series provides a comprehensive overview of processes, key endpoints, and computational approaches in reproductive and developmental toxicology.

At the time of going to press, another book is in the pipeline. It deals with the different *in silico* strategies that are used to find new substances active on mosquitoes and their diseases.

I gratefully acknowledge Hilary LaFoe for her willingness to assist me in the development of this series.

James Devillers
Series Editor

Preface

This contribution exploring computational methods in reproductive and developmental toxicology provides a unique series of contributions from a talented group of authors. The volume begins with descriptions of two pressing needs—safety assessment of pharmaceutical and industrial chemicals followed by a discussion of the role of REACH and environmental chemicals. Both, individually and jointly, lay out the challenges facing toxicology, regulatory agencies, industry, and population health.

The next section summarizes the molecular, biochemical, and cellular processes necessary for reproduction and development, describing the biological processes and endpoints that can be utilized in modeling and simulation. These themes identified in the first and second sections of the volume represent the foundation for discussion in the third section, which describes in greater detail biological and computational approaches for reproduction and development.

The third section begins with a discussion of animal study protocols for characterizing reproductive and developmental toxicology. This is followed by a description of the resources available at the National Library of Medicine for computational toxicology. The next four chapters describe unique and fascinating case studies dealing with

- Menstrual cycle modeling
- Chemical characteristics that influence partitioning into breast milk
- Pediatric drug-induced liver injury
- Use of adverse outcome pathways in describing reproductive and developmental toxicology

The pathway from data needs through biological processes and endpoints to computational methods represents an important resource for improving our understanding of and response to reproductive and developmental toxicology.

Editor

Dr. Donald Mattison continues a distinguished career in medicine and public health. In 2012, he was concurrently appointed chief medical officer and senior vice president of Risk Sciences International and associate director of the McLaughlin Centre for Population Health Risk Assessment at the University of Ottawa (Ottawa, Ontario, Canada). In 2013, he was also appointed medical advisor to QuarterWatch: Monitoring FDA MedWatch Reports, Institute for Safe Medication Practices (Horsham, Pennsylvania), and in 2014, he was appointed as a senior research fellow at the International Prevention Research Institute (Lyon, France). From 2002 to 2012, he was a senior advisor to the director of the Eunice Kennedy Shriver National Institute of Child Health and Human Development, National Institutes of Health. He has also been the medical director of the March of Dimes, the dean of the Graduate School of Public Health at the University of Pittsburgh, the director of Human Risk Assessment at the FDA National Center for Toxicological Research, and on the faculty of the University of Arkansas for Medical Sciences, University of Pittsburgh, and Columbia University. During this time, he has also served in the US Public Health Service, with deployments for medical and public health support.

In his research, Dr. Mattison has led drug development in pediatrics and obstetrics, including safety signal identification and evaluation. He has also led the development of methods in risk assessment for reproductive and developmental endpoints. Currently his research explores approaches to understand drug use, effectiveness, safety and risk–benefit communications, as well as environmental impacts on population health.

Dr. Mattison earned a BA degree (chemistry and mathematics) from Augsburg College (Minneapolis, Minnesota), an MS degree (chemistry) from the Massachusetts Institute of Technology (Cambridge, Massachusetts), and an MD degree from the College of Physicians and Surgeons, Columbia University (New York, New York). His clinical training in obstetrics and gynecology was at the Sloane Hospital for Women in the Columbia Presbyterian Medical Center in New York. His training in pharmacology and toxicology was at the National Institutes of Health, Bethesda, Maryland.

He has published more than 200 peer-reviewed articles. In 1997, he was elected a fellow of the American Association for the Advancement of Science; in 1999, a fellow of the New York Academy of Medicine; in 2000, a member of the Institute of Medicine; in 2005, Distinguished Alumni of Augsburg College; and in 2009, a fellow of the Royal Society of Medicine.

Contributors

Snezana Agatonovic-Kustrin
School of Pharmacy and Applied
   Science
Faculty of Pharmacy
MARA University of Technology
Selangor, Malaysia

Karine Audouze
Molecules Therapeutiques in silico
   (MTi)
Inserm UMR-S 973
Universite Paris Diderot
Paris, France

Karen A. Augustine-Rauch
Discovery Toxicology
Bristol-Myers Squibb Research and
   Development
Pennington, New Jersey

Søren Brunak
Center for Biological Sequence Analysis
Department of Systems Biology
Technical University of Denmark
Lyngby, Denmark

Pertti J. Hakkinen
Office of Clinical Toxicology
Specialized Information Services
National Library of Medicine
National Institutes of Health
Bethesda, Maryland

Leona A. Harris
Department of Mathematics and
   Statistics
The College of New Jersey
Ewing, New Jersey

Philip N. Judson
Lhasa Limited
Leeds, United Kingdom

Mariangela Maluf
Department of Pathology
University of São Paulo School
   of Medicine
and
Division of Reproductive Medicine
CEERH—Specialized Center
   for Human Reproduction
São Paulo, Brazil

Katherine L. Marlow
National Institute for Occupational
   Safety and Health
Cincinnati, Ohio

Donald R. Mattison
National Institutes of Health
Bethesda, Maryland

and

Risk Sciences International
Ottawa, Ontario, Canada

David W. Morton
School of Pharmacy and Applied
   Science
La Trobe Institute of Molecular
   Sciences
La Trobe University
Bendigo, Australia

Paulo Marcelo Perin
Department of Pathology
University of São Paulo School
   of Medicine
and
Division of Reproductive Medicine
CEERH—Specialized Center
   for Human Reproduction
São Paulo, Brazil

Aldert H. Piersma
Laboratory for Health Protection
   Research
National Institute for Public Health
   and the Environment
Bilthoven, the Netherlands

Damiano Portinari
Lhasa Limited
Leeds, United Kingdom

Susan Reutman
National Institute for Occupational
   Safety and Health
Cincinnati, Ohio

William F. Salminen
Division of Systems Biology
National Center for Toxicological
   Research
Food and Drug Administration
Jefferson, Arkansas

Steven M. Schrader
National Institute for Occupational
   Safety and Health
Cincinnati, Ohio

James F. Selgrade
Department of Mathematics and
   Biomathematics Program
North Carolina State University
Raleigh, North Carolina

Olivier Taboureau
Molecules Therapeutiques in silico
   (MTi)
Inserm UMR-S 973
Université Paris Diderot
Paris, France

Xi Yang
Division of Systems Biology
National Center for Toxicological
   Research
Food and Drug Administration
Jefferson, Arkansas

Section I

Introduction

1 Introduction

Donald R. Mattison

Over the past several decades, driven by the need for more effective, accurate, and efficient methods to understand hazards and risks for reproduction and development, there have been incredible advances in both our understanding of these complex processes and approaches to characterize toxicity. These advancements have been stimulated by application of new approaches recommended by the National Research Council, the Organization for Economic Cooperation and Development, and many different regulatory agencies.

This work has improved our knowledge concerning the structure and consequent biological activity of individual chemicals through improvements in structure–activity relationships and use of high-throughput screening. Additionally, improved understanding of the pathways and targets of individual chemicals through the use of adverse outcome pathways and systems biology has provided new knowledge about biological consequences. Additionally, advancements in informatics and databases have provided resources for hypothesis generation and testing, further stimulating our understanding.

These and other approaches are given attention in this collection of work in three general areas. Section I provides reviews of methods and approaches to meet needs for safety assessment in product development and regulatory approaches for environmental chemicals. Section II reviews the biological processes and endpoints involved in reproduction and development. Section III (the largest) summarizes protocols for evaluating biological processes and endpoints within reproduction and development as well as informatics resources and computational methods.

I would like to thank the authors for their willingness to participate in this volume and for the work they have done in preparing their chapters.

2 Visions and Advancements for Meeting the Needs of Safety Assessment of Pharmaceutical and Industrial Chemicals

Karen A. Augustine-Rauch

CONTENTS

ABSTRACT

Characterization of toxicity has remained unchanged for many years; assessment has focused on apical endpoints related to observed adverse effects in laboratory animals treated with high doses of agents. This approach, while the core to safety assessment of pharmaceuticals and industrial chemicals, is costly (e.g., test article, animals, and time). Advancements in high-throughput screening, in vitro model systems, and computational biology provide an opportunity to revolutionize toxicology. Such applications are expected to provide more mechanistic insight into the basis of toxicity. Chemical genetics and target deconvolution strategies, used since the 1990s in drug discovery work, have been adapted by toxicologists to enable toxicological characterization to be conducted proactively. To this end, cell culture and simple organismal systems are currently being integrated into discovery-phase screening assays to characterize various types of toxicity, providing a toxicity or teratogenic liability profile that can facilitate optimization of lead selection for the drug development phase. In response to recent legislation in Europe, the chemical industry is faced with the daunting task of assessing the toxicity of all chemicals sold in Europe in quantities of more than 1 ton/year. Standard toxicological assessment is not feasible since ≥68,000 compounds will require testing, and the task is required to be complete within the next decade. The National Research Council has provided a vision for toxicological assessment that requires a complete paradigm shift in testing approaches, eventually minimizing or eliminating whole-animal testing. Instead, assessment will rely upon a mechanistic systems-based approach integrating biological/molecular response from panels of cell-based assays that will determine whether the compound has potential to alter toxicity pathways. If successful, it is envisioned that this form of testing will affect how compounds are assessed in both the pharmaceutical and chemical industries. This chapter will review the needs set by the industry for advancements in toxicity testing, provide an overview of the efforts that are being undertaken to meet these needs, and discuss the potential challenges that are expected to be encountered during this process.

KEYWORDS

Computational modeling, developmental toxicology assays, high-throughput screening, in vitro models, NRC Toxicity Testing in the 21st Century report, systems biology, teratogenicity

2.1   INTRODUCTION

The pharmaceutical and chemical industries are undergoing changes in their approach to toxicological assessment of compounds. There are several factors that drive this change. In the pharmaceutical industry, functional characterization of gene targets has directed chemistry toward specific molecular targets associated with disease. This has increased emergence of promising drug candidates from the discovery pipeline. However, the extensive compound requirements and labor/time associated with in vivo toxicity studies dampen the impact of the expanding pipeline. Proactive toxicology assessment at the discovery phase allows selection of candidates for development with reduced toxicity potential. This approach is anticipated to optimize toxicological profiles of drug candidates early, thus reducing toxicity-based attrition in later stages of development.

Increasing animal rights pressures and regulatory legislations such as Registration, Evaluation, Authorization, and Restriction of Chemicals (REACH) have influenced both the pharmaceutical and chemical industries. In 1981, the United States and European Union (EU) introduced comprehensive safety evaluations of novel chemicals coming on to the market. However, chemicals in use that have been characterized for toxicity only represent a small fraction of the total population. The 10-year goal of REACH is to assess the toxicity of all chemicals sold in Europe in quantities of more than 1 ton/year. A projected 68,000 substances will require toxicity characterization, using approximately 54 million research animals [1]. Together, these projections highlight the lack of feasibility associated with conducting standard toxicology assessment on the vast number of compounds over a limited period. Furthermore, in vivo reproductive toxicology studies (mostly related to studies profiling teratogenicity and postnatal development liabilities) represent approximately 70% of the animals and 90% of the costs associated with these estimates. With the need for industry to better align with the 3 R's (reduction, refinement, and replacement) of animals with alternative methods, in vitro screens and in silico modeling have become a strategic focus.

Several years ago, the National Research Council (NRC), tasked by the US Environmental Protection Agency (EPA) and US National Institute of Environmental Health Sciences (NIEHS), provided recommendations for toxicity testing. In 2007, the NRC published a seminal report ("Toxicity Testing in the 21st Century: A Vision and a Strategy") that harnesses current capabilities in high-throughput screening technology and computational biology for meeting these needs [2,3]. The proposal defines an overhaul of the current practices conducted in safety assessment. In lieu of assessing compound toxicity by reviewing apical in vivo findings in animal models, the approach is mechanistic based and primarily uses a suite of in vitro models for testing. To this end, compounds will be screened through a battery of high-throughput in vitro assays and a systems biology approach will be applied for interpreting data for signatures indicative of toxicity. Most of the in vitro assays are anticipated to be derived from human sources and together will profile more expansive concentration ranges, enabling the data to be interpreted for human risk assessment [4].

This chapter will review the needs set by industry for advancements in toxicity testing and will provide an overview of the NRC vision, efforts that organizations are taking to meet this vision, and the anticipated challenges of this initiative. Based upon the significant need for streamlining developmental toxicity testing, this chapter will emphasize advancements made in this particular area.

2.2 SETTING THE STAGE: NEEDS FOR ADVANCEMENT POSED BY THE PHARMACEUTICAL AND CHEMICAL INDUSTRIES

2.2.1 DRUG DISCOVERY BY CHEMICAL GENETICS AND TARGET DECONVOLUTION

In the 19th century when the pharmaceutical industry was still in its infancy, natural medicines were used to treat illnesses. Frequently, these medicines were derived from natural sources found to have efficacy against disease. "Testing" was conducted on the level of traditional medicine practitioner who produced extracts of plants and tested them in organisms (i.e., animals or people) for a change in phenotype (i.e., behavior, relief in symptom, healing, and disease efficacy). The pharmaceutical industry adapted this approach and applied chemical means to extract and isolate specific alkaloids that could be empirically tested in complex living systems for changes in phenotype. Throughout the 20th century, chemistry became more sophisticated and synthetic analogs and unique chemotypes were generated and many efficacious drugs were discovered. However, the basic approach of testing chemicals and evaluating phenotypic change in biological systems has remained generally unchanged.

At the end of the 20th century, drug discovery underwent a sea of change. Advances in molecular biology and human genome sequence characterization led to the industry taking a more reductionist approach to drug discovery. In this regard, drug discovery could be directed toward a specific molecular target speculated to regulate aberrant cellular processes associated with a disease. Newly approved drugs will have well-characterized target biology.

These advancements simplify approaches to characterize compound-related phenotypes in biological systems. Whole-animal models are no longer the sole biological system. Instead, cell lines and simple biological models like nematodes, *Drosophila*, yeast, and zebrafish are being utilized. A systems-based approach provides an opportunity to interpret molecular responses at multiple targets to define the compound's effect on signaling pathways. This "chemical genetics" approach helps classical forward genetics, where the target is known and the phenotype is characterized subsequently [5].

Another change in drug discovery utilizes target deconvolution approaches. This involves the retrospective identification of molecular targets that underlie an observed phenotype of interest [5]. The zebrafish developmental biology field has applied this concept to link phenotype back to genotype. For instance, N-ethyl-N-nitrosourea mutagenesis has been used to generate large populations of mutant zebrafish, where an investigator can select a mutant phenotype of interest and then use a series of breeding and genomic linkage marker and mapping strategies to identify the mutated target [6,7]. Similar linkage and chromosomal mapping approaches can identify mutations in mammals including humans. Target deconvolution has improved our understanding of mechanisms of disease. Discovery chemistry adapted this concept in generating approaches to characterize small molecule–target protein interactions; modifying a small molecule enables tracking by tagging/labeling or binding to a substrate for biochemical assays. Affinity chromatography, three-hybrid systems, phage and mRNA display, and biochemical suppression identify proteins that bind with high affinity to the compound (reviewed by Terstappen et al. [5]).

Target-based drug design benefits greatly from these applications that critically support mechanism-based safety assessment.

2.2.2 OVERCOMING THE PRECLINICAL TESTING BOTTLENECK

Target-based drug design expanded the drug candidate pipeline by providing more flexibility in designing small molecules. However, once the candidate enters formal drug development, its progression slows down considerably. This is because regulatory animal toxicity test batteries have not changed significantly over the last two to three decades. The standard toxicity testing approach requires relatively rigid timelines against associated milestones. This has led to a strategic quandary. How can an expanded pipeline continue to maintain a steady clip or an accelerated pace when financial resources and regulatory study design logistics limit pipeline capacity? This problem can be addressed by profiling compounds for toxicity proactively at the discovery phase to facilitate drug candidate selection.

Safety-related failures comprise approximately 35% of attrition in clinical development [8]. By identifying toxicity liabilities in the discovery phase, an opportunity exists to attrite unacceptable molecules earlier and optimize better compounds faster. The early attrition translates into savings from costly clinical development, and faster optimization may translate into lower late-stage attrition; even a small improvement in drug marketing approvals will translate into significant sales and revenue increases. The pharmaceutical industry recognized the value of early toxicity evaluation and developed appropriate testing strategies.

In contrast, standard timelines for regulatory reproductive toxicology testing typically occur two or more years after a compound's entry into drug development. Regulatory in vivo teratogenicity studies are usually not undertaken until approximately a year before the milestone in drug development where women of childbearing potential (WoCBP) enter into clinical trials. Teratogenicity is one of the most prevalent forms of toxicity-based attrition, ranking third behind cardiovascular and hepatic toxicity, respectively [9]. The impact of an in vivo teratogenic finding may be very significant considering (1) the substantial investment (i.e., money, time, resources) in discovery and development made before observing it and (2) the delayed progression or termination of the compound because of inability to enroll WoCBP in clinical trials or a drug label potentially leading to marketing disadvantages for the therapeutic indication.

Because advancements in forward genetics, chemical genetics, and target deconvolution lead to extensive biological characterization of drug targets and compounds, a significant opportunity exists to proactively identify target-based teratogenic liability in the discovery phase. The strategy is based on assessments that would address the two origins of teratogenicity: (1) the therapeutic target being essential in embryonic development and (2) off-target effects associated with the chemical structure. Either origin can be characterized proactively using a series of evaluations (profiling the known developmental biology of the target, in vitro screens, and approaches to understand whether the effect is on- or off-target). A systematic approach to profiling provides a process outlining subsequent activities (e.g., adjusting timing of in vivo teratogenicity studies relative to clinical trial initiation) (Figure 2.1).

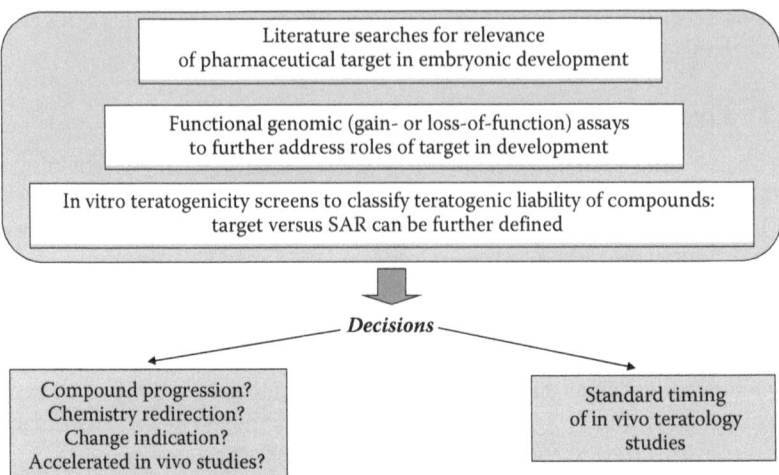

FIGURE 2.1 Proactive approaches for assessing teratogenic liability. This diagram illustrates the integrated process of how drug targets and compounds are evaluated at the discovery phase for teratogenic potential. Literature searches to obtain an understanding of what is known about the relevance of the target in embryogenesis combined with functional genomic assays and teratogenicity screens can better define on- or off-target teratogenicity concerns and possible structure–activity relationships of chemotypes. The positive or negative outcomes of these activities can influence proactive decisions regarding timing of in vivo studies, lead selection, or indication-based risk–benefit analysis.

Characterizing embryo–fetal developmental effects of a compound in discovery is advantageous from the business perspective because it enables proactive risk–benefit analysis. To this end, this information helps define therapeutic and commercial potential for the drug against any teratogenic liabilities that have been revealed by profiling. Some indications have more tolerance for teratogenic liability than others. For instance, the combination of a life-threatening outcome and unmet medical need associated with cancer typically makes cancer therapeutics more tolerant against teratogenic liability. In contrast, virtually no tolerance exists for teratogenic liability with drug treatments for obesity because WoCBP comprise a large segment of the clinical population and alternative efficacious therapies exist without teratogenic liability (e.g., diet and exercise). These analyses extend beyond a "yes/no" decision of whether a compound should progress into development by considering the timing of regulatory in vivo testing for teratogenicity relative to the start of clinical trials and chronic toxicology studies. Early detection of teratogenic liability may save millions of dollars in preclinical and clinical study costs.

Currently, the battery of developmental toxicology assays recommended by the European Centre for the Validation of Alternative Methods (ECVAM) provides in vitro screens (e.g., rat whole embryo culture, rat midbrain micromass culture, and mouse embryonic stem cells) for teratogenic potential. These assays will be described in more detail in Section 2.4.2. Researchers in the pharmaceutical and chemical industries either use the validated assays or have developed their own versions that provide improved throughput and performance. Some researchers also use rabbit whole embryo culture as a means to evaluate teratogenic potential in a second/ nonrodent species [9]. In addition, various groups use zebrafish embryos in developmental toxicology; this model has gained acceptance as a screening tool [10–12].

2.2.3 CHALLENGES POSED BY REACH

REACH legislation went into effect in 2006 and aims to assess toxicity of all chemicals sold in Europe in quantities of more than 1 ton/year within the next decade [1]. This is an ambitious goal since approximately 97% of all chemicals on the market currently lack safety data. As the draft legislation neared ratification, testing synthetic intermediates was added and guidance on testing was changed. This substantially increased the number of compounds for testing. Considering the costs of in vivo studies, current regulatory approaches for evaluating teratogenicity would not be feasible for the number of compounds being considered. The official EU estimates of chemicals requiring filing was approximately 29,000. However, 144,000 have been preregistered for testing. One assumes that some increase in filing relates to multiple companies filing the same compound and additional filing increase relates to expansion of the EU membership. The Transatlantic Think Tank for Toxicology reevaluated these figures to model worst- and best-case estimates of chemicals requiring testing (68,000 to 101,0000), which would require between 54 and 141 million animals for this evaluation. Originally, only 2.6 million animals were projected to be required for this testing [13].

Altogether, current projections suggest a 20-fold increase in animals and a 6-fold increase in costs as was previously estimated, which makes the goal potentially impossible since regulatory toxicology does not include high-throughput strategies or in vitro alternatives to manage these extensive numbers. Reproductive toxicology testing incurs the largest use in animals (90%) and costs (70%). In the short term, a moratorium of reproductive toxicology testing has been recommended until a less costly testing approach has been identified [13]. One near-term approach replaces the two-generation postnatal development study with an extended one-generation study, under development by the Organisation for Economic Co-operation and Development. The test extends the observation period for the first-generation offspring with additional testing on developmental neuro- and immunotoxicity if triggered by positive findings [14]. This in vivo test replacement is estimated to reduce animal use by 40%–60% in the postnatal development evaluation and a 15% reduction in overall animal use by REACH. However, different approaches will eventually be needed to make a more substantial impact in reducing costs/animals as well as increase throughput. Such changes will require the regulatory toxicology field to take a large step into the 21st century by applying existing and emerging high-throughput screens (HTSs) and a mechanistic-based systems approach to interpret biological/phenotypic changes.

2.3 SUMMARY OF THE NRC VISION FOR TOXICITY TESTING IN THE 21ST CENTURY

Current toxicity testing focuses primarily on the observed adverse effects in laboratory animals at high exposure of test articles. Human risk assessment of chemicals requires extrapolation to environmental levels that are usually orders of magnitude lower than exposures in animals with further extrapolated risk from animals to humans. This paradigm dates back a number of decades, when mechanisms of cellular and molecular toxicity were largely unknown. Extensive advances in cellular and molecular biology and computational sciences still have not been fully leveraged by toxicologists. The EPA and NIEHS requested the NRC to address this gap by providing guidance on new directions in toxicity testing that incorporate these advancements. In the final NRC report, design criteria for a state-of-the-art approach to toxicity testing were outlined. This testing paradigm provides (1) broad coverage of chemicals, chemical mixtures, outcomes, and life stages; (2) reduced cost and time required for toxicity testing; (3) a more robust scientific basis for assessing health effects of environmental chemicals; and (4) minimal use of animals in testing [3,4].

The key assessments include (1) chemical characterization, (2) toxicity pathways and targeted testing, (3) dose–response and extrapolation modeling, and (4) population-based and human exposure data (Figure 2.2). To achieve this vision, expanded in silico methods combined with in vitro biological models will generate a systems biology approach for characterizing toxicity of chemicals (Table 2.1). Using primarily human cells, cell lines, or tissues enables better risk assessment. A suite of tests identifies the range of potential significant perturbations of human biology from chemical exposure. Biological perturbations, in this context, are changes in defined toxicity pathways, previously identified using many known in vivo toxicants. The

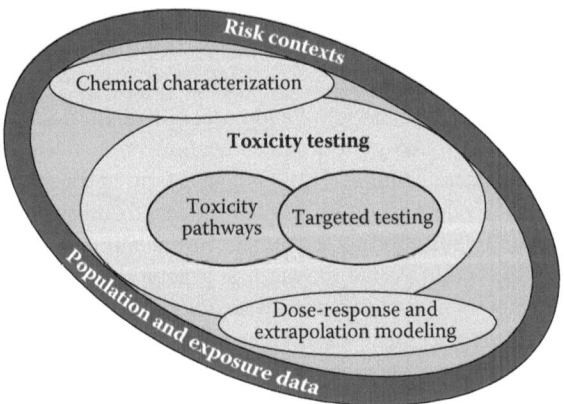

FIGURE 2.2 Schematic for the components required in the vision for toxicity testing in the 21st century. This schematic illustrates the key elements that comprise the NRC proposal for toxicity testing. Core in the proposal are in vitro tests and short-term in vivo tests to evaluate perturbations in toxicology pathways. The in vitro testing approach includes testing over an expansive concentration range, which is anticipated to improve extrapolation modeling. Computational approaches will also be used to integrate these data with chemical structure information to enhance understanding of structure–activity relationships. Altogether, these components prove the requisite tools for interpreting toxicity test results for assessing human risk assessment. (Reproduced with permission from the National Research Council [NRC]. 2007. *Toxicity Testing in the 21st Century: A Vision and Strategy.* Washington, D.C.: National Academy Press.)

TABLE 2.1
Toxicity Testing Tools and Their Application in Risk Assessment

Tool	Application
High-throughput screens	Efficiently identify critical toxicity pathway perturbations across a range of doses and molecular and cellular targets
Stem cell biology	Develop in vitro toxicity pathway assays using human cells produced from directed stem cell differentiation
Functional genomics	Identify the structure of cellular circuits involved in toxicity pathway responses to assist computational dose–response modeling
Bioinformatics	Interpret complex multivariable data from HTS and genomic assays in relation to target identification and effects of sustained perturbations on organs and tissues
Systems biology	Organize information from multiple cellular response pathways to understand integrated cellular and tissue responses
Computational systems biology	Describe dose–response relationships on the basis of perturbations of cell circuitry underlying toxicity pathway responses giving rise to thresholds, dose–dependent transitions, and other dose-related biological behaviors
PBPK models	Identify human exposure situations likely to provide tissue concentrations equivalent to in vitro activation of toxicity pathways
Structure–activity relationships	Predict toxicological responses and metabolic pathways on the basis of the chemical properties of environmental agents and comparison to other active structures
Biomarkers	Establish biomarkers of biological change representing critical toxicity pathway perturbations

Source: Reproduced from M.E. Andersen and D. Krewski, *Toxicological Sciences* 107 (2009), pp. 324–330. With permission.

population-based component focuses on continuation of human health surveillance and linking population studies to advancements in toxicity pathways with biomonitoring and biomarker surveillance. Together, the strategy is predicted to reduce the need for whole-animal testing and provide mechanistic-based data for human health risk assessment. However, from the perspective of pharmaceutical evaluation, some degree of animal testing will likely need to be retained to provide in vivo toxicokinetic and toxicity data before supporting human dose selection for clinical studies.

Low-dose and interspecies extrapolations may be less challenging because high-throughput in vitro screens allow evaluation from high concentrations to environmentally relevant concentrations or below. Thus, thresholds for altered biological function could be defined. However, these efforts will require better interpretation of in vitro data for human health risk assessment. Research in this area is particularly active in Europe where scientists working on alternatives to animal tests are developing physiologically based pharmacokinetic (PBPK) models that link in vitro concentrations to those expected in vivo [15].

The NRC considered four options for toxicity testing to predict human health risks [16]. Option I is essentially the current toxicity testing paradigm and relies primarily on in vivo animal toxicity tests with apical endpoints. As such, the approach incurs the greatest costs in animals, time, and resources with low throughput. Option II involves tiered testing starting initially with in vitro and in silico approaches to prioritize a reduced number of in vivo studies, which makes the current testing strategy more efficient. With this approach, mechanisms of action, intended use, and estimated exposures of a chemical (or class) are considered. This allows priority for compounds of greatest concern early with subsequent advanced testing, as needed [16]. The third and fourth options require HTSs and a systems biology approach, a transformational shift from Options I and II. Option III includes some degree of animal testing combined with high- and medium-throughput in vitro screening. Option III combines an emphasis on human biology and focus on data as it relates to perturbations of toxicity pathways. Option IV is similar to Option III except the assessment methods are envisioned to be completely high throughput in nature and with virtually no animal use. In addition, the overall database and computational algorithms are anticipated to be sufficiently robust to support in silico screens. Until toxicological signatures have been fully characterized by extensive validation efforts, Option III is the most realistic option to apply the NRC vision at present. To this end, the EPA has already started an effort called "ToxCast" that uses in vitro profiling to prioritize compounds for in vivo toxicity testing [17]. This approach will be described in more detail in Section 2.4. Option IV defines the complete shift to the NRC vision, high-throughput in vitro testing, and stronger, mechanistically based predictive tools for human health risk assessment that eliminates whole-animal testing [16].

2.4   MEETING THE NRC VISION: ENVIRONMENTAL HEALTH PROTECTION ACTIVITIES, ECVAM, AND ADDITIONAL EFFORTS

2.4.1   ENVIRONMENTAL HEALTH PROTECTION ACTIVITIES

The EPA and NIEHS originally funded the NRC project to develop a vision for toxicity testing and a strategic plan for implementation. The EPA established the National Center for Computational Toxicology (NCCT), which supports efforts from the EPA and the National Toxicology Program (NTP). As a joint venture, EPA, NTP, and NCCT collaborate to transform toxicology from a predominantly observational science of whole-animal models to a predominantly predictive science focused on biological and molecular effects observed in in vitro systems [18].

Using existing in vitro HTSs, these agencies are evaluating >2800 NTP and EPA compounds in more than 50 biochemical and cell-based assays. The compounds are being evaluated for response across ≤15 concentrations between ~0.005 and ~100 μM. This approach may provide reproducible results with acceptable false-positive and false-negative rates. The NCCT will be charged with building an informatics platform that compares results among the HTSs and with historical toxicological NTP and EPA data. The HTS data will also be linked to historical toxicological test results from these agencies that will create relational databases with controlled ontologies and chemical annotation enriching interpretation of NTP data (e.g., linking chemical structure to physicochemical and toxicological data) [18,19].

Approximately 2800 toxicants are being explored by the National Institutes of Health (NIH) Chemical Genomics Center (NCGC) for tissue-related sensitivity using human cell lines from the HapMap Project. In addition, the NTP established a Host Susceptibility Program to investigate the genetic basis for disease response in various mouse strains. Cell lines derived from these animals will be evaluated at the NCGC for differential sensitivity to the compounds tested in the HapMap Project. This work may identify biological pathways that, when perturbed, lead to toxicity, and ultimately, in vitro signatures of in vivo toxicity [18,20].

The EPA launched ToxCast in 2007 to evaluate HTS for prioritizing compounds for standard in vivo toxicology testing [17,21]. The first phase of this effort will evaluate 300 compounds, previously tested in vivo. These compounds will be evaluated through a series of biological models and cell-based assays, including zebrafish embryos. More than 400 biological endpoints will be collected for the relational database.

2.4.2   ECVAM

In anticipation of REACH legislation going into effect, ECVAM performed an extensive intralaboratory validation of several in vitro teratogenicity assays to assess general concordance and provide guidance on assays acceptable for regulatory testing in the EU [22–24]. Three rodent-based developmental toxicity assays (the mouse embryonic stem cell test [EST], the rat micromass assay, and the rat whole embryo culture assay) were evaluated. These assays are reviewed in detail by Augustine-Rauch et al. [25] and summarized in Sections 2.4.2.1 through 2.4.2.3.

2.4.2.1   Rat Whole Embryo Culture Assay

ECVAM developed a novel rat whole embryo culture assay for classifying the teratogenic potential of compounds. Compounds are first characterized for cytotoxicity along a concentration range in NIH3T3 cells. The concentration causing 50% loss in cellular viability (EC_{50}) was determined and represents a surrogate value for adult toxicity. Developmental toxicity was determined using the rat whole embryo culture model. Early somite-stage (gestational day 9) rat embryos were treated with various concentrations of test compounds for 2 days, after which, embryos were scored for growth/developmental staging parameters using the Brown and Fabro morphological score system to determine a Total Morphological Score (TMS) [26]. A statistical prediction model incorporating both TMS and NIH3T3 EC_{50} values was generated and tested at several laboratories. The concordance in correct classification of in vivo teratogens and nonteratogens was 80%, which was the most robust outcome among the three developmental toxicology assays evaluated [23].

Although the ECVAM rat whole embryo culture assay presented the highest concordance, it is not conducive for HTS. However, the model has potential application for systems biology studies supporting the longer-term NRC strategy.

2.4.2.2   Rat Micromass Assay

In this assay, limb buds are harvested from gestational day 12 rat embryos and the disassociated cells are treated with test compounds and cultured for 5 days. The cells are stained with alcian blue to identify foci undergoing cartilaginous differentiation. The concentration at which each compound inhibits the formation of differentiated foci 50% of the control value (IC_{50}) is determined. The rat limb bud micromass assay was assessed in the ECVAM validation project and determined to have reasonable concordance (70%) in correctly classifying in vivo teratogens and nonteratogens [23,24]. The limb bud micromass assay has potential for HTS given the relative ease of dissecting midgestation limbs and the substantial cell yield. However, developing HTS assays using embryonic stem cells (described below) are taking precedence because (1) total concordance was superior at 78%, (2) no animals are used to support the assay, and (3) all aspects of the culture and assay can be conducted automated [9,23,25,27].

2.4.2.3 Mouse EST

The mouse EST uses D3 embryonic stem cells cultured with test compounds in hanging drops to initiate differentiation into embryoid bodies (EBs). By the end of the 10-day culture, morphological evaluation of each EB for the presence of beating cardiomyocytes is assessed and an ID_{50} is calculated (the test concentration at which 50% of differentiation into cardiomyocytes is inhibited). In addition, the prediction model for the EST requires determining the compound's IC_{50} (50% inhibition in viability) in ES and NIH3T3 cells. ECVAM validated the EST assay and found it to be 78% concordant in correct classification of in vivo teratogens and nonteratogens.

2.4.3 ADDITIONAL EFFORTS

Embryonic cells and simple organismal systems were identified by the NRC as promising biological systems for characterizing in silico systems biology of toxicological response. As such, ToxCast included these models in their characterization efforts. Efforts related to refining the EST and zebrafish model for HTS are summarized in Sections 2.4.3.1 and 2.4.3.2.

2.4.3.1 Optimizing the EST for High-Throughput Screening

Efforts are continuing to streamline the EST so that the assay could be outfitted for HTS. Large numbers of EBs can now be grown in low-adherence plates in lieu of manual hanging drop cultures. Progress in identifying molecular endpoints and fluorescent markers of differentiation are expected to shorten the duration of the assay and enhance predictivity [28,29]. For instance, gene expression changes representing specific cell types or organ systems affected by certain teratogens have been reported for ≤10-day EB cultures [30–32]. Adding statistical algorithms will simplify the assay and make it conducive for HTS.

2.4.3.2 Zebrafish Teratogenicity Assays

The zebrafish embryo–larva model is used for characterizing mechanisms of teratogenicity and has also been integrated into screening assays for developmental toxicity (reviewed by Chapin et al. [33] and Augustine-Rauch et al. [25]). Recently, we described a zebrafish teratogenicity assay using dechorionated zebrafish embryos [12]. Evaluation of 30 pharmaceutical compounds with this assay yielded promising results with approximately 87% total concordance and ≤15% error rate in misclassification of in vivo teratogens and nonteratogens [12]. However, the assay requires manual removal of the chorion and an extensive morphological scoring procedure that renders it incompatable for automated screening. Using the basic framework of the dechorionated assay, a number of pharmaceutical and biotechnology companies are working as a consortium to develop and evaluate a chorion-on developmental toxicology assay. A subsequent objective is to streamline the protocol so it could be adapted to HTS [25]. Approximately 60 compounds will be evaluated in multiple wild-type strains by various laboratories. This consortium will share current assay experiences and results, assess variability between laboratories and wild-type strains, establish a harmonized protocol, and develop a statistical predication model to support using fewer morphological endpoints. The major obstacle in integrating the zebrafish assay into a high-throughput platform is the morphological assessment. Recent advancements of an automated platform for zebrafish manipulation and imaging were reported recently [34]. Functional attributes of this system include automated fish larvae loading with placement in correct orientation for imaging, high-speed confocal imaging, and laser manipulation for superficial and deep organ imaging. With this technology available, the feasibility of a future HTS zebrafish assay is promising.

2.5 CHALLENGES

Although the NRC vision for toxicity testing is revolutionary and theoretically feasible, challenges lay ahead to achieve this. From the scientific level, characterization of pathway perturbation to support a systems biology approach will take years and the expense will be considerable. The NRC estimated approximately $1 billion cost in implementation over a period of 10–20 years [4]. The collective budget across the NTP, EPA, and NCGC is not formally established and future budgets will depend on demonstrated success of initial efforts [18]. Furthermore, additional public and private partners may be needed to support this effort. The efforts will also need to rely upon advancement in other fields such as computational biology. Advancements in regenerative medicine using human stem cells is an example. For instance, fetal amniotic stem cells show promise for these efforts; they are not tumorigenic, do not require feeder cocultures, and have a short (36 h) doubling time while maintaining long teleomers and normal karyotype [35].

Another challenge is appropriately interpreting the new data sets for hazard identification versus risk assessment. This challenge affects the toxicology field from the ground up. Active toxicologists will need to retool from an intellectual or technical level to be able to interpret toxicity by a systems biology approach. Furthermore, there will need to be considerable changes in the curricula used to train students for careers in toxicology. This paradigm change emphasizes translational toxicology: an area focusing on how testing results and interpretative tools are used to make human risk assessment decisions regarding hazardous environmental agents [4]. Integrating functional genomics, computational biology, and bioinformatics will foster better data interpretation.

Scientific and technical challenges will occur during the transformation from vision to established testing platform. Examples include establishing baseline data for cellular response to various vehicles and clearly interpreting data from chemical mixtures. Strain differences in animal cells/tissues and normal human variations/polymorphisms in cells and cell lines will be important in defining systems biology applied to risk assessment. It will be important to establish methods that more accurately model in vivo exposure from in vitro data. In addition, the reliance on cell lines in the screening batteries raises questions regarding how well they reflect molecular pathways in normal whole tissue in vivo. For instance, cellular physiology may be altered when removed from its endogeneous environment, where it may lack paracrine mechanisms, cell matrix interactions, and intercellular communication that exist in the intact organism. Evaluating epigenetic change was not addressed in detail in the NRC. Substantial research already suggests that epigenetic mechanisms may be quite relevant to toxicological mechanisms and possibly transgenerational effects. For instance, there are various reports implicating an epigenetic toxicological mechanism of estrogenic and endocrine disrupting environmental agents [36,37]. Addressing epigenetic change could increase the challenge considerably given the multitude of mechanisms (methylation, histone deacetylation, non-coding RNAs, etc.) and their respective interplay, requiring a systems biology approach to deconvolute adaptive versus adverse response [38].

Finally, regulatory acceptance will be a challenge. An overarching ethical question to be addressed is whether the new data are sufficient for human risk assessment. To this end, if the paradigm went through a complete shift to Option IV, and the pharmaceutical industry adapted the approach, this would imply that the initial in vivo treatment may be directed to the human, since animal models have been removed from the test scheme. Furthermore, use of human stem cells in the platform leads to ethical considerations if the platform includes human embryonic stem cells. From a technical end, the screening platform will need to undergo an extensive validation and harmonization process to ensure good laboratory practice standards. Such practices are extremely labor intensive and time consuming and could take 10–12 years to accomplish [39]. There may be reluctance to accept the investment required for these changes and proactive efforts may be needed to optimize time- and cost-effectiveness as it relates to scientific development of the testing platform. Regulatory authorities will need to consider whether current risk assessment practices can be adapted to in vitro–based systems biology data sets. In addition, lawmakers will need to consider how to interpret or possibly update regulatory statues, such as the Toxic Substances Control Act, since interpretation of toxicity may eventually emphasize perturbations of toxicity pathways rather than adverse effects in experimental animals [16].

2.6 CONCLUSION

Needs from the pharmaceutical and chemical sectors for higher-throughput toxicological approaches have converged with advancements in HTS and computational biology, making in vitro screening and the NRC vision timely and feasible. In vitro developmental toxicology screening approaches are already underway in the pharmaceutical and chemical industries to address an expanding pipeline and legislation such as REACH. Concurrently, combined efforts of US environmental health agencies are progressing the NRC vision, a strategy that may eventually be adapted by the drug and chemical industries for safety evaluation. Once scientific and technical challenges have been addressed, the testing approach is expected to provide an improved basis for human health risk assessment and will drastically reduce the amount of animal testing. Committed, broad involvement of the scientific community will be needed to progress the paradigm shift in toxicity testing and address the ethics and politics associated with the effort.

REFERENCES

1. T. Hartung and C. Rovida, *Chemical regulators have overreached*, Nature 460 (2009), pp. 1080–1081.
2. National Research Council (NRC), *Toxicity Testing for Assessment of Environmental Agents: Interim Report*, National Academy Press, Washington, DC, 2006.
3. National Research Council (NRC), *Toxicity Testing in the 21st Century: A Vision and Strategy*, National Academy Press, Washington, DC, 2007.
4. M.E. Andersen and D. Krewski, *Toxicity testing in the 21st century: Bringing the vision to life*, Toxicol. Sci. 107 (2009), pp. 324–330.
5. G.C. Terstappen, C. Schlupen, R. Raggiaschi, and G. Gaviraghi, *Target deconvolution strategies in drug discovery*, Nat. Rev. Drug Discov. 6 (2007), pp. 891–903.
6. F. Pelegri, *Mutagenesis*, in *Zebrafish*, C. Nusslein-Volhard and R. Dahm, eds., Oxford University Press, Oxford, UK, 2002.
7. R. Geisler, *Mapping and cloning*, in *Zebrafish*, C. Nusslein-Volhard and R. Dahm, eds., Oxford University Press, Oxford, UK, 2002.
8. J.R. Empfield and P.D. Leeson, *Lessons learned from candidate drug attrition*, IDrugs 13 (2010), pp. 869–873.
9. K. Augustine-Rauch, *Predictive teratology: Teratogenic risk-hazard identification partnered in the discovery process*, Curr. Drug Metabol. 9 (2008), pp. 971–977.
10. R. Nagel, *DarT: The embryo test with the Zebrafish Danio rerio—A general model in ecotoxicology and toxicology*, Altex 19 (2002), pp. 38–48.
11. P. McGrath and C.Q. Li, *Zebrafish: A predictive model for assessing drug-induced toxicity*, Drug Discov. Today 13 (2008), pp. 394–401.
12. K.C. Brannen, J. Panzica-Kelly, T. Danberry, and K. Augustine-Rauch, *Development of a zebrafish embryo teratogenicity assay and a quantitative prediction model*, Birth Defects Res. B 89 (2010), pp. 66–77.
13. C. Rovida and T. Hartung, Re-evaluation of animal numbers and costs for in vivo tests to accomplish REACH legislation requirements for chemicals. Transatlantic Think Tank for Toxicology, 2009. Available at http://www.altex.ch or http://altweb.jhsph.edu.
14. OECD guideline for the testing of chemicals. Draft version 28. Draft proposal for an extended one-generation reproductive toxicology study. Available at http://www.oecd.org/dataoecd/55/24/43965303.pdf, accessed October 2009.
15. H. DeJongh, A. Forsby, J.B. Houston, M. Beckman, R. Combes, and B.J. Blaauboer, *An integrated approach to the prediction of systemic toxicity using computer-based biokinetic models and biological in vitro test methods: Overview of a prevalidation study based on the ECITTS*, Toxicol. In Vitro 3 (1999), pp. 549–554.
16. D. Krewski, M.E. Andersen, E. Mantus, and L. Zeise, *Toxicity testing in the 21st century: Implications for human health risk assessment*, Risk Anal. 29 (2009), pp. 474–479.
17. D.J. Dix, K.A. Houck, M.T. Martin, A.M. Richard, R.W. Setzer, and R.J. Kavlock, *The ToxCast program for prioritizing toxicity testing of environmental chemicals*, Toxicol. Sci. 95 (2007), pp. 5–12.

18. F.S. Collins, G.M. Gray, and J.R. Bucher, *Transforming environmental health protection*, Science 319 (2008), pp. 906–907.
19. DSSTox: Distributed Structure Searchable Toxicology. Available at http://www.epa.gov/ncct/dsstox//.
20. International HapMap Project. Available at http://www.hapmap.org/.
21. ToxCast. Available at http://www.epa.gov/ncct/toxcast.
22. W. Lilienblum, W. Dekant, H. Foth, T. Gebel, J.G. Hengstler, R. Kahl, P.J. Kramer, H. Schweinfurth, and K.M. Wollin, *Alternative methods to safety studies in experimental animals: Role in the risk assessment of chemicals under the new European Chemicals Legislation (REACH)*, Arch. Toxicol. 82 (2008), pp. 211–236.
23. E. Genschow, H. Spielmann, G. Scholz, A. Seiler, N. Brown, A. Piersma, M. Brady, N. Clemann, H. Huuskonen, F. Paillard, S. Bremer, and K. Becker, *The ECVAM international validation study on in vitro embryotoxicity tests: Results of the definitive phase and evaluation of prediction models. European Centre for the Validation of Alternative Methods*, Altern. Lab. Anim. 30 (2002), pp. 151–176.
24. H. Spielmann and M. Liebsch, *Lessons learned from validation of in vitro toxicity test: From failure to acceptance into regulatory practice*, Toxicol. In Vitro 15 (2001), pp. 585–590.
25. K. Augustine-Rauch, C.X. Zhang, and J. Panzica-Kelly, *In vitro developmental toxicology assays: A review of the state of the science of rodent and zebrafish embryo culture and embryonic stem cell assays*, Birth Defects Res. C 90 (2010), pp. 89–98.
26. N.A. Brown and S. Fabro, *Quantitation of rat embryonic development in vitro: A morphological scoring system*, Teratology 24 (1981), pp. 65–78.
27. E. Genschow, H. Spielmann, G. Scholz, I. Pohl, A. Seiler, N. Clemann, S. Bremer, and K. Becker, *Validation of the embryonic stem cell test in the international ECVAM validation study on three in vitro embryotoxicity tests*, Altern. Lab. Anim. 32 (2004), pp. 209–244.
28. R. Buesen, A. Visan, E. Genschow, B. Slawik, H. Spielmann, and A. Seiler, *Trends in improving the embryonic stem cell test (EST): An overview*, Altex 21 (2004), pp. 15–22.
29. R. Buesen, E. Genschow, B. Slawik, A. Visan, H. Spielmann, A. Luch, and A. Seiler, *Embryonic stem cell test remastered: Comparison between the validated EST and the new molecular FACS-EST for assessing developmental toxicity in vitro*, Toxicol. Sci. 108 (2009), pp. 389–400.
30. N.I. zur Nieden, G. Kempka, and H.J. Ahr, *Molecular multiple endpoint embryonic stem cell test-a possible approach to test for the teratogenic potential of compounds*, Toxicol. Appl. Pharmacol. 194 (2004), pp. 257–269.
31. M. Festag, B. Viertel, P. Steinberg, and C. Sehner, *An in vitro embryotoxicity assay based on the disturbance of the differentiation of murine embryonic stem cells into endothelial cells. II. Testing of compounds*, Toxicol. In Vitro 21 (2007), pp. 1631–1640.
32. D.A. van Dartel, J.L. Pennings, P.J. Hendriksen, F.J. van Schooten, and A.H. Piersma, *Early gene expression changes during embryonic stem cell differentiation into cardiomyocytes and their modulation by monobutyl phthalate*, Reprod. Toxicol. 27 (2009), pp. 93–102.
33. R. Chapin, K. Augustine-Rauch, B. Beyer, G. Daston, R. Finnell, T. Flynn, S. Hunter, P. Mirkes, K.S. O'Shea, A. Piersma, D. Sandler, P. Vanparys, and G. Van Maele-Fabry, *State of the art in developmental toxicity screening methods and a way forward: A meeting report addressing embryonic stem cells, whole embryo culture, and zebrafish*, Birth Defects Res. B Dev. Reprod. Toxicol. 83 (2008), pp. 446–456.
34. C. Pardo-Martin, T.Y. Chang, B.K. Koo, C.L. Gilleland, S.C. Wasserman, and M.F. Yanik, *High-throughput in vivo vertebrate screening*, Nat. Methods 7 (2010), pp. 634–636.

35. P. DeCoppi, G. Bartsch Jr., M.M. Siddiqui, T. Xu, C.C. Santos, L. Perin, G. Mostoslavsky, A.C. Serre, E.Y. Snyder, and J.J. Yoo, *Isolation of amniotic stem cell lines with potential for therapy*, Nat. Biotechnol. 26 (2007), pp. 100–106.
36. D.C. Dolinoy, J.R. Weidman, and R.L. Jirtle, *Epigenetic gene regulation: Linking early developmental environment to adult disease*, Reprod. Toxicol. 23 (2007), pp. 297–307.
37. J.J. Heindel, *Role of exposure to environmental chemicals in the developmental basis of disease and dysfunction*, Reprod. Toxicol. 23 (2007), pp. 257–259.
38. J.I. Goodman, K.A. Augustine, M.L. Cunnningham, D. Dixon, Y.P. Dragan, J.G. Falls, R.J. Rasoulpour, R.C. Sills, R.D. Storer, D.C. Wolf, and S.D. Pettit, *What do we need to know prior to thinking about incorporating an epigenetic evaluation into safety assessments?*, Toxicol. Sci. 116 (2010), pp. 375–381.
39. T. Hartung, *A toxicology for the 21st century—Mapping the road ahead*, Toxicol. Sci. 109 (2009), pp. 18–23.

3 REACH and Environmental Chemicals

Olivier Taboureau, Karine Audouze, and Søren Brunak

CONTENTS

ABSTRACT

Which chemicals are we exposed to? What are their biological effects? And what may be the risk on human health? Pharmaceutical products, personal care products, nutritional ingredients, and industrial chemicals are all potentially dangerous and need to be assessed and documented. Only recently, in 2007, the European Union (EU) promulgated a regulatory initiative for the Registration, Evaluation, Authorization, and Restriction of Chemicals (REACH). It is expected that all chemicals manufactured or imported into EU will be registered into REACH with physicochemical characteristics, as well as toxicological properties. Here, we will present the implementation, the need, and the challenge of REACH to evaluate the chemical safety and risk assessment. We will focus on the legislation for reproduction and developmental toxicity and discuss alternative methods and emerging strategies to supplement in vivo studies required by REACH in chemical risk assessment.

KEYWORDS

REACH, QSAR, environmental chemicals, systems biology

3.1 INTRODUCTION

Both the general public and regulators have become increasingly concerned about the possible threat of chemical substances in our environment, the effectiveness of chemical safety assessment, and the number of substances to be tested. More than 50 million chemicals are currently registered in the Chemical Abstract System. However, different industrial sectors including pharmaceutical, cosmetic, and chemistry areas are constantly generating new chemical entities. To assess the safety of all these available chemicals on the market, an accurate, affordable risk evaluation is required.

Unfortunately, traditional tools cannot cope with chemical safety assessment on this large scale for various reasons: (1) the cost of standard testing is too high, (2) the time for testing is too long, (3) the ethical reasons to reduce animal testing, (4) the lack of experimental testing resources, and (5) that current tests tend to have unsatisfactory accuracy. Instead, novel high-throughput screening–based tools for testing chemical safety are needed to remedy these problems.

Therefore, the poor efficacy of the current risk assessment process and the limited information obtained on the hazard properties of chemicals have driven the need for new regulatory dispositions, although the need for a core set of data necessary for prioritization and risk assessment of chemicals was recognized already in the 1970s. Only recently, in 2007, a systematic, chemical management system, known as REACH (Registration, Evaluation, and Authorization of Chemicals) was introduced in Europe [1].

3.2 THE GOAL OF REACH

Human diseases may occur in response to chemical exposures. As chemicals are present everywhere in our surroundings, at the workplace, in ambient environments, and through consumer products, this multiple exposure to chemicals causes an increasing fear in our society. To allow proper regulatory decision making on the usage and availability of individual chemicals, the generation of relevant and accurate documentation related to toxicity and risk assessment is the major priority of REACH.

REACH aims to assess all chemicals marketed at more than 1 ton/year in Europe in order to characterize their toxicological properties. For this purpose, three main goals are considered, that is, (1) improve knowledge about properties and uses of individual chemical substances to ensure a high level of protection of human health and the environment, (2) increase the speed and efficiency of the risk assessment process, and (3) explore the free movement of chemical substances, on their own, in preparations and in products while enhancing competitiveness and innovation [2].

Under the term *chemicals*, REACH includes any organic or inorganic compounds, substances, and additives in the natural state or obtained by any manufacturing process. Interestingly, pharmaceutical drugs are regulated under other laws and are excluded from the REACH regulation [3]. Approximately 100,000 existing substances are already registered in the European Inventory of Existing Chemical Substances, of which approximately 30,000 are expected to be registered in REACH [4]. Potential financial burdens for societies might result from the use of industrial chemicals in the marketplace including the need for purification of drinking water, maintenance of sewage treatment plants, and disposal of dredged sediment. Therefore, substances classified as dangerous or PBT (persistent, bioaccumulative, and toxic) or vPvB (very persistent and very bioaccumulative) will need a complete risk assessment (including an exposure analysis) [5].

Diverse toxicological information is required for substances to be mandated by REACH, namely, skin/eye irritation or corrosion and respiratory irritation, skin or respiratory sensitization, acute toxicity, repeated dose toxicity, mutagenicity and carcinogenicity, aquatic toxicity, toxicokinetics, and reproductive or developmental

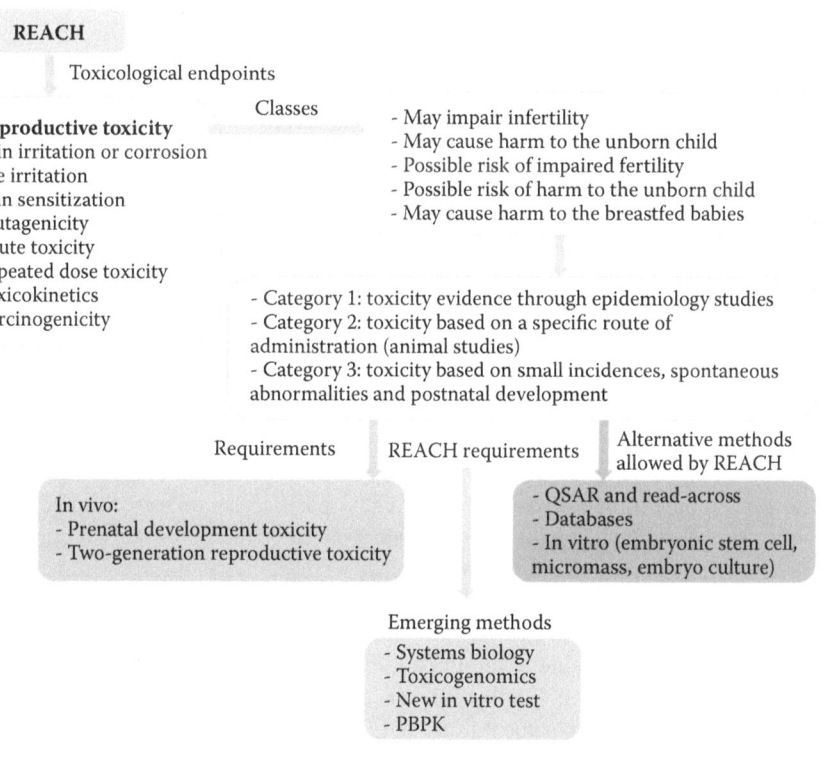

FIGURE 3.1 Schema about the toxicological endpoints assessed by REACH and the different requirements or alternative methods relevant for toxicity testing in chemical risk assessment.

toxicity, as well as degradation, biodegradation, bioconcentration, and bioaccumulation (Figure 3.1) [6].

One of the issues is the absence of epidemiological dose–response functions for these endpoints, especially for reproductive and developmental toxicity. Many health science professionals will be very interested in European epidemiological studies and case reports of these diseases, to see whether a measurable reduction in overall diseases or even certain diseases can be linked to reduction in exposure to specific chemicals [7]. Therefore, considering the scope and amount of information required when completing a registration dossier, REACH legislation has accepted that data may be obtained through several avenues including public and commercial databases, the Organisation for Economic Co-operation and Development (OECD) high production volume (HPV) Chemicals Program, and peer-reviewed literature. Of course, critical assessment of the available information must be considered for the chemical registration including relevance, reliability, and adequacy.

3.3 REACH LEGISLATION FOR REPRODUCTIVE AND DEVELOPMENTAL TOXICOLOGY

Reproductive and developmental toxicology is estimated to be one of the most difficult endpoints to assess. It is known from mouse and human genetics that more than a thousand genes are necessary for the reproductive cycle to be successful. In addition, understanding the process is even more complicated than just knowing the function of genes as the control of mRNA stability, protein translation, protein modification, and protection of germline play an important role in the regulation of the genes in the reproductive process [8]. Therefore, REACH establishes some rules for the assessment of reproductive and developmental toxicology.

Depending on the metric tons per year, several reproductive and developmental toxicology data are required by REACH, including prenatal developmental toxicity studies in one or two species and two-generation reproductive toxicity studies in one species. Such studies are quite expensive and may reach half a million Euros for testing one chemical on rats. As it is estimated that REACH could require the use of almost 22 million experimental animals for reproductive and developmental toxicity testing, alternative methods are necessary and will be discussed in Section 3.4 [9,10].

Reproductive and developmental toxicology effects are categorized into five classes: (a) may impair fertility, (b) may cause harm to the unborn child, (c) possible risk of impaired fertility, (d) possible risk of harm to the unborn child, and (e) may cause harm to breastfed babies. Within each of these five classes, substances known to produce reproductive and developmental toxicity based on epidemiology studies are defined as category 1. Toxicity data for substances obtained from animal studies using an appropriate route of administration and not to be attributed to generalized toxicity are in category 2. Finally, substances showing adverse reproductive or developmental effects in experimental animal studies that are attributed to generalized toxicity, including small incidences of spontaneous abnormalities or in postnatal developmental assessments, are in category 3. Hence, all substances that impair fertility, cause developmental toxicity in humans, or generate concern in relation to the healthy reproduction of a child (substances absorbed by women that may interfere with lactation or may be present in milk in a sufficient amount to be harmful) should be reported as toxic substances. Studies have shown that persistent chemicals give rise to exposure of newborn babies through breastfeeding after birth and by transfer across the placenta [11–13].

REACH assumes that a substance that does not produce reproductive or developmental toxicity at the level of 1000 mg/kg body weight can be assumed to be without important damaging reproductive or developmental effects. Application of the Threshold of Toxicological Concern and Exposure-Based Waiving may be useful to obtain reliable information on the toxicological properties of chemicals. However, all stages in the life cycle of a chemical must be taken into account for a valid justification of withdrawing, especially for long-term reproduction and carcinogenicity.

Thus, for chemicals with incomplete toxicological information, data from in vitro experiments are acceptable by REACH if they are derived from a validated test method such as those approved by the European Centre for the Validation of Alternative Methods (ECVAM). Adaptation of the standard testing routine also allows the use of computational approaches, including Quantitative Structure–Activity Relationships (QSARs), under certain conditions, such as the applicability domain of the model if the model has been appropriately validated [4]. A summary of the REACH legislation for reproductive and developmental toxicity as well as alternative methods for risk assessment are presented in Figure 3.1.

3.4 ALTERNATIVE METHODS

The American National Research Council recently published a seminal report entitled "Toxicity Testing in the 21st Century: A Vision and a Strategy" [14]. The report reviewed established toxicology methodologies and discussed the use of alternative approaches and emerging technologies as a vision and strategy to increase efficiency and relevance of toxicity testing in chemical risk assessment. Among the emerging strategies, the use of toxicity databases, high-throughput in vitro screening, computational toxicology, and systems biology were identified as promising tools to fulfill the needs for hazard identification and risk assessment. These tools are separately described below.

3.4.1 Toxicity Databases

REACH promotes the use of toxicity databases in gathering information on chemical safety profiles in order to speed up the risk assessment process [15]. Toxicity databases serve as a library that can be mined to recover information on undesirable effects of chemicals. The main advantage of using existing databases is that they allow collecting valuable information and high scientific quality data from prior toxicity studies in order to build an electronic resource that may be searchable by different features, that is, chemical name, chemical structure, toxicology endpoint, gene, and so on. In addition, such databases can provide predictive information using diverse computational modeling in silico toxicology methods, that is, QSAR, chemoinformatics, and "read across" strategies in order to fill gaps in the hazard and risk assessment [16]. The reproducibility of the results, preferably referenced in scientific journals or by regulatory agencies, with an explicit and controlled vocabulary should facilitate integration between the different sources and help ensure quality of the databases as well as the assessments [17,18].

A large amount of chemical toxicity information is publicly available through various databases, for example, Aggregated Computational Toxicology Resource (ACToR), Adverse Event Reporting System (AERS), Distributed Structure-Searchable Toxicity (DSSTox), Comparative Toxicogenomics Database (CTD), and TOXicology Data NETwork (TOXNET). These databases have been expanded in recent years with the aim of including and describing effects of substances on human health—information that is very useful for risk assessment. For a more complete list, a recent review of available toxicity databases and in silico toxicology tools with their advantages and limitations has been published recently [19].

3.4.2 IN VITRO TESTS

Significant efforts have been undertaken to develop in vitro tests as alternative methods to assess reproductive toxicity. However, the majority of these tests have not yet gained regulatory acceptance, essentially because of a lack of understanding of the mode of action of reproductive toxicants. Actually, three in vitro tests have been validated as acceptable alternative tests according to the ECVAM: (1) embryonic stem cell test, (2) limb bud micromass culture, and (3) whole postimplantation embryo culture. These tests are considered to be useful for screening of chemicals rather than a replacement for whole-animal toxicity tests, as negative in vitro tests cannot be interpreted with confidence as nontoxic whereas positive results will provide justification for further testing [4].

1. The embryonic stem cells test has the potential to assess adverse effects and cytotoxicity on all cell types of the mammalian organism that might be relevant for in vivo embryotoxicity [20]. Although such tests allow an easy monitoring of toxic effects in medium-throughput applications, some limitations have been discussed, especially its applicability domain and its predictive capacity. Therefore, this approach is still in its infant status and current scientific efforts are ongoing to stabilize stem cell differentiation [21].
2. The micromass test makes use of cell cultures of limb bud and of neuronal cells. The cells are isolated from the limb or the cephalic tissues of midorganogenesis embryos. Then, the cells undergo differentiation into chondrocytes and neurons. The differentiation after exposure to test chemicals is analyzed by using defined toxicological endpoints [22,23].
3. The whole postimplantation embryo culture is widely used to understand the mechanism of action and as a screening test for developmental toxicants. Embryo cultures from rodent, zebrafish, frog, and chicken are the most common cultures used. The protocol has been standardized and validated according to the ECVAM criteria. However, the predictability and applicability domains of the whole postimplantation are not sufficiently defined yet to allow regulatory implementation [24].

3.4.3 QSAR AND STRUCTURAL SIMILARITY

It has been reported by the pharmaceutical industry that computational toxicology tools are useful for early assessment of the toxic potential of lead molecules in drug discovery. Such methods like chemoinformatics, predictive computer-based modeling, structural similarity search, and QSARs are common techniques considered as preliminary screening methods to assess the degree of diverse toxicity endpoints [25].

QSARs for endpoints within reproductive toxicity have been developed by a wide variety of approaches ranging from regression analysis to multivariate analyses. A broad range of reproductive effects has been studied by QSARs. A general application is to predict the Absorption, Distribution, Metabolism and Excretion (ADME) properties of chemicals in order to determine their bioavailability [26]. QSARs have been also developed for the passive diffusion of chemicals across the placenta, blood–testis diffusion, and other relevant barriers [27,28].

Specific models have been generated to receptor binding associated with reproductive hazard and toxicology. Therefore, a panel of QSAR models for chemical binding to estrogen receptors, androgen receptors, and thyroid receptors is available.

Finally, structural similarity (defined also as grouping of substances) is another method, based on the concept that compound toxicity, structurally related to well-studied substances about toxicological properties, can be extrapolated or read across. A good example is the description of a category approach for a group of 10 ortho-phthalate esters with side-chain lengths C4 to C6 that are commonly known to cause reproductive effects [29].

However, the accuracy of such models is limited and the models should therefore be applied with caution. Most of the QSAR models can be performed only for obvious structural analogs and are seldom able to differentiate between agonist and antagonist. Therefore, the performance of these models in external assessments is usually poor for predicting diverse toxicity endpoints. Especially, reproductive effects are among the most difficult endpoints to predict in silico because of a lack of knowledge of mechanisms of action, oversimplification of the systems biology, and the limited number of data associated to the toxicants. REACH seems to be aware of the limitation of these predictive models as it is indicated that a battery of QSAR models, on the basis of current knowledge, cannot adequately cover a large number of potential targets/mechanisms associated with reproductive toxicity. For example, a reproductive and developmental toxicity model based on 2134 chemicals for seven endpoints showed an unbalanced and poor accuracy (specificity, >80%; sensitivity, 25%–50%) [19].

3.4.4 SYSTEMS BIOLOGY

To explore environmental toxicity hazard, systems biology appears as a new emerging alternative method and a powerful scientific approach. Systems biology is the study of an organism, viewed as an integrated, dynamic, and interacting network of genes, proteins, and biochemical reactions that gives rise to life. Instead of analyzing individual components or aspects of the organism, systems biologists focus on all the components and the interactions among them, all as part of one system. These interactions are ultimately responsible for an organism's form and functions. Accordingly, systems biology may help elucidate complex networks of genetic interactions that lead to toxicity of chemicals. Powerful support to such approaches for chemical safety assessment using in vitro methods comes currently from diverse studies aimed for drug discovery and the establishment of the "connectivity map" [30,31]; the latter serves as a large reference catalog of gene expression data from cultured human cells perturbed with more than 1300 chemicals and genetic reagents.

Systems biology is possible because of recent advances within the "omics revolution." These advances provide the tools to study biological systems comprehensively, for example, by transcriptomics, toxicogenomics, or metabolomics. These techniques enable detection of a large number of components on biological systems in parallel, which is required for the adoption of the systems approach. While the technologies keep improving, they alone are not sufficient for systems biology. The large amounts of data being generated by various omics technologies need to be analyzed and studied in the biological context. Computational modeling of biological systems is thus an essential component of the systems approach, which, in turn, can generate testable hypotheses based on analyzing and modeling large amounts of data. Systems biology can be viewed as an iterative cycle that involves studies of biological systems, measuring them, handling the data, and developing/refining the models. The modeling leads to new hypotheses, which further can be validated, with new experiments.

Existing approaches are largely based on bioinformatics analysis and data integration. Recently, various pharmacology network-based approaches have been developed with the aim of understanding the molecular dysfunction that is caused by small molecules [32,33]. An application of network toxicology addresses the fact that small molecules whose activity was predicted by specific target binding assays or gene expression profiling may have more than one interacting partner such that its activity is determined by its multiple interactions leading to unwanted toxic effects. Moreover, chemicals could alter the activity of proteins that are situated in the neighborhood of the disease module. The promise of systems biology by generated network-based methods is illustrated in the area of toxicogenomics and human health [34]. Toxicogenomics provides a gene expression profile of a substance that is associated to some adverse effects and toxicological effects in response to its exposure. The use of toxicogenomics for identifying the mechanism of action of genotoxic and nongenotoxic carcinogens has been increasing over the past few years and there are now training sets for these endpoints. Implementation of toxicogenomics has not yet been fully investigated for reproductive and developmental toxicology but is clearly an alternative method to animal testing, which would have a significant impact on reducing the use of animal testing.

Patel and Butte [34] implemented a method to predict a list of environmental chemicals associated with differentially expressed genes. This toxicogenomics approach allows predicting chemicals linked with various types of cancer, for example, cadmium-prostate cancer, doxorubicin-lung cancer, and bisphenol A-breast cancer. Recently, another knowledge- and data-driven method has been developed using toxicogenomics information, protein–protein interactions, and disease annotations with the aim of predicting chemical–disease connections and chemical–protein associations [35]. The latest point is very important in order to identify potential novel molecular mechanisms of action of chemicals, meaning deciphering unknown links between a chemical and a protein, which may lead to a toxic effect. For example, bis(2-ethylhexyl)phthalate has been predicted to potentially affect some receptors in the human brain called gamma-aminobutyric acid (GABA) A receptors. These receptors control the particular release of gonadotropin hormone networks, which play an important role in reproduction. It should be emphasized that the contexts for which these systems biology methods identify hypotheses need to be verified experimentally before they can be confirmed, but compared to most current approaches in toxicology, which have focused on a small number of chemicals and their influences on single or small groups of gene/protein, toxicogenomics-based models can concentrate on genome-wide responses for chemicals. Different from QSAR methods, which are based on the chemical structure of the compounds, systems biology offers a new vision, which is also more sophisticated as it includes complex biological data, protein–protein interaction data, and pathway information, to mention but a few data types.

3.4.5 DOSE–RESPONSE MODELING

Dose–response modeling using systems biology models for the perturbed biological pathways in combination with physiologically based pharmacokinetic (PBPK) models is also an alternative method to explore the toxicity of chemicals. Such methods can provide the information required for in vitro to in vivo extrapolation from relevant doses used in the in vitro tests rather than from high doses typically used in animal studies. PBPK (physiologically based toxicokinetics) is a process that enables hypothesis generation and creates model-driven experimentation, taking into account physiological and biological changes that might be related to disease and environmental exposure [36].

3.4.6 CHEMICAL MIXTURES

Finally, we have to take into account that, in reality, humans are widely exposed over the same period to not only one environmental toxicant but to complex mixtures. Recent animal studies show that such mixtures can have profound effects on male reproductive development at concentrations at which the individual chemical exposure has no effect [37]. A recent ecological study comparing the presence of endocrine disruptors in human breast milk samples in Denmark and Finland showed that polychlorinated biphenyls, organochlorine pesticides, and polychlorinated dibenzo-p-dioxins were found in significantly higher concentrations in Danish samples, although still at low concentrations [38]. Nevertheless, these classes of chemicals have been implicated in impairment of fetal testis development or testis cancer and could lead to the question whether the evaluation of the effect of chemicals on human health should include as many as possible of the agents constituting the total pollution cocktail to estimate the combined effects [39].

3.5 WHAT IS NEW ON REPRODUCTIVE AND DEVELOPMENTAL TOXICOLOGY?

In the United States, the ongoing National Health and Nutrition Examination Survey conducted by the US National Center for Health Statistics has provided a unique documentation of the presence of a wide variety of chemicals in members of the general population, giving rise to concern about their possible effects in human health. The Centers for Disease Control and Prevention has measured 212 chemicals in people's blood or urine and detected widespread exposure to polybrominated diphenyl ether (BDE-47), bisphenol A, perfluorooctanoic acid, and acrylamide [40].

In Europe, the European Commission was conscious of the potential threats from endocrine disruptors to human health and the environment. Therefore, a strategy was developed to prioritize and track "suspected endocrine disrupters." A list of 553 chemicals identified to be potentially harmful to humans, because of their production by industry at high volumes or highly persistent in the environment, has been evaluated. From this study, evidence of endocrine disruption was noted for 147 chemicals. Among them, 129 chemicals (essentially heavy metals, persistent organic compounds, polybrominated diphenyl ethers, polyfluorinated substances, polychlorinated dioxins, and furans) were chemical classes already subject to ban or restriction [41].

The US Environmental Protection Agency's ToxCast program has recently tested 309 environmental chemicals in 467 assays across diverse technologies and cell-based assays in order to prioritize chemicals for potential human toxicity. Such in vitro screening in association with advanced computational approaches provides meaningful information on the understanding of the complex biological systems targeted by environmental contaminants and in the prioritization of potential toxicant to be tested [42].

3.6 CONCLUSION

A lot of debate is ongoing regarding the challenge and the future impact of REACH. Although some scientists believe that the characterization of hundreds of chemicals will require a huge number of animal tests, it will for sure encourage scientists to develop and validate rapidly new alternative methods such as in vitro tests or complex computer models to limit animal testing. In addition, the designation of a substance as a reproductive or developmental toxicant follows criteria that, until now, do not consider the dose level of the substance at which reproductive or developmental effects occur, as long as excessive generalized toxicity does not occur. This method of labeling substances without consideration of effective dose level does not provide information on the actual risk of the chemical and might be an important issue.

Finally, reducing the overall uncertainty rather than reducing the uncertainty for a few individual chemicals assessed might be an option to consider if we want to speed up the regulatory acceptance process. To achieve these challenges, involvement and concern from pharmaceutical, chemical, and cosmetic industries are also expected by REACH to ensure a suitable safety assessment for the human health protection.

ACKNOWLEDGMENT

The authors would like to acknowledge the support of the Developmental Effects of Environment on Reproductive health EU project (DEER), the Innovative Medicines Initiative Joint Undertaking (eTOX), and the Villum Foundation.

REFERENCES

1. L. Michielan and S. Moro, *Pharmaceutical perspectives of nonlinear QSAR strategies*, J. Chem. Inf. Model. 50 (2010), pp. 961–978.
2. C.J. Van Leeuwen, B.G. Hansen, and J.H.M. de Bruijn, *Management of industrial chemicals in the European Union (REACH)*, in *Risk Assessment of Chemicals. An Introduction*, 2nd ed., C.J. Van Leeuwen and T.G. Vermeire, eds., Springer Publishers, Dordrecht, The Netherlands, 2007, pp. 511–551.
3. EChA, *Guidance for Identification and Naming of Substances under REACH*, European Chemicals Agency, Helsinki, 2007.
4. A.R. Scialli, *The challenge of reproductive and developmental toxicology under REACH*, Reg. Toxicol. Pharmacol. 51 (2008), pp. 244–250.
5. DHI, *The Impact of REACH on the Environment and Human Health (DGEnvironment)*, 2005. Available at http://ec-europa.eu/environment/chemicals/reach/background/docs/impact_on_environment_report.pdf.
6. E.S. Williams, J. Panko, and D.J. Paustenbach, *The European union's REACH regulation: A review of its history and requirements*, Crit. Rev. Toxicol. 39 (2009), pp. 553–575.
7. C. Ruden and S.O. Hansson, *Improving REACH*, Regul. Toxicol. Pharmacol. 44 (2006), pp. 33–42.
8. M.M. Matzuk and D.J. Lamb, *The biology of infertility: Research advances and clinical challenges*, Nat. Med. 14 (2008), pp. 1197–1213.
9. A.H. Piersma, *Alternative methods for developmental toxicity testing*, Basic Clin. Pharmacol. Toxicol. 98 (2006), pp. 427–431.
10. T. Höfer, I. Gerner, U. Gunder-Remy, M. Liebsch, A. Schulte, H. Spielmann, R. Vogel, and K. Wettig, *Animal testing and alternative approaches for the human health risk assessment under the proposed new European chemicals regulation*, Arch. Toxicol. 78 (2004), pp. 549–564.
11. R.Y. Wang and L.L. Needham, *Environmental chemical: From the environment to food, to breast milk, to the infant*, J. Toxicol. Environ. Health B. Crit. Rev. 10 (2007), pp. 597–609.
12. I.N. Damgaard, N.E. Skakkebaek, J. Toppari, H.E. Virtanen, H. Shen, K.W. Schramm, J.H. Petersen, T.K. Jensen, and K.M. Main, *Nordic cryptorchidism study group. Persistent pesticides in human breast milk and cryptorchidism*, Environ. Health Perspect. 114 (2006), pp. 1133–1138.
13. K.M. Main, G.K. Mortensen, M.M. Kaleva, K.A. Boisen, I.N. Damgaard, M. Chellakooty, I.M. Schmidt, A.M. Suomi, H.E. Virtanen, D.V. Petersen, A.M. Anderson, J. Toppari, and N.E. Skakkebaek, *Human breast milk contamination with phthalates and alterations of endogenous reproductive hormones in infants three months of age*, Environ. Health Perspect. 114 (2006), pp. 270–276.
14. NRC, *Tools and Technologies Chapter 4. Toxicity Testing in the 21st Century: A Vision and a Strategy*, National Academies Press, Washington, DC, 2007, pp. 98–119.
15. G. Schaafsma, E.D. Kroese, E.L. Tielemanns, J.J. Van de Sandt, and C.J. Van Leeuwen, *REACH, non testing approaches and the urgent need for a change in mind set*, Regul. Toxicol. Pharmacol. 53 (2009), pp. 70–80.

16. OECD, *Approaches to Data Gap Filling in Chemical Categories. Chapter 3: Guidance on Grouping of Chemicals, Environment Directorate, Joint Meeting of the Chemical Committee and the Working Party on Chemicals, Pesticides and Biotechnology, Series on Testing and Assessment*, Vol. 80, 2007, pp. 30–41. Available at http://www.oecd.org /officialdocuments/publicdisplaydocumentpdf/?doclanguage=en&cote=env/jm/mono %282007%2928.

17. A.M. Richard, C. Yang, and R.S. Judson, *Toxicity data informatics: Supporting a new paradigm for toxicity prediction*, Toxicol. Mech. Methods 18 (2008), pp. 103–118.

18. C. Yang, R.D. Benz, and M.A. Cheeseman, *Landscape of current toxicity databases and database standards*, Curr. Opin. Drug. Discov. Dev. 9 (2006), pp. 124–133.

19. L.G. Valerio Jr., *In silico toxicology for the pharmaceutical sciences*, Toxicol. Appl. Pharmacol. 241 (2009), pp. 356–370.

20. M. Balls and E. Hellsten, *Statement of the scientific validity of the embryonic stem cell test (EST)—An in vitro test for embryotoxicity*, Altern. Lab. Anim. 30 (2002), pp. 265–268.

21. P. Marx-Stoelting, E. Adriaens, H.J. Ahr, S. Bremer, B. Garthoff, H.P. Gelbke, A. Piersma, C. Pellizer, U. Reuter, V. Rogiers, B. Schenk, S. Schwengberg, A. Seiler, H. Spielmann, M. Steemans, D.B. Stedman, P. Vanparys, J.A. Vericat, M. Verwei, F. van der Water, M. Weimer, and M. Schwarz, *A review of implementation of the embryonic stem cell test (EST). The report and recommendations of an ECVAM/reProTect work-shop*, Altern. Lab. Anim. 37 (2009), pp. 313–328.

22. O.P. Flint, *A micromass culture method for rat embryonic neural cells*, J. Cell. Sci. 61 (1983), pp. 247–262.

23. N.A. Brown, H. Spielmann, R. Bechter, O.P. Flint, S.J. Freeman, R.J. Jeläinek, E. Koch, H. Nau, D.R. Newall, A.K. Palmer, J.Y. Renault, M.F. Repetto, R. Vogel, and R. Wiger, *Screening chemicals for reproductive toxicity: The current alternatives; The report and recommendations of an ECVAM/ETS workshop (ECVAM workshop 12)*, Altern. Lab. Anim. 23 (1995), pp. 868–882.

24. E. Genschow, H. Spielmann, G. Scholz, A. Seiler, N. Brown, A. Piersma, M. Brady, N. Clemann, H. Huuskonen, F. Paillard, S. Bremer, and K. Becker, *The ECVAM interna-tional validation study on in vitro embryotoxicity tests: Results of the definitive phase and evaluation of prediction models. European Centre for the Validation of Alternative Methods*, Altern. Lab. Anim. 30 (2002), pp. 151–176.

25. N.L. Kruhlak, J.F. Contrera, R.D. Benz, and E.J. Matthews, *Progress in QSAR toxicity screening of pharmaceutical impurities and other FDA regulated product*, Adv. Drug. Deliv. Rev. 59 (2007), pp. 43–55.

26. J.C. Madden, *In silico approaches for predicting ADME properties*, in *Recent Advances in QSAR Studies—Methods and Applications*, T. Puzyn, J. Leszczynski, and M.T.D. Cronin, eds., Springer, Dordrecht, 2010, pp. 283–304.

27. M. Hewitt, J.C. Madden, P.H. Rowe, and M.T.D. Cronin, *Structure-based modeling in reproductive toxicology: (Q)SAR for the placenta barrier*, SAR QSAR Environ. Res. 18 (2007), pp. 57–76.

28. M.T.D. Cronin and M. Hewitt, *In silico models to predict passage through the skin and other barriers*, in *Comprehensive Medicinal Chemistry II*, Vol. 5, B. Testa and H. van de Waterbeemd, eds., Elsevier, Oxford, 2007, pp. 725–744.

29. E. Fabjan, E. Hulzebos, W. Mennes, and A.H. Piersma, *A category approach for repro-ductive effects of phthalates*, Crit. Rev. Toxicol. 36 (2006), pp. 695–726.

30. J. Lamb, E.D. Crawford, D. Peck, J.W. Modell, I.C. Blat, M.J. Wrobel, J. Lerner, J.P. Brunet, A. Subramanian, K.N. Ross, M. Reich, H. Hieronymus, G. Wei, S.A. Amstrong, S.J. Haggarty, P.A. Clermons, R. Wei, S.A. Carr, E.S. Lander, and T.R. Golub, *The con-nectivity map: Using gene-expression signatures to connect small molecules, genes, and disease*, Science 313 (2006), pp. 1929–1935.

31. J. Lamb, *The connectivity map: A new tool for biomedical research*, Nat. Rev. Cancer 7 (2007), pp. 54–60.
32. M. Kuhn, M. Campillos, I. Letunic, L.J. Jensen, and P. Bork, *A side effect resource to capture phenotypic effect of drugs*, Mol. Syst. Biol. 6 (2010), p. 343.
33. O. Taboureau, S.K. Nielsen, K. Audouze, N. Weinhold, D. Edsgärd, F.S. Roque, I. Kouskoumvekaki, A. Bora, R. Curpan, T.S. Jensen, S. Brunak, and T.I. Oprea, *ChemProt: A disease chemical biology database*, Nucleic Acids Res. 39 (2011), pp. 367–372.
34. C.J. Patel and A.J. Butte, *Predicting environmental chemical factors associated with disease-related gene expression data*, BMC Med. Genomics 3 (2010), pp. 1–17.
35. K. Audouze, A.S. Juncker, F.J. Roque, K. Krysiak-Baltyn, N. Weinhold, O. Taboureau, T.S. Jensen, and S. Brunak, *Deciphering diseases and biological targets for environmental chemicals using toxicogenomics networks*, PLoS Comput. Biol. 6 (2010), p. e10000788.
36. H.J. Clewell, Y.M. Tan, J.L. Campbell, and M.E. Andersen, *Quantitative interpretation of human biomonitoring data*, Toxicol. Appl. Pharmacol. 231 (2008), pp. 122–133.
37. C.V. Rider, V.S. Wilson, K.L. Howdeshell, A.K. Hotchkiss, J.R. Furr, C.R. Lambright, and L.E. Gray Jr., *Cumulative effects of in utero administration of mixtures of antiandrogens on male rat reproductive development*, Toxicol. Pathol. 37 (2009), pp. 100–113.
38. K. Krysiak-Baltyn, J. Toppari, N.E. Skakkebaek, T.S. Jensen, H.E. Virtanen, K.W. Schramm, H. Shen, T. Vartiainen, H. Kiviranta, O. Taboureau, S. Brunak, and K.M. Main, *Country-specific chemical signatures of persistent environmental compounds in breast milk*, Int. J. Androl. 33 (2010), pp. 270–278.
39. S. Christiansen, M. Scholze, M. Axelstad, J. Boberg, A. Kortenkamp, and U. Hass, *Combined exposure to anti-androgens causes markedly increased frequencies of hypospadias in the rat*, Int. J. Androl. 31 (2008), pp. 241–248.
40. Available at http://www.cdc.gov/ExposureReport/.
41. Available at http://ec.europa.eu/environment/endocrine/strategy/substances_en.htm.
42. R.S. Judson, K.R. Houck, R.J. Kavlock, T.B. Knudsen, M.T. Martin, H.M. Mortensen, D.M. Reif, D.M. Rotroff, I. Shah, A.M. Richard, and D.J. Dix, *In vitro screening of environmental chemicals for targeted testing prioritization: The ToxCast project*, Environ. Health Perspect. 118 (2010), pp. 485–491.

Section II

Reproduction and Development

Biological Processes and Endpoints

4 Female Reproductive Physiology

Mariangela Maluf and Paulo Marcelo Perin

CONTENTS

ABSTRACT

Female reproductive development and function are controlled by timely and coordinated endocrine, paracrine, and autocrine signals starting at the embryonic stage. Steroid hormones, proteins, growth factors, and other signaling molecules, which affect gene expression and protein synthesis in target cells, are involved in these complex biological processes. The ovary is composed of germ (oocytes) and somatic (granulosa, theca, stromal) cells whose interactions are essential not only for the development, maturation, and release of a fertilizable oocyte capable of developing into a viable embryo, but also in the maintenance of luteal cell function essential for successful implantation and early pregnancy development. The formation of the primordial follicles, which occurs when oocytes are surrounded with squamous pre-granulosa cells and remain quiescent in the ovary until recruited into the growing pool throughout reproductive life, represents the first stage of folliculogenesis. From the limited pool of primordial follicles, some are stimulated to growth and develop into primary and secondary preantral follicles through a precise interaction of multiple genes. At the antral stage, the few follicles that survive atresia reach the preovulatory stage under the cyclic gonadotropin stimulation that occurs after puberty. One of the follicles attains dominance over the rest of the cohort and, in response to the preovulatory surge of gonadotropins, releases the mature oocyte for fertilization. The residual follicle undergoes luteinization to become the corpus luteum, a transient hormone-regulated ovarian organ that secretes progesterone to support pregnancy. The exposure to reproductive toxicants during the process of folliculogenesis can affect ovarian function in distinct ways depending on the type of follicle affected. An immediate but reversible loss of reproductive function is observed when the target is the antral or preovulatory follicle. On the other hand, ovarian function can be impaired by the exposure to toxicants that destroy primordial and/or preantral follicles resulting in the disruption of endocrine balance that will be manifested only several years after exposure as subfertility and finally infertility. The depletion of the pool of resting follicles ultimately leads to irreversible ovarian failure. This chapter reviews current knowledge about key molecular, cellular, and endocrine events involved in ovarian function including follicular development, ovulation, and corpus luteum formation. The effects of the exposure to reproductive toxicants on ovarian function and reserve are briefly discussed.

KEYWORDS

Folliculogenesis, primordial follicles, oocyte, corpus luteum, reproductive toxicants, ovarian reserve

4.1   INTRODUCTION

Female reproductive development and function are controlled by timely and coordinated endocrine, paracrine, and autocrine signals starting at the embryonic stage. Steroid hormones, proteins, growth factors, and other signaling molecules that affect gene expression and protein synthesis in target cells are involved in these complex biological processes. The ovary is basically composed of two cell lines, the germ cells (oocytes) and the somatic cells (granulosa, theca, and stromal cells), whose interactions are essential for oocyte growth and the regulation of meiotic maturation, which plays a key role during the menacme not only in the development, maturation, and release of a fertilizable oocyte capable of developing into a viable embryo but also in the maintenance of luteal cell function essential for successful implantation and early pregnancy development [1].

The first stage of folliculogenesis is represented by the formation of the primordial follicles, which occurs when oocytes that survive the process of germ cell cluster breakdown are individually surrounded with squamous pregranulosa cells and remain quiescent in the ovary until recruited into the growing pool throughout reproductive life. From the limited pool of primordial follicles, some are stimulated to growth and develop into primary and secondary preantral follicles before acquiring an antral cavity through a precise spatiotemporal expression and interaction of multiple genes. At the antral stage, most follicles undergo atresia; the few that survive reach the preovulatory stage under the cyclic gonadotropin stimulation that occurs after puberty. One of the follicles attains dominance over the rest of the cohort and, in response to the preovulatory surge of gonadotropins, releases the mature oocyte for fertilization. The residual follicle undergoes luteinization to become the corpus luteum (*CL*), a transient hormone-regulated ovarian organ that secretes progesterone (P) to support pregnancy [2–4].

Over the past few years, an increasing amount of evidence has shown the negative effects of environmental contaminants on human reproductive health. On a daily basis, women are unavoidably exposed to these ubiquitous contaminants through various routes, and the effects of this exposure may remain unnoticed over a long period [5]. The exposure to reproductive toxicants during the process of folliculogenesis can affect ovarian function in distinct ways depending on the type of follicle affected. An immediate but reversible loss of reproductive function is observed when the target is the antral or preovulatory follicle. On the other hand, ovarian function can be impaired by the exposure to toxicants that destroy primordial and preantral follicles, resulting in the disruption of endocrine balance that will be manifested only several years after exposure as subfertility and finally infertility. The depletion of the pool of resting follicles ultimately leads to irreversible ovarian failure [6,7].

This chapter reviews current knowledge about key molecular, cellular, and endocrine events involved in ovarian function including follicular development, ovulation, and *CL* formation. The effects of the exposure to reproductive toxicants on ovarian function and reserve are briefly discussed.

4.2 OVARIAN CYCLE

The development and release of a mature oocyte capable of being fertilized to produce an embryo depends on a delicate balance between the ovaries, pituitary, and hypothalamus. The follicle represents the basic functional unit of the ovary and consists of an oocyte surrounded by one or more layers of granulosa cells. The process of folliculogenesis begins when the follicles at rest within the ovarian cortex (primordial follicles), in response to signals not completely known and independent of gonadotropins, are recruited and become activated for entry into the cohort of developing follicles and culminates with the production of a single dominant follicle during each menstrual cycle [4].

The first stage of folliculogenesis, which occurs in the second half of pregnancy, is represented by the formation of the primordial follicles in which oocytes that are surrounded by an adequate number of pregranulosa cells survive the process of germ cell cluster breakdown [3,4]. At the time of birth, both human ovaries contain approximately one to two million primordial follicles arrested in prophase of meiosis I, representing the finite pool of female gametes available throughout the reproductive life span [8]. During childhood, most ovarian follicles become atretic and their number decreases to around 400,000 at puberty. During this period, follicles cannot attain ovulatory sizes or produce a significant amount of estrogen in the absence of follicle-stimulating hormone (FSH) and luteinizing hormone (LH) stimulation. Puberty marks the onset of the cyclic gonadotropin stimulation of the oocyte to complete maturation. A small group of primordial follicles (~1000) is recruited from the ovarian cohort and undergoes a primordial-to-primary follicle transition [2]. The rate at which primordial follicles join the growing pool is directly related to the size of the ovarian reserve, suggesting that intraovarian paracrine signaling between growing follicles, resting follicles, and ovarian stroma may be responsible for this event [9].

4.2.1 ACTIVATION OF RESTING PRIMORDIAL FOLLICLES

Primordial follicles remain in a quiescent state with the oocyte arrested at the diplotene stage of first meiotic prophase until they gradually leave the arrested pool undergoing the primordial-to-primary follicle transition. This transition, a gonadotropin-independent phase, is a nonreversible process in which follicles continue growth until their inevitable fate: apoptosis/atresia or ovulation. The selective loss of oocytes in this transition has been attributed to classic apoptotic mechanisms involving the actions of the B cell lymphoma/leukemia-2 (*BCL-2*) family of proteins and acts as a quality control mechanism in which deficient nuclei are lost and healthy oocytes are preferably encapsulated into primordial follicles (Figure 4.1) [10,11]. Since the initial pool of primordial follicles at birth is fixed in women, representing the only source of follicles throughout the reproductive life span, follicular assembly rate and primordial-to-primary follicle transition are of critical importance to female reproduction. When the supply of follicles is finally depleted, reproduction ceases and women enter menopause.

During follicle transition, epithelial-derived granulosa cells change from a flattened to a cuboidal shape and start to proliferate, beginning the process of folliculogenesis. The oocyte increases in diameter and starts the synthesis of its unique extracellular glycoprotein matrix, the zona pellucida (*ZP*), which is deposited between the oocyte and the granulosa cells. Progenitor theca cells are recruited from the surrounding stromal/mesenchymal-derived cell population by activated follicles and start to proliferate. The undifferentiated theca cells do not express LH receptors (*LHr*) and are, as granulosa cells, steroidogenically inactive. These cells are separated from the outermost layer of mural granulosa cells by a pronounced basement membrane [12]. Both granulosa and theca cells provide essential nutrients, information molecules, metabolic precursors, growth factors, and hormones necessary to support oocyte development throughout folliculogenesis and integrate ovarian function with the body by directing gonadal steroidogenesis. The oocyte itself synthesizes factors that control the fate of granulosa and theca cells playing an active role in directing follicle growth [13]. Transzonal projections established at the onset of folliculogenesis are small finger-like processes that extend from granulosa cells, traverse the *ZP*, and directly interact with the microvilli of the oocyte plasma membrane. These projections allow for transfer of nutrients and small molecules such as biosynthetic substrates and meiosis-arresting signals as well as communication between the two cell compartments. Additionally, these adhesive contact sites may act as signaling domains for the interaction of receptor kinases with growth factors allowing the processing, activation, and delivery of certain oocyte- and granulosa cell–derived paracrine factors to specific receptor targets [13].

The expression of a variety of activating proteins in both the oocyte and somatic cells of follicles and the release from active repression are involved in the activation of the primordial-to-primary follicle transition. Several transcription factors have a critical role during the transition from primordial to primary follicles. The factor in the germline alpha (*FIGLA*) regulates the expression of three genes encoding proteins (*ZP1, ZP2, ZP3*) that form the *ZP* [14]. The newborn ovary

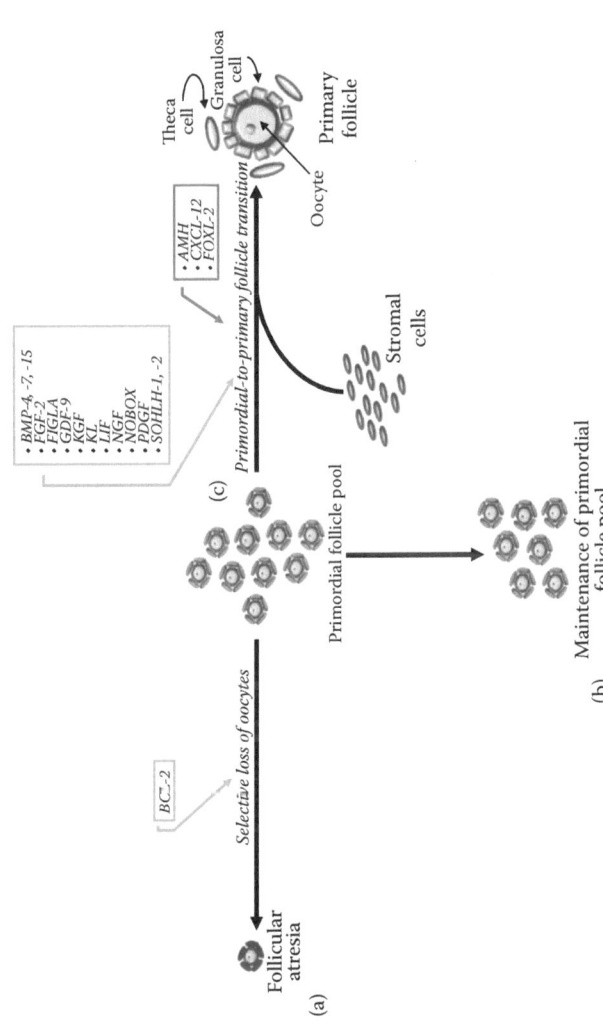

FIGURE 4.1 (**See color insert.**) Developmental fate of ovarian primordial follicles: (a) follicular atresia, (b) maintenance of the primordial follicle pool, and (c) primordial-to-primary follicle transition. Genes inside green and red boxes are involved in the activation or suppression of processes that determine primordial follicle fate.

homeobox (*NOBOX*) is necessary for expression of several key oocyte-specific genes, including growth-differentiation factor 9 (*GDF-9*) and the transcription factor octamer-binding protein 4 (*OCT-4*) [15]. Spermatogenesis and oogenesis helix-loop-helix 1 (*SOHLH-1*) is another germ cell–specific transcription factor that is upstream of the LIM homeobox protein 8 (*LHX-8*) gene, which has a critical role in early follicle formation and oocyte differentiation and functions in part by regulating the *NOBOX* and *FIGLA* pathways [2,16]. Spermatogenesis and oogenesis helix-loop-helix 2 (*SOHLH-2*), which has an expression pattern that mimics that of *SOHLH-1*, is also critical in early follicle formation. *SOHLH-2* regulates the differentiation of pregranulosa cells surrounding the oocyte of primordial follicles into cuboidal and multilayered granulosa cells [2]. *SOHLH-1* and *SOHLH-2* are independently required for successful oocyte differentiation [15]. Forkhead box L2 (*FOXL-2*) and nerve growth factor (*NGF*) are also crucial in the transition from squamous to cuboidal granulosa cells that occur during primordial follicle activation. Leukemia inhibitory factor (*LIF*) found in pregranulosa and somatic cells promotes primordial-to-primary follicle transition by up-regulation of kit ligand (*KL*) in granulosa cells [3]. Anti-Mullerian hormone (*AMH*) expressed by granulosa cells of growing follicles has been shown to suppress primordial follicle recruitment functioning as an inhibitory growth factor in the ovary during the early stages of folliculogenesis. This effect probably results from a paracrine pregranulosa cell–derived effect of *AMH* on the primordial follicle. Chemoattractive chemokine (C-X-C motif) ligand 12 (*CXCL-12*) has been identified as a second inhibitor of primordial follicle transition. *FOXL-2* is an additional inhibitory signal that maintains primordial follicles in the dormant state in humans (Figure 4.1) [3,17,18].

Two oocyte-specific growth factors, *GDF-9* and bone morphogenetic protein 15 (*BMP-15*), both members of the transforming growth factor β (*TGF-β*) superfamily and expressed throughout folliculogenesis, are required for progression beyond the primary stage of development. *GDF-9* has an important role in both granulosa cell proliferation/survival and theca cell recruitment and function modulation while limiting oocyte growth through the suppression of *KL* expression in granulosa cells. *BMP-15* is also required for granulosa cell proliferation and follicle progression [12,19,20]. Nevertheless, *BMP-15* promotes *KL* expression, which, in turn, down-regulates *BMP-15* expression, establishing a paracrine negative feedback loop between granulosa cells and the oocyte [13]. Basic fibroblast growth factor (*bFGF*), also produced by the oocyte in the primordial and early-stage follicles, stimulates granulosa cell mitosis and both theca and stromal cell growth, influencing primordial follicle development [21]. Moreover, *bFGF* may mediate follicular activation through enhancement of *KL* production by granulosa cells [2].

Granulosa cells produce *KL* that can control oocyte development and influence theca cell recruitment from the stromal–interstitial cell population, regulate their proliferation, and induce the primordial-to-primary follicle transition although the mechanisms by which *KL* signaling contribute to this transition are not entirely known [21]. Evidence suggests that the *KL* pathway induces the phosphatidylinositol 3-kinase (*PI3K*)/protein kinase B (*AKT*) pathway leading to phosphorylation and inactivation of transcription factor forkhead box O3 (*FOXO-3*), an inhibitor of primordial follicle activation. *KL* produced by granulosa cells activates oocyte surface cognate tyrosine kinase receptor (*c-KIT*), which later activates the *PI3K* pathway leading to oocyte growth and the production of oocyte factors, which, in turn, stimulate the proliferation and differentiation of the surrounding granulosa cells. Additionally, the phosphorylation and functional suppression of *FOXO-3* induced by *KL* release the oocytes from their quiescent state [4]. Insulin-like growth factor (*IGF*) produced by granulosa cells promotes thecal differentiation [12].

Theca cells produce bone morphogenetic protein 4 (*BMP-4*), keratinocyte growth factor (*KGF*), transforming growth factor α (*TGF-α*), and hepatocyte growth factor (*HGF*) that can regulate granulosa cells. The developing theca cells produce *BMP-4*, a growth factor that promotes the primordial-to-primary follicle transition and acts on granulosa cells to sustain oocyte survival [22]. *KGF*, a fibroblast growth factor (*FGF-7*) that mediates mesenchymal–epithelial interactions and is produced by the recruited precursor theca cells, has also been found to stimulate the primordial-to-primary follicle transition regulating the expression of factors from the adjacent granulosa cells [23]. *TGF-α*, a mitogenic factor, regulates granulosa cell proliferation via paracrine mechanisms acting as a survival factor [24]. *HGF*, another mitogenic factor, stimulates *KL* gene expression in granulosa cells, which, in turn, stimulates *HGF* gene expression in theca cells, suggesting a positive feedback loop between theca and granulosa cells mediated by *HGF* and *KL* [25].

4.2.2 PROGRESSION OF PRIMARY FOLLICLES TO THE EARLY ANTRAL STAGE

The transition of a primary follicle into a preantral follicle (a gonadotropin-responsive phase) is characterized by complex bidirectional communication between the oocyte and the surrounding somatic cells. Throughout this process, granulosa cells have an established role in supporting oocyte growth, the acquisition of meiotic competence, and modulating the global transcriptional activity in the oocyte genome [26]. On the other hand, the rate of follicular development is critically dependent on oocyte-secreted factors that regulate several aspects of granulosa cell development including proliferation, differentiation, and extracellular matrix (*ECM*) and steroid hormone production, thereby controlling the development of a healthy oocyte [27]. At this stage, the selection of follicles for further development results from a delicate balance between endocrine and intraovarian (autocrine and paracrine) regulatory signals necessary for survival, and follicular atresia is a consequence of an inadequate growth support.

Follicle progression in this stage of development is characterized by a growing oocyte surrounded by several granulosa cell layers, granulosa cell proliferation, *ZP* formation, and the acquisition of an additional outermost somatic cell layer (the theca) from mesenchymal precursor cells present in the adjacent ovarian stroma through a gonadotropin-independent process (Figure 4.2). This layer is composed of a well-developed theca interna and a less well-defined theca externa. The theca interna layer that surrounds the basement membrane found around granulosa cells is formed by cells with characteristic features including numerous mitochondria with vesicular cristae, smooth endoplasmic reticulum, and abundant lipid vesicles that correspond

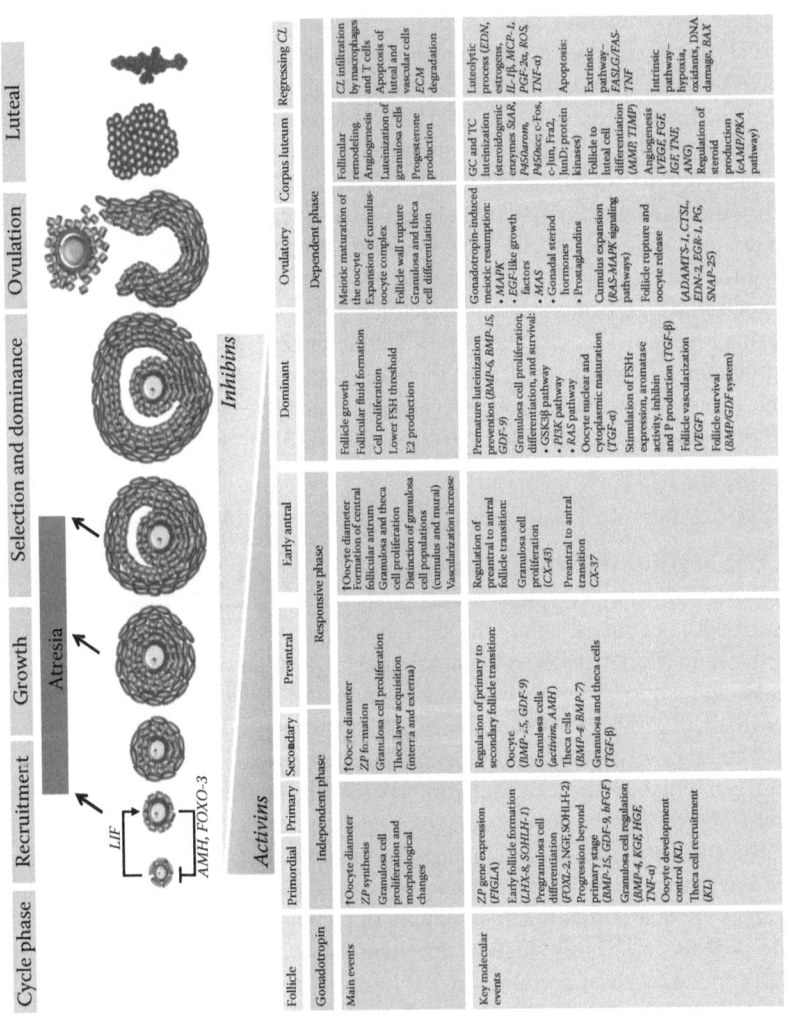

FIGURE 4.2 (**See color insert.**) Factors involved in ovarian follicle development.

with their main function as a source of androgens for neighboring granulosa cells to convert to estrogens, crucial to the pituitary–gonadal axis and endocrine control of reproduction [3]. The theca externa (an ill-defined layer that surrounds the theca interna) is composed of fibroblasts, smooth muscle–like cells, and macrophages, as well as circumferential collagen bundles, blood, and lymphatic vessels [12].

Several members of the *TGF-β* superfamily, locally expressed by the oocyte (*GDF-9* and *BMP-15*), granulosa cells (activins, *AMH*), and theca cells (*BMP-4* and bone morphogenetic protein 7 [*BMP-7*]) or both (*TGF-β*), are involved in an intricate process of positive and negative feedback to regulate the primary-to-secondary transition and subsequent follicle growth to the late preantral and early antral stages of development [18]. *GDF-9* and *BMP-15* (two oocyte-specific growth factors) act as positive regulators of follicle growth into preantral and antral stages, stimulating granulosa cell mitosis in a gonadotropic-independent manner [13]. *GDF-9* has an important role as a granulosa cell survival factor during preantral-to-early antral transition and is required to maintain FSH receptor (*FSHr*) expression in the preantral follicles [28]. Activin βA, βB, and βAB subunits and follistatin (activin-binding protein) are primarily expressed by granulosa cells, while their receptors are expressed by theca cells, granulosa cells, and the oocyte. As another positive regulator, activin promotes granulosa cell proliferation/differentiation, enhancing growth and survival of preantral follicles [29]. In contrast, granulosa cells of growing follicles secrete *AMH*, which plays a role in the inhibition of recruitment of primordial follicles into the pool of growing follicles and decreasing responsiveness of growing preantral and small antral follicles to FSH [30]. Theca-derived *BMP-4* and *BMP-7* have a positive paracrine action on granulosa cells of growing preantral follicles, promoting follicle growth beyond the primary stage. Additionally, *BMP-4* and *BMP-7* modulate FSH signaling, promoting estradiol (E2) production via aromatization while inhibiting P synthesis, acting as a luteinization inhibitor [18]. *TGF-β* (isoforms *TGF-β1*, *TGF-β2*, and *TGF-β3*) expression, by both granulosa and theca cells, is first observed during preantral follicular growth and intensifies as the follicle matures. Expression of 17α-hydroxylase (a key protein controlling androgen production), which catalyzes the conversion of progestogens to androgens, and steroidogenic acute regulatory protein (*StAR*), which facilitates the transport of cholesterol into the mitochondria for steroid synthesis, is down-regulated by *TGF-β* in thecal cells, showing the suppression of steroidogenesis in these cells [31].

The basal lamina acts as a blood–follicle barrier separating the oocyte and granulosa cells from the vascularized thecal cell layer. At the onset of follicle growth, gap junctions already connect oocytes with granulosa cells and granulosa cells with each other, integrating them as a functional syncytium. This process allows all cells inside the basal lamina not only to have effective communication but also to effectively share metabolites and signaling molecules between them and is required for oocyte growth as well as meiotic maturation [26]. Connexins (*CX*) are the core proteins that make up gap junctions and at least two (*CX-43* and *CX-37*) have essential and distinct roles during folliculogenesis to support proper oocyte development. *CX-43* are required for granulosa cell proliferation earlier in folliculogenesis to form multilayered follicles, whereas *CX-37* oocyte–granulosa cell localized gap junctions are essential for the preantral-to-antral follicle transition [3].

4.2.3 Antral Follicle Growth and Follicle Selection Mechanism

During antral folliculogenesis, follicle growth is characterized by further oocyte growth, proliferation of granulosa and theca cells, increased vascularization, and formation of a central follicular antrum or cavity (a fluid-filled space that separates two anatomically and functionally distinct granulosa cell populations). Mural granulosa cells line the follicle walls and are critical for steroidogenesis and ovulation, whereas cumulus granulosa cells that surround the oocyte promote its growth and developmental competence (Figure 4.2) [3]. The differentiation of these distinct cell populations is regulated by the oocyte, which establishes a heterogeneous pattern of gene expression by granulosa cells. The oocyte promotes the cumulus cell phenotype in those granulosa cells immediately adjacent to it, stimulates their proliferation, enhances their E2 production, and induces cumulus cells to provide it with the necessary metabolic support. The expression of *LHr* and *KL*, as well as P synthesis by cumulus cells, is suppressed by paracrine factors secreted by the oocyte [31]. In contrast, mural granulosa cells express *LHr* and the highest expression is observed in cells that are in close apposition to the basal lamina, which has components that enhance gonadotropin-induced expression of these receptors [19]. In addition to the ovarian steroids and regulatory growth factors, follicular growth is controlled by endocrine factors such as pituitary gonadotropins (FSH and LH).

4.2.4 Role of Gonadotropins and Estrogen

Preantral-to-antral transition is marked by the change from intraovarian to extraovarian regulatory processes of folliculogenesis as the hypothalamic–pituitary–gonadal (*HPG*) axis starts functioning. FSH becomes a critical determinant of follicle growth beyond the late-preantral/small-antral stage, stimulating granulosa cell mitosis, E2 production, and *LHr* expression. This prevents granulosa cell apoptosis and follicular atresia at this stage. Throughout childhood, blood concentrations of FSH and LH remain too low to stimulate full preovulatory follicular development. At menarche, cyclic increases in pituitary FSH secretion rescue a cohort of antral follicles from atresia for further growth, leading to the emergence of the preovulatory follicle. In addition to the various positive and negative feedback loops in the *HPG* axis involved in follicle maturation and selection, the ovary also produces growth factors such as activins, inhibins (functional antagonists of activins), and follistatins that not only modulate pituitary FSH secretion but also act locally to regulate follicular development [3].

FSH plays a fundamental role in the growth and differentiation of the leading follicle through its ability to promote follicular fluid formation, cell proliferation, E2 production, and *LHr* expression. During the late luteal and early follicular phases of the menstrual cycle, the increase in circulating FSH levels allows a cohort of antral follicles to escape apoptosis (Figure 4.3). One of the leading follicles in this group dominates by producing higher levels of estrogens and inhibins to suppress pituitary FSH release during the midfollicular phase. The dominant follicle that emerges is more sensitive to FSH (lower FSH threshold) because of enhanced expression of FSH/LH receptors or increase in FSH responsiveness resultant from the increase of local growth factors and vasculature (positive selection). The granulosa cells of the

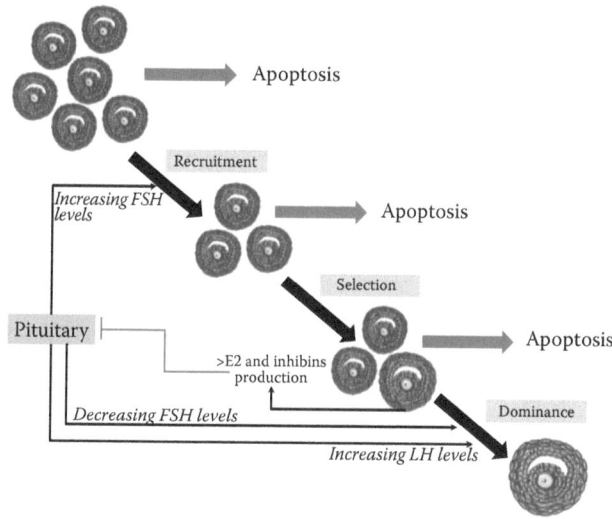

FIGURE 4.3 **(See color insert.)** Cyclic recruitment, selection, and dominance of early antral follicles. FSH increasing levels rescue a cohort of antral follicles from atresia for further growth, leading to the emergence of the preovulatory follicle. The dominant follicle that emerges from this cohort produces higher levels of estrogens and inhibins to suppress pituitary FSH release during the midfollicular phase and is more sensitive to FSH (lower FSH threshold) because of enhanced expression of FSH/LH receptors or increase in FSH responsiveness resultant from the increase of local growth factors and vasculature (positive selection). The decline in FSH level to a concentration insufficient to sustain the growth of the remaining antral follicles (higher FSH threshold) leads to their loss as they begin to undergo atresia (negative selection).

dominant follicle are not only sensitized to FSH but now express *LHr* responding directly to LH as well as FSH. The decline of the FSH level to a concentration insufficient to sustain the growth of the remaining antral follicles (higher FSH threshold) leads to their loss as they begin to undergo atresia (negative selection) [32–35].

Ovarian E2 production depends on a delicate interplay between granulosa and theca cells and represents the hallmark of successful ovulatory follicles. The intra-ovarian effects of E2 on folliculogenesis (growth, differentiation, survival) are mediated by two estrogen receptors (*ER*), *ERα* (*Esr-1*) predominantly expressed in thecal and interstitial cells and *ERβ* (*Esr-2*) expressed in granulosa cells of growing follicles and regulated by gonadotropins. E2 modulates granulosa cell differentiation by enhancing the ability of FSH to induce the expression of *LHr*, which, at the time of follicular selection, promotes follicular growth and differentiation. Although E2 is not essential for antral follicle formation, it has a critical role in granulosa cell growth and differentiation to maintain antral follicles and promote ovulation [3].

As the midcycle approaches, the dominant preovulatory follicle increases E2 production through endocrine (LH) and paracrine (inhibin) signaling. Inhibin secretion by the dominant follicle increases LH-induced androgen secretion from theca cells. This in turn provides granulosa cells with a sufficient supply of androgen for conversion into estrogens by the aromatase enzyme as the demand for estrogen synthesis increases significantly during the periovulatory period. Cell proliferation and follicle growth are maintained during this period while premature luteinization is prevented by the secretion of oocyte-derived growth factors bone morphogenetic protein 6 (*BMP-6*), *BMP-15*, and *GDF-9* that act within the follicle suppressing gonadotropin-driven P synthesis [18,31]. The midcycle LH surge initiates the ovulatory cascade through paracrine signaling involving growth factors *GDF-9* and *BMP-15* that results in oocyte maturation and maturation and expansion of the cumulus cell mass [36]. The midcycle LH surge initiates the ovulatory cascade through paracrine signaling involving growth factors *GDF-9* and *BMP-15* that results in oocyte maturation and maturation and expansion of the cumulus cell mass [37,38]. Cumulus expansion facilitates the release of the oocyte into the abdominal cavity, capture of the oocyte by the oviductal fimbria, sperm penetration, and fertilization [39]. After the release of the oocyte at ovulation, the effect of oocyte-derived luteinization inhibitors is lost and luteinization begins [18].

4.2.5   ROLE OF INTRAOVARIAN FACTORS

Differential exposure to paracrine oocyte-derived (*GDF-9, BMP-6, BMP-15*), and to autocrine/paracrine granulosa-derived (activin, *BMP-6*) and theca-derived (*BMP-4, BMP-7*) factors, which promotes granulosa cell proliferation, and FSH-dependent follicle function modulation sensitize some follicles to FSH for further growth to the preovulatory stage. On the other hand, granulosa-derived *AMH* has a negative role in the cyclic recruitment of follicles and the dominant follicle selection process by reducing the responsiveness of preantral and small antral follicles to FSH, thereby exerting a controlling influence on the rate at which follicles become available for preovulatory development [9,36].

FSH is required for granulosa cell differentiation, and these cells in turn rely on this gonadotropin to facilitate follicular growth. The classical signaling cascade activated by FSH is the adenylyl cyclase (*AC*)/adenosine $3',5'$-cyclic monophosphate(*cAMP*)/ protein kinase A (*PKA*) pathway. This results in the production of *cAMP* and activation of *PKA*, thus regulating a number of target genes including aromatase, inhibin α- and β-subunits, and *LHr* among others. FSH activates other *cAMP*-independent signaling cascades including *PI3K* (also activated by *IGF-I*), rat sarcoma (*RAS*), and glycogen synthase kinase 3β (*GSK3β*) pathways [1,3]. The activation of the *PI3K* pathway by FSH and *IGF-I* increases phosphorylation and degradation of forkhead box O1 (*FOXO-1*), which affects proliferation, differentiation, and survival of granulosa cells. Insulin-like growth factor 1 (*IGF-1*) enhances granulosa cell responsiveness to FSH by increasing their quantity of *FSHr*. Thus, follicles that possess higher levels of insulin-like growth factor (*IGF*) survive in the face of declining FSH levels. In conjunction with FSH and *IGF-1*, the E2 signaling cascade controls granulosa cell proliferation through cell cycle modulation (Figure 4.4) [1,3,40].

The actions of FSH are enhanced by activins and E2 during the later stages of follicular development. Systemically, activins promote FSH release from the anterior pituitary. Within the ovary, activins facilitate proliferation of granulosa cells and promote their expression of *FSHr* and are involved in the regulation of steroidogenic enzyme aromatase activity. This activity converts theca cell–derived androgens to

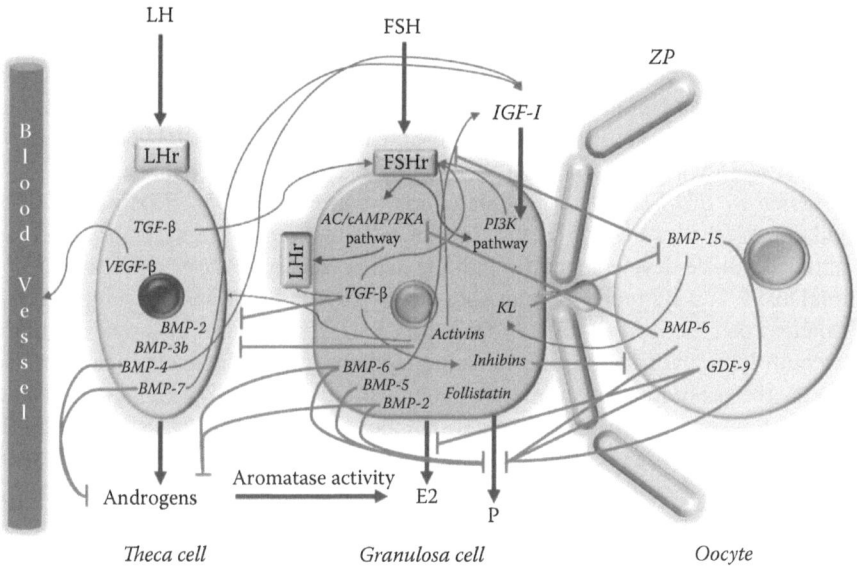

FIGURE 4.4 (**See color insert.**) Paracrine and autocrine pathways involved in bidirectional communication between the oocyte and cumulus/granulosa and theca cells. Green lines ending with arrows represent stimulatory effects and red lines ending with bars represent inhibitory effects.

E2, increasing E2 production and enhancing FSH actions. Small follicles secrete more activin relative to inhibin, whereas larger selected antral follicles approaching preovulatory status produce more inhibin, which leads to an increase in LH-induced androgen production. Consequently, granulosa cells are able to maintain an adequate supply of the thecal androgen required for their increased estrogen synthesis during the preovulatory period. Additionally, activin has an important role in oocyte development of growing antral follicles, accelerating maturation and improving developmental competence of the oocyte. Conversely, inhibin acts as a meiotic inhibitor negatively affecting both oocyte maturation and developmental competence [9,18].

   TGF-α, which is structurally analogous to epidermal growth factor (*EGF*), and *TGF-β* are autocrine growth regulators. *TGF-α* stimulates oocyte nuclear and cytoplasmic maturation. *TGF-β* produced by both theca and granulosa cells stimulates *FSHr* expression, aromatase activity, inhibin and *P* production, and *LHr* induction, and in a similar manner to activin A, it suppresses androgen production in theca cells and is involved in recruitment and development of the follicle [9,41]. Follicle vascularization is influenced by peptides present in the follicular fluid, especially vascular endothelial growth factor (*VEGF*). The induction of *VEGF* thecal cell expression in growing ovarian follicles by gonadotropins during folliculogenesis increases the vascular density in the thecal cell layer. Additionally, *VEGF* increases vascular permeability, not only facilitating extravasation of plasma and accumulation of antral fluid in the growing follicles but also enhancing the delivery of mediators such as lipids that are used as precursors for androgen synthesis in thecal cells [41].

BMP and *GDF-9* expressed within different compartments of the antral follicle have a critical role in folliculogenesis and oocyte development. At this stage of development, the oocyte expresses *BMP-6*, *BMP-15*, and *GDF-9*; granulosa cells bone morphogenetic protein 2 (*BMP-2*), bone morphogenetic protein 5 (*BMP-5*), and *BMP-6*; and thecal cells *BMP-2*, bone morphogenetic protein 3b (*BMP-3b*), *BMP-4*, and *BMP-7*. Down-regulation of *AC* activity by *BMP-6* and the suppression of *FSHr* expression by *BMP-15* attenuate FSH action on granulosa cells. FSH itself down-regulates *BMP-6* expression at the time of follicle selection, a necessary step for continued follicle development through the action of FSH. *GDF-9* suppresses FSH-stimulated P and E2 production and attenuates FSH-induced *LHr* formation [9,31]. Granulosa cell–derived *BMP-2*, *BMP-5*, and *BMP-6* interact in a complex manner with both *IGF-* and FSH-dependent signaling pathways to promote follicle survival through the prevention of premature luteinization or atresia and maintenance of cell proliferation. *BMP-6* increases cell number and basal and *IGF*-stimulated E2, inhibin A, activin A, and follistatin secretion, and as *BMP-2* and *BMP-5*, suppresses FSH-stimulated P production [42]. Similarly, theca cell–derived *BMP-4* and *BMP-7* also act as paracrine regulators of granulosa cell function, increasing cell number and basal and *IGF*-stimulated E2, inhibin A, activin A, and follistatin secretion, while *P* secretion is suppressed [9,42]. Basal and LH-induced androgen secretion by theca cells is suppressed by *BMP-6* of granulosa cell origin and by theca cell–derived *BMP-4* and *BMP-7* [18]. *BMP-2* also suppresses androgen secretion by theca cells. The suppression of androgen secretion by theca cells followed by the decrease of estrogen synthesis impairs the development of the nondominant follicles. *BMP*-binding proteins (follistatin, noggin, chordin, and gremlin) regulate the bioavailability of *BMP* and are responsible for the modulation of the intrafollicular *BMP*/growth differentiation factor (*GDF*) system (Figure 4.4) [9].

4.2.6 Ovulation, Luteinization, and CL Formation

The ovarian process of ovulation is based on a positive hypothalamic–pituitary ovarian interaction that triggers the estrogen-mediated preovulatory LH surge. The surge terminates follicle growth and initiates a sequence of events that results in meiotic maturation of the oocyte, expansion of the cumulus–oocyte complex (*COC*), follicle wall rupture, and the differentiation of the remaining granulosa and theca cells to create a new structure, the *CL* (Figure 4.5). The integration of endocrine, paracrine, and autocrine signaling pathways between the oocyte and the surrounding granulosa and cumulus cells is required for the resumption of meiosis during oocyte maturation. This is a process morphologically characterized by the dissolution of the nuclear envelope referred to as germinal vesicle breakdown (*GVBD*) that follows the preovulatory gonadotropin surge.

The concentration of *cAMP* in the oocyte plays a critical role in the regulation of meiotic resumption. The presence of high levels of *cAMP* within the oocyte resulting from endogenous production by the stimulation of the guanosine triphosphate-binding (g) protein-coupled receptors/G proteins/*AC* pathway, *cAMP* transport from adjacent cumulus cells through gap junctions, or the inhibition of phosphodiesterase 3A (*PDE3A*) by guanosine 3′,5′-cyclic monophosphate (*cGMP*) transported from

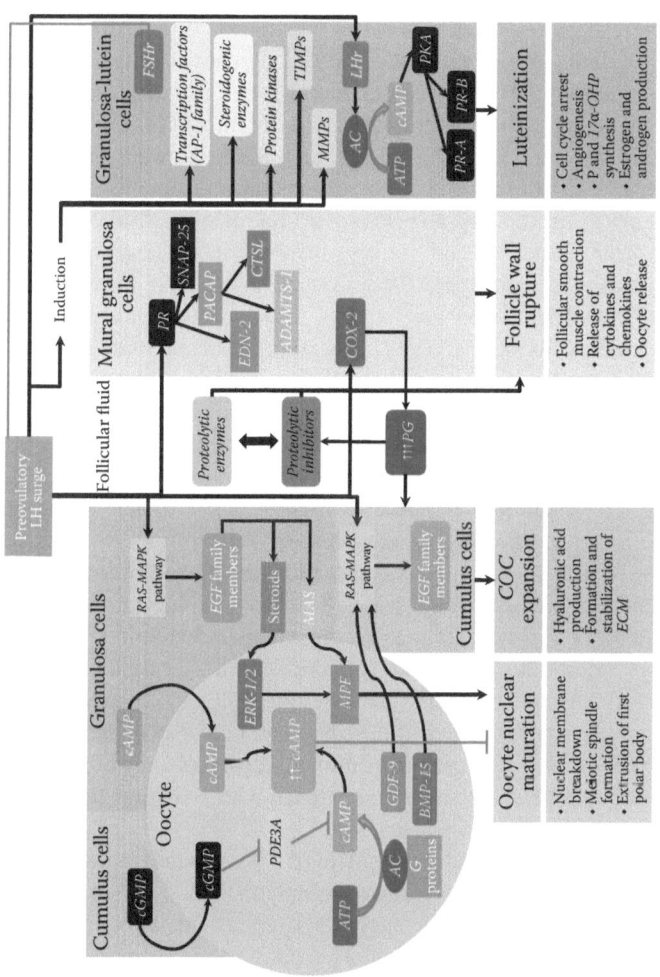

FIGURE 4.5 (See color insert.) LH-mediated pathways involved in ovulation and *CL* formation.

the somatic compartment prevent the oocyte from resuming meiosis [38]. Mitogen-activated protein kinase (*MAPK*) activation in follicular somatic cells is required for gonadotropin-induced meiotic resumption. Gonadotropin-induced *EGF*-like growth factors (epiregulin [*EREG*], amphiregulin [*AREG*], and β-cellulin [*BTC*]), meiosis-activating sterol (*MAS*), and gonadal steroid hormones (probably through *cAMP*-dependent protein kinase A II [*PKAII*] and protein kinase C [*PKC*] pathways) are involved in the activation of *MAPK*. *MAPK* mediates LH-induced oocyte matura-tion not only by inducing the synthesis of meiosis resumption–inducing factors but also by interrupting communication between oocyte and surrounding somatic cells through phosphorylation of *CX-43*. This prevents the meiosis-inhibiting signals from entering the oocyte [43]. LH activation of mural granulosa cells stimulates *cAMP* signaling, which, in turn, induces the expression of the *EGF*-like growth factors that activate *EGF* receptor and *RAS–MAPK* signaling pathways that act as intra-follicular mediators to stimulate cumulus expansion and oocyte maturation [38]. Prostaglandins (*PG*) mimicking LH action up-regulate *EGF*-like growth factor biosynthesis in granulosa cells and are also involved in the resumption of oocyte meiosis. Additionally, LH causes *MAPK* activation in follicular somatic cells, which phosphorylates *CX-43*, decreasing gap junction permeability between the oocyte and surrounding somatic cells before *GVBD*. Interruption of communication ends the supply of *cAMP* from the somatic cells to the oocyte, decreasing the intraoocyte concentration of this cyclic nucleotide. This interruption determines a rapid increase of *cAMP* in cumulus cells, which activates *PDE3A* and decreases the *cAMP* level in the oocyte through the *PI3K/AKT* pathways. The release of signals that trigger meiotic resumption despite the presence of high levels of *cAMP* in the oocyte is induced by the increase in *cAMP* in cumulus cells determined by gonadotropins [43]. In response to decreased *cAMP* concentration, the oocyte resumes meiosis and progresses through a precisely synchronized process of nuclear and cytoplasmic maturation to achieve full developmental competence. Oocyte nuclear maturation involves nuclear membrane breakdown, meiotic spindle formation, and extrusion of half the oocyte's chromosomes into the first polar body to complete metaphase I. After the LH surge, progression of the oocyte through metaphase I involves the accumulation of *MAS* (a meiotic progression mediator) that may contribute to syn-chronization of oocyte developmental competence with follicular development and ovulation [44].

The preovulatory LH surge induces cumulus cells surrounding the oocyte to pro-duce hyaluronic acid (a high-molecular-weight unbranched glycosaminoglycan) that binds to these cells and expands the spaces between them, a process required for normal ovulation and fertilization. Cumulus expansion in response to LH is medi-ated by *EGF*-like factors and the *MAPK* signaling pathway. The rapid increase of *EGF*-like factors in mural granulosa cells of the preovulatory follicle up-regulates the expression of prostaglandin synthase 2 (*PGS-2*), hyaluronan synthase 2 (*HAS-2*), and *TNF*-α-induced protein 6 (*TNFAIP-6*) genes, whose products are essential for the formation and stabilization of the *ECM* of the *COC*. Another important require-ment for cumulus matrix formation or stabilization is the entrance of serum-derived protein inter-α-inhibitor (*IαI*) during ovulation into the follicle after disintegra-tion of the basal lamina [45]. The LH surge leads to the activation of the *MAPK* signaling pathway, another important mediator of cumulus expansion. On the other hand, paracrine factors from the oocyte are also required for this process. *GDF-9* promotes the expansion of cumulus cells through mothers against decapentaplegic homolog (*SMAD*) family members 2/3 (*SMAD-2/3*)–dependent and –independent pathways, up-regulating prostaglandin-endoperoxide synthase 2 (*PTGS-2*), *HAS-2*, *TNFAIP-6*, prostaglandin E receptor 2 (*PGER-2*), and pentraxin 3 (*PTX-3*) genes, and activating *MAPK* pathways, respectively. *BMP-15*, another oocyte-secreted factor, acts synergistically with *GDF-9* through *SMAD-1/5/8* pathways as a cumulus expansion-enabling factor [3,19].

Structural remodeling of the follicle wall is an important step for oocyte release and *CL* formation, two events that are functionally dissociated. After the LH surge, the induction of different transcriptional regulators by *LHr* activation is necessary for ovulation. LH acts on the dominant preovulatory follicle, terminating the program of gene expression associated with folliculogenesis and steroidogenesis and inducing genes involved in ovulation and luteinization such as the progesterone receptor (*PR*), cyclooxygenase-2 (*COX-2*), CCAAT enhancer-binding protein β (*C/EBPβ*), early growth response factor 1 (*EGR-1*), pituitary adenylyl cyclase-activating peptide (*PACAP*), cell cycle inhibitors ($p21^{CIP1}$ and $p27^{KIP1}$), steroidogenic enzymes (*StAR*; *P450* aromatase and cholesterol side-chain cleavage—*P450arom* and *P450scc*, respectively), specific members of the activator protein-1 (*AP-1*) family of transcription factors (FBJ murine osteosarcoma viral oncogene homolog [*c-Fos*], jun proto-oncogene [*c-Jun*], FOS-like antigen 2 [*Fra2*], and jun D proto-oncogene [*JunD*]), and protein kinases, respectively [1,3,45].

In the ovary, LH rapidly and selectively induces *PR* in mural granulosa cells of the preovulatory follicle. Molecular targets of *PR* action such as a disintegrin-like and metallopeptidase with thrombospondin type 1 motif (*ADAMTS-1*), cathepsin L (*CTSL*), endothelin 2 (*EDN-2*), and synaptosomal-associated protein 25 (*SNAP-25*) are involved in the control of follicular rupture [3]. Ovulation requires the action of proteases (*ADAMTS-1*, *CTSL*, and *ADAM* metallopeptidase domain 8 [*ADAM-8*]) directly or indirectly regulated by *PR* to facilitate the release of the oocyte from the follicle. In particular, *ADAMTS-1* has a critical role in mediating the *PR*-regulated ovarian activity that culminates in the rupture of the follicle and in controlling the amount and cellular location of various proteoglycans that regulate the activity of specific growth factors such as *GDF-9*, *FGF-2/-7*, *EFG*, and *TGF-α* [44,46]. *EDN-2*, a potent vasoconstrictor produced by mural granulosa cells in a *PR*-dependent manner, induces follicular smooth muscle contraction that leads to follicular rupture [47]. *SNAP*-25, another *PR*-targeted gene, regulates the release of vesicle-contained factors such as cytokines and chemokines from the *COC* and granulosa cells as part of the inflammatory and immune-like response of ovulation [48].

The LH surge selectively induces the biosynthesis of *COX-2* in granulosa cells, which, in turn, produces a rapid rise in follicular fluid *PG* (predominantly prostaglandin E2 [*PGE-2*]) that has an important role in follicular rupture. *PGE-2* acting through G-protein coupled *PG* receptors stimulates ovulatory events such as cumulus expansion and enhanced expression of proteases associated with the rupture of the follicle [49]. Another transcriptional regulator that mediates the ovulatory response to the LH surge is *C/EBPβ*, a critical downstream mediator of *MAPK1/3* (also known as extracellular signal–regulated kinases 1 and 2 [*ERK-1/2*]). LH rapidly induces the expression of *EGF*-like growth factors *AREG*, *BTC*, and *EREG*, which, in turn, bind their receptors (present on granulosa and cumulus cells), activating the *RAS* signaling cascade. In response to *AREG*, *C/EBPβ* induces the expression of downstream target genes such as *HAS-2*, *PTGS-2*, *StAR*, and *TNFAIP-6*, which are up-regulated in the periovulatory period [1,3]. The onset of the LH surge activates a transient periovulatory increase in *EGR-1*, a transcription factor that serves as a master switch to promote the expression of various genes involved in inflammation, vascular hyperpermeability, coagulation, and other events associated with tissue damage that were implicated as mediators of the ovulatory process [44,50,51]. The expression of *PACAP* is induced in granulosa and cumulus cells of the preovulatory follicle through LH- and *PR*-mediated mechanisms. *PR*-regulated expression of proteases *ADAMTS-1* and *CTSL* mediated by *PACAP* is required to facilitate release of the oocyte from within the follicular structure [46,52].

Folliculogenesis culminates with the release of the *COC* and the terminal differentiation of the remaining granulosa and theca cells into a *CL* through a process termed *luteinization* that occurs within a few hours after the LH surge. The newly formed *CL* synthesizes and secretes P, a steroid hormone essential for the establishment and maintenance of pregnancy. After the LH surge, granulosa cells of the preovulatory follicle exit the cell cycle and are mainly arrested at the G_0/G_1 phase. LH, through *MAPK1/3* pathway activation, determines the up-regulation of cyclin-dependent kinase (*CDK*) inhibitors of cell cycle progression such as *p27^{KIP1}* and *p21^{CIP1}*, which bind to cyclin/CDK complexes to inhibit their activity, thus initiating cell cycle arrest. The up-regulation of these *CDK* inhibitors associated with the progressive loss of positive cell cycle regulators (cyclins and *CDK-2*) interrupts the proliferation of luteinized cells [3,53].

The P and 17α-hydroxyprogesterone (*17α-OHP*) synthesis that occurs after the LH surge marks the beginning of granulosa and theca cell luteinization. This process involves the increased expression of *LHr* and *PR* and the induction of steroidogenic enzymes (*StAR*, *P450arom*, and *P450scc*), transcription factors of the *AP-1* family (*c-Fos*, *c-Jun*, *Fra2*, and *JunD*), and protein kinases in the granulosa cells [54]. During luteinization, the responsiveness of luteal cells to external signals is changed, allowing these cells to respond to a new set of hormones. Briefly, the LH surge down-regulates the expression of *FSHr* not only in granulosa cells during luteinization but also throughout the life span of the *CL*. *LHr* in luteinized cells are first activated, then undergo a transient desensitization, and thereafter have their expression increased becoming abundant in the *CL*. *PR* expression is rapidly increased in response to LH stimulation and maintained throughout the luteal phase of the menstrual cycle [53,54]. *LHr* activation on granulosa cells in response to the LH surge stimulates *AC* to increase intracellular *cAMP*, thereby activating *cAMP*-dependent *PKA* that directly or through *MAPK* pathway activation induces within the *PR* gene mRNA expression, giving rise to two protein isoforms, *PR-A* and *PR-B*. *PR-B* in particular has an important role in preventing granulosa cell apoptosis and in steroid synthesis during luteinization, while *PR-A* is involved in ovulation control [46].

CL formation, a process absolutely dependent on the LH surge, involves the remodeling of the entire follicular structure through intense angiogenesis and infiltration of the antral space by endothelial cells, fibroblasts, and theca and immune cells occurring in conjunction with luteinization of granulosa cells. The *CL* is a complex, heterogeneous ovarian structure with a limited life span composed of several cell types including steroidogenic (granulosa and theca) luteal cells and nonsteroidogenic (endothelial, fibroblast, pericyte, and immune) cells with distinct morphological, endocrine, and biochemical features [44,53,54]. During the luteal phase after acquiring a high concentration of *LHr* granulosa-lutein, cells respond to LH undergoing extensive hypertrophy and synthesizing P and E2. Theca-lutein cells do not undergo hypertrophy and produce androgen precursors that are aromatized by granulosa-lutein cells, indicating that the two-cell model of estrogen biosynthesis used to explain follicular estrogen production is preserved in the *CL*. These cells are also the site of *17α-OHP* production [53,54].

In addition to luteal cell differentiation, extensive changes in the *ECM* (a complex system formed by a network of collagens associated with proteoglycans and glycoproteins that allow cell migration and neovascularization of the newly formed *CL*) occur. The LH surge induces the expression of various matrix metalloproteinases (*MMPs*) and tissue inhibitors of metalloproteinases (*TIMPs*). An adequate ratio between *MMP* and *TIMP* is responsible for the maintenance of an *ECM* microenvironment favorable to the differentiation of follicular-derived cells into luteal cells. In addition to its role as a protein scaffold, *ECM* is also responsible for luteal function modulation through the presence of cell surface receptors in the luteal cells [53].

During the formation of the *CL*, endothelial cells proliferate extensively to establish a dense capillary network in which each luteal cell is in close contact with various capillaries. This allows an efficient supply of oxygen, nutrients, hormones, and cholesterol to the luteal cells as well as the removal of secretory products, mainly steroid hormones. After ovulation, perivascular cells (pericytes), similar to vascular smooth muscle cells and derived from the theca compartment, are recruited to the outer wall of the newly formed blood vessels to afford their stabilization and to guide them toward the former antrum [53,55]. Several factors including *VEGF*, endocrine gland–derived *VEGF* (*EG-VEGF*), acidic and basic *FGF* (*FGF-1* and *FGF-2*, respectively), *IGF*, tumor necrosis factor (*TNF*), and angiopoietins (*ANG*) are involved in the angiogenesis within the *CL*. *VEGF* expression is up-regulated by gonadotropins in luteinized granulosa cells during the early luteal phase when the angiogenic process is more intensive. *VEGF* acting through its tyrosine kinase receptors (*VEGFR-1* and *VEGFR-2*) is the major regulatory molecule of luteal endothelial cell proliferation and plays a critical role in the maintenance of the viability and functionality of luteal blood vessels during pregnancy [56,57]. *EG-VEGF*, a steroidogenic endocrine gland–specific angiogenic regulator, is up-regulated during the mid- to late luteal phase and has similar biological actions to *VEGF* inducing proliferation, migration, and fenestration of endothelial cells. Fenestration has an important role in the high permeability of ovarian endothelial cells, facilitating the large exchange of materials between interstitial fluid and plasma associated with the *CL* [53,55].

Steroidogenic and endothelial cells of the *CL* produce *FGF*, an angiogenic factor that stimulates proliferation and motility of luteal endothelial cells. *FAS* ligand (*FASLG*) is a cytokine mainly produced by immune cells that bind to cell surface *TNF* receptor superfamily member 6 (*FAS*), triggering structural luteolysis through apoptosis. *IGF*, a survival factor, exerts a luteotropic action in the *CL* by inhibiting *FAS*-mediated cell death. Macrophages and endothelial cells that infiltrate the newly formed *CL* are the source of *TNF*, which is a potent stimulator of luteal *PGs* including *PGE2*, *PGF2α*, and *PGI2*. *TNF* and *TNF*-induced *PGE2* acting as autocrine or paracrine regulators of vascular angiogenesis partly promote *CL* formation [53,58]. Also critical for angiogenesis and vessel integrity are *ANG-1* and *ANG-2* growth factors that act through the same tyrosine kinase receptor (*TIE-2*) but have opposite roles in vascular regulation of the *CL* (quiescence and angiogenesis). *ANG-1* stimulates sprouting and maturation of blood vessels and enhances their stability by recruiting pericytes to enclose the capillaries, playing a crucial role in the interaction between endothelial cells and surrounding matrix and in the control of vessel function. *ANG-2* causes destabilization and remodeling of *CL* blood vessels in collaboration with *VEGF* by blocking the function of *ANG-1* [59].

Regulation of steroid production by the *CL* is largely dependent on LH, which acts through the *cAMP/PKA* pathway. Progesterone is the primary steroid hormone produced by the *CL* in which an increased expression of enzymes necessary for conversion of cholesterol to progesterone (*P450scc* and 3β-hydroxysteroid dehydrogenase/Δ^5,Δ^4 isomerase [*3β-HSD*]) and a decreased expression of the enzymes that convert P to estrogen (cytochrome P450 17α-hydroxylase/C17–20 lyase [*CYP-17*]; cytochrome P450 aromatase [*CYP-19*]) occur [60]. Although steroidogenic luteal cells can produce cholesterol by de novo synthesis, plasma low-density lipoprotein (*LDL*) is the major source of cholesterol for steroid production taken up by these cells through *LDL* receptor–mediated endocytosis. Esterified cholesterol stored within luteal cells binds to sterol carrier protein 2 (*SCP-2*) and is moved to the outer mitochondrial membrane. Cholesterol transport from the outer to the inner mitochondrial membrane through the aqueous intermembrane space involves several proteins including *StAR*, which acts as an intermitochondrial shuttle and is up-regulated by LH/*cAMP*, peripheral-type benzodiazepine receptor (*PBR*) that functions as a pore, and hormone-sensitive lipase [53,61]. The conversion of cholesterol to pregnenolone (*P5*) is catalyzed by *P450scc* located on the inner mitochondrial membrane and then later converted to P through an enzymatic reaction catalyzed by *3β-HSD* present in the smooth endoplasmic reticulum [54,62]. Overall expression of these enzymes remains elevated and relatively constant throughout the luteal phase. In addition to P, androgens and estrogens are also produced by the *CL*, a heterogeneous structure made up of small and large steroidogenic luteal cells. Conversion of P to androstenedione, the major androgen produced by the ovary, is mediated by *CYP-17*, an enzyme that is mainly expressed in small/theca luteal cells that synthesize aromatizable androgens. LH-dependent E2 biosynthesis results from selective stimulation of *CYP-17* and occurs in large/granulosa luteal cells. In the *CL*, granulosa luteal cells express high levels of *CYP-19* responsible for the conversion of androstenedione to estrone and also 17β-hydroxysteroid dehydrogenase type 7 (*17β-HSD-7*), responsible for the conversion of estrone to E2 [53,54].

The *CL* has a limited life span and it degenerates within approximately 13 days after ovulation if fertilization and implantation do not occur. The function of the *CL* is maintained by LH during the second phase of the menstrual cycle and luteal regression occurs when locally produced luteolytic agents inhibit LH action. Luteal functional regression, characterized by the reduced production of P associated with a gradual decrease in the expression of the *StAR* gene, precedes the morphological regression of the *CL* [54]. The expression of several factors including prostaglandin F2α (*PGF-2α*), tumor necrosis factor α (*TNF-α*), interleukin 1β (*IL-1β*), endothelin (*EDN*), monocyte chemoattractant protein 1 (*MCP-1*), reactive oxygen species (*ROS*), and estrogens is involved in the luteolytic process. Intraluteal *PGF-2α* production is crucial not only for the initiation and the amplification of the luteolytic cascade, decreasing luteal P production by the suppression of *StAR* expression in steroidogenic luteal cells and increasing luteal *MCP-1* and *EDN* production, but also for the structural demise of the *CL* [63,64]. *PGF-2α* induces apoptosis in luteal capillary endothelial cells resulting in a marked reduction in capillary density, thereby decreasing blood flow to the luteal parenchyma and thus depriving the *CL* of nutrients, substrates for steroidogenesis, and luteotropic support [60].

During luteolysis, macrophages and T lymphocytes infiltrate the *CL* in response to chemotactic factors including *MCP-1*. Macrophages are involved in the phagocytosis of degenerative luteal cells, cytokine-mediated inhibition of steroidogenesis, and stimulation of *CL* secretion of *PGF-2α*. Secretion of pro-inflammatory cytokines (*IL-1β* and *TNF-α*) by luteal macrophages stimulates *PGF-2α* production and inhibits basal P secretion. Activated T lymphocytes infiltrate the *CL* and produce interferon γ (*IFN-γ*) and *TNF-α*, which increase expression of class I and class II major histocompatibility complex (*MHC*) molecules on luteal cells. By this means, a positive feedback loop of antigenic peptide presentation and T cell activation predispose luteal cells to an autoimmune-type *MHC*-mediated response during luteal regression, facilitating the rapid demise of the tissue that occurs during luteolysis [65]. Additionally, *ROS* produced by macrophages including hydrogen peroxide (H_2O_2), superoxide anion $\left(O_2^-\right)$, oxygen (O_2), and nitric oxide (*NO*) contribute to the functional and structural luteolysis [66].

Structural luteolysis is associated with cell death by apoptosis of luteal and vascular cells and an increased expression of connective tissue growth factor (*CTGF*) and *MMPs* (*MMP-2* and *MMP-9*) in luteal fibroblasts. Cell-to-cell interactions regulate apoptosis and *CL* regression is associated with the loss of cell–cell adhesion sites. Two major apoptotic signaling pathways, the death receptor–mediated (extrinsic) and the mitochondrial (intrinsic), are involved in *CL* regression. In the extrinsic pathway, the interaction between *FASLG* and cell death receptors (*FAS* and *TNF*) initiates the activation of caspases, which cleave a variety of intracellular polypeptides including major structural elements of the cytoplasm such as actin, components of the DNA repair machinery, protein kinases, and the inhibitor of caspase-activated deoxyribonuclease (*ICAD*) [53]. *PGF-2α* initiates luteal regression by increasing the bioactivity or bioavailability of cytokines such as *FASLG* and by inducing 20α-hydroxysteroid dehydrogenase (*20α-HSD*) expression in luteal cells, which, in turn, lose their capacity to secrete P, facilitating the expression of *FAS* on their surface and the invasion of the *CL* by immune cells. *FAS* receptors present on luteal cells bind to *FASLG* expressed by immune cells activating the apoptotic pathway through the caspase cascade [53,67]. Stress-inducing stimuli such as hypoxia, oxidants, deviation of cell cycle, and DNA damage among others activate the intrinsic apoptotic pathway, which is regulated by the proteins of the *BCL-2* family. *PGF-2α* determines the increase in the expression of pro-apoptotic *BCL-2* family protein B cell lymphoma/leukemia-2-associated X protein (*BAX*) leading to permeabilization of the mitochondria. This results in the release of cytochrome c, which, in turn, binds to apoptotic protease-activating factor-1 (*APAF-1*)/caspase-9 complex initiating the apoptotic cascade. Finally, *MMP*-mediated *ECM* degradation observed in the structural regression of the *CL* is also attributed to *PGF-2α* [34,53].

4.3   CRITICAL WINDOWS OF SUSCEPTIBILITY

Female reproductive health can be compromised by exposure to contaminants such as metals and chemical substances that are increasingly released into the environment in modern society. Exposure to these contaminants is sometimes unavoidable and occurs through various routes including ingestion of water, food consumption, absorption through the skin, and inhalation. Unlike the male who continues to produce sperm cells throughout life, women are born with a finite number of primordial follicles. The current trend in many populations to delay pregnancy until a later age may increase the impact of environmental toxicants on female reproductive system function as a result of the exposure to these toxicants over a longer period during the reproductive life span in addition to the age-associated decline in oocyte number and quality.

Four major regulatory events are involved in folliculogenesis including recruitment, preantral follicle development, selection, and atresia. The different types of follicles present in the ovary (primordial, growing, and preovulatory) have distinct sensitivities to environmental toxicants that could affect each step of follicular development through mechanisms such as the reduction in the primordial follicle pool, altered initial recruitment of primordial follicles, defective follicular maturation, and increased follicular atresia. The damage caused to the ovary by the toxicant(s) and implications on female reproductive health depend therefore on the type of follicle affected (Figure 4.6). A toxicant that destroys the primordial follicle pool can lead to permanent infertility because of the depletion of this nonrenewable pool. This type of exposure is of concern because it can go unnoticed as normal menstrual cycles may be present until the preovulatory and growing follicle pools are exhausted through ovulation or atresia. On the other hand, a toxicant that damages primary or preovulatory follicles may lead to transient infertility, delayed or immediate, since the pool of primordial follicles is unaffected. Within the oocyte, ooplasmic organelles such as mitochondria and cortical granules, the spindle apparatus, and chromosomes are potential targets for environmental toxicants that can directly impair oocyte viability [68].

In addition to the oocyte itself, granulosa and cumulus cells, the basal lamina, and the theca interna and externa may be targets for contaminants (Figure 4.6). Gap junctions and membrane or intracellular hormone receptors may serve as targets for reproductive toxicants in granulosa and cumulus cells. Granulosa cells have a central role in the exchange of intra- and extraovarian communication and secretion of steroid hormones. Connexins, the core proteins that make up gap junctions, have essential and distinct roles during folliculogenesis to support proper oocyte development. In response to environmental toxicants, altered connexin expression may represent one mechanism responsible for ovarian dysfunction. Endocrine disruptors

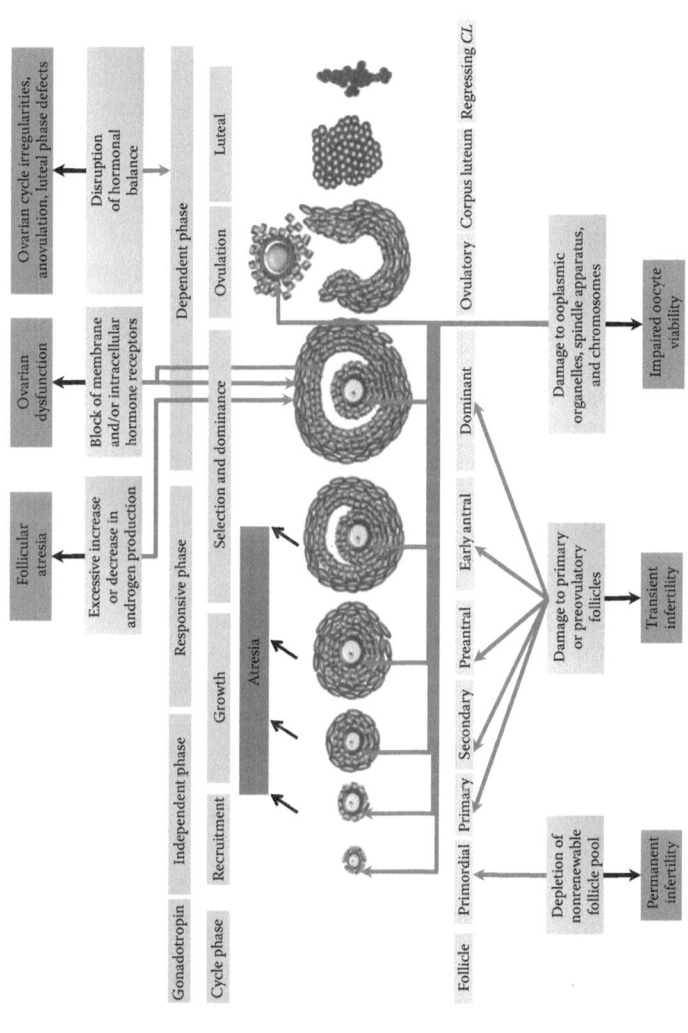

FIGURE 4.6 (See color insert.) Major regulatory events involved in folliculogenesis and critical windows of susceptibility.

(*ED*), including environmental toxicants, chemicals, drugs, and hormones, may lead to altered ovarian function by altering endocrine signaling within the ovary. They may exert their effect directly by binding to transcription factor receptors by altering the expression of hormone receptors and their ability to bind their endogenous ligands, by affecting the expression or activity of enzymes required for synthesis or catabolism of ovarian sex steroids, and by interfering with signal transduction of various ovarian growth regulators such as gonadotropins, steroids, and growth factors. *ED* may negatively affect endocrine function of granulosa cells by directly blocking the mitotic signal or indirectly through the interaction of protein tyrosine kinase/microtubule-associated protein 2 (*MAP-2*) kinase and protein kinase signaling or by impairing spindle microtubule dynamics at the centrosome, which results in metaphase arrest and abnormal chromosome organization [69].

Although the basal lamina acts as a blood–follicle barrier separating the oocyte and granulosa cells from the vascularized thecal cell layer, it may be permeable to xenobiotics, which may accumulate in the follicular fluid influencing oocyte quality by changing the follicular fluid environment. Thecal cells, which provide the necessary precursors for steroid synthesis by granulosa cells, are recruited from ovarian stroma cells during folliculogenesis. Xenobiotics that affect cell proliferation, migration, or communication will have a negative impact on theca cell function. The theca cells' excessive increase or decrease in androgen production induced by environmental toxicants will lead to follicular atresia or impairment of estrogen production by granulosa cells, respectively, negatively affecting female reproductive health [70].

Follicular development and ovulation depend on a delicate hormonal balance between the ovary and the pituitary gland. The final fate of a growing follicle depends on factors whose expression and actions promote follicular cell proliferation, growth, differentiation, and atresia, controlled by gonadotropins, paracrine oocyte-derived (*GDF-9*, *BMP-15*, *BMP-6*), autocrine/paracrine granulosa-derived (activin, *BMP-6*), and theca-derived (*BMP-4*, *BMP-7*) factors. A positive hypothalamic–pituitary ovarian interaction (triggering the estrogen-mediated preovulatory LH surge, which terminates follicle growth and initiates a sequence of events that result in meiotic maturation of the oocyte, expansion of the *COC*, follicle wall rupture, and differentiation of the remaining granulosa and theca cells to form the *CL*) is involved in the process of ovulation. Exposure to environmental toxicants may lead to disruption of the hormonal balance necessary for ovulation through different mechanisms including interference with hormone synthesis, storage, release, transport and clearance, interference with hormone receptor recognition and binding, interference with hormone postreceptor activation, and interference with the central nervous system. The resultant hormonal imbalance causes ovarian cycle irregularities, anovulation, or luteal phase defects, all of them compromising fertility (Figure 4.6) [71].

GLOSSARY

3β-HSD: 3β-hydroxysteroid dehydrogenase/Δ^5,Δ^4 isomerase
17α-OHP: 17α-hydroxyprogesterone
17β-HSD-7: 17β-hydroxysteroid dehydrogenase type 7

20α-HSD: 20α-hydroxysteroid dehydrogenase
AC: adenylyl cyclase
ADAM: a disintegrin and metalloprotease proteins
ADAM-8: *ADAM* metallopeptidase domain 8
ADAMTS-1: disintegrin-like and metallopeptidase with thrombospondin type 1 motif
AKT: protein kinase B
AMH: anti-Mullerian hormone
ANG: angiopoietins
AP-1: activator protein-1
APAF-1: apoptotic protease-activating factor 1
AREG: amphiregulin
BAX: B cell lymphoma/leukemia-2-associated X protein
BCL-2: B cell lymphoma/leukemia-2
bFGF: basic fibroblast growth factor
BMP-2: bone morphogenetic protein 2
BMP-3b: bone morphogenetic protein 3b
BMP-4: bone morphogenetic protein 4
BMP-5: bone morphogenetic protein 5
BMP-6: bone morphogenetic protein 6
BMP-7: bone morphogenetic protein 7
BMP-15: bone morphogenetic protein 15
BTC: β-cellulin
cAMP: adenosine 3′,5′-cyclic monophosphate
CDK: cyclin-dependent kinases
C/EBPβ: CCAAT enhancer-binding protein β
cGMP: guanosine 3′,5′-cyclic monophosphate
c-Fos: FBJ murine osteosarcoma viral oncogene homolog
c-Jun: jun proto-oncogene
c-KIT: cognate tyrosine kinase receptor
CL: corpus luteum
COC: cumulus–oocyte complex
COX-2: cyclooxygenase-2
CTGF: connective tissue growth factor
CTSL: cathepsin L
CX: connexins
CXCL-12: chemokine (C-X-C motif) ligand 12
CYP-17: cytochrome P450 17α-hydroxylase/C17–20 lyase
CYP-19: cytochrome P450 aromatase
E2: estradiol
ECM: extracellular matrix
ED: endocrine disruptors
EDN: endothelin
EDN-2: endothelin 2
EGF: epidermal growth factor

EGR-1: early growth response factor 1
EG-VEGF: endocrine gland–derived *VEGF*
ER: estrogen receptors
EREG: epiregulin
ERK-1/2: extracellular signal–regulated kinases 1 and 2
FAS: Fas (*TNF* receptor superfamily, member 6)
FASLG: *FAS* ligand
FGF: fibroblast growth factor
FIGLA: factor in the germline alpha
FOXL-2: forkhead box L2
FOXO-1: forkhead box O1
FOXO-3: forkhead box O3
Fra2: FOS-like antigen 2
FSH: follicle-stimulating hormone
FSHr: follicle-stimulating hormone receptors
GDF: growth-differentiation factor
GDF-9: growth-differentiation factor 9
GSK3β: glycogen synthase kinase 3β
GVBD: germinal vesicle breakdown
H₂O₂: hydrogen peroxide
HAS-2: hyaluronan synthase 2
HGF: hepatocyte growth factor
HPG: hypothalamic–pituitary–gonadal axis
IαI: inter-α-inhibitor
ICAD: inhibitor of caspase-activated deoxyribonuclease
IFN-γ: interferon γ
IGF: insulin-like growth factor
IGF-1: insulin-like growth factor 1
IL-1β: interleukin 1β
JunD: jun D proto-oncogene
KGF: keratinocyte growth factor
KL: kit ligand
LDL: low-density lipoprotein
LH: luteinizing hormone
LHr: luteinizing hormone receptors
LHX-8: LIM homeobox protein 8
LIF: leukemia inhibitory factor
MAP-2: microtubule-associated protein 2
MAPK: mitogen-activated protein kinase
MAS: meiosis-activating sterol
MCP-1: monocyte chemoattractant protein 1
MHC: major histocompatibility complex
MMPs: matrix metalloproteinases
MPF: maturation promoting factor
NGF: nerve growth factor
NO: nitric oxide

NOBOX: newborn ovary homeobox
O$_2$: oxygen
O$_2^-$: superoxide anion
OCT-4: octamer-binding protein 4
P: progesterone
P5: pregnenolone
P450arom: P450 aromatase
P450scc: P450 cholesterol side-chain cleavage
PACAP: pituitary adenylyl cyclase-activating peptide
PBR: peripheral-type benzodiazepine receptor
PDE3A: phosphodiesterase 3A
PDGF: platelet-derived growth factor
PG: prostaglandins
PGE-2: prostaglandin E2
PGF-2α: prostaglandin F2α
PGER-2: prostaglandin E receptor 2
PGS-2: prostaglandin synthase 2
PI3K: phosphatidylinositol 3-kinase
PKA: protein kinase A
PKAII: protein kinase A II
PKC: protein kinase C
PR: progesterone receptor
PTGS-2: prostaglandin-endoperoxide synthase 2
PTX-3: pentraxin 3
RAS: rat sarcoma
ROS: reactive oxygen species
SCP-2: sterol carrier protein 2
SMAD: mothers against decapentaplegic homolog
SMAD-1: SMAD family member 1
SMAD-2: SMAD family member 2
SMAD-3: SMAD family member 3
SMAD-5: SMAD family member 5
SMAD-8: SMAD family member 8
SNAP-25: synaptosomal-associated protein 25
SOHLH-1: spermatogenesis and oogenesis helix-loop-helix 1
SOHLH-2: spermatogenesis and oogenesis helix-loop-helix 2
StAR: steroidogenic acute regulatory protein
TGF-α: transforming growth factor α
TGF-β: transforming growth factor β
TIE-2: tyrosine kinase receptor
TIMPs: tissue inhibitors of metalloproteinases
TNF: tumor necrosis factor
TNF-α: tumor necrosis factor α
TNFAIP-6: TNF-α-induced protein 6
VEGF: vascular endothelial growth factor
ZP: zona pellucida

REFERENCES

1. J.S. Richards and S.A. Pangas, *The ovary: Basic biology and clinical implications*, J. Clin. Invest. 120 (2010), pp. 963–972.
2. D. Adhikari and K. Liu, *Molecular mechanisms underlying the activation of mammalian primordial follicles*, Endocr. Rev. 30 (2009), pp. 438–464.
3. M.A. Edson, A.K. Nagaraja, and M.M. Matzuk, *The mammalian ovary from genesis to revelation*, Endocr. Rev. 30 (2009), pp. 624–712.
4. E.A. McLaughlin and S.C. McIver, *Awakening the oocyte: Controlling primordial follicle development*, Reproduction 137 (2009), pp. 1–11.
5. D.R. Mattison, D.R. Plowchalk, M.J. Meadows, A.Z. Al-Juburi, J. Gandy, and A. Malek, *Reproductive toxicity: Male and female reproductive systems as targets for chemical injury*, Med. Clin. North Am. 74 (1990), pp. 391–411.
6. R.G. Cortvrindt and J.E. Smitz, *Follicle culture in reproductive toxicology: A tool for in-vitro testing of ovarian function?* Hum. Reprod. Update 8 (2002), pp. 243–254.
7. P.B. Hoyer, *Ovarian toxicity in small pre-antral follicles*, in *Ovarian Toxicology*, P. Hoyer, ed., CRC Press, Boca Raton, FL, 2004, pp. 17–39.
8. O. Oktem and K. Oktay, *The ovary: Anatomy and function throughout human life*, Ann. N. Y. Acad. Sci. 1127 (2008), pp. 1–9.
9. P.G. Knight and C. Glister, *TGF-beta superfamily members and ovarian follicle development*, Reproduction 132 (2006), pp. 191–206.
10. S.K. Bristol-Gould, P.K. Kreeger, C.G. Selkirk, S.M. Kilen, R.W. Cook, J.L. Kipp, L.D. Shea, K.E. Mayo, and T.K. Woodruff, *Postnatal regulation of germ cells by activin: The establishment of the initial follicle pool*, Dev. Biol. 298 (2006), pp. 132–148.
11. C. Tingen, A. Kim, and T.K. Woodruff, *The primordial pool of follicles and nest breakdown in mammalian ovaries*, Mol. Hum. Reprod. 15 (2009), pp. 795–803.
12. J.M. Young and A.S. McNeilly, *Theca: The forgotten cell of the ovarian follicle*, Reproduction 140 (2010), pp. 489–504.
13. K.J. Hutt and D.F. Albertini, *An oocentric view of folliculogenesis and embryogenesis*, Reprod. Biomed. Online 14 (2007), pp. 758–764.
14. S. Joshi, H. Davies, L.P. Sims, S.E. Levy, and J. Dean, *Ovarian gene expression in the absence of FIGLA, an oocyte-specific transcription factor*, BMC Dev. Biol. 7 (2007), p. 67.
15. H.S. Choi, S.H. Lee, H. Kim, and Y. Lee, *Germ cell-specific gene 1 targets testis-specific poly(A) polymerase to the endoplasmic reticulum through protein-protein interactions*, FEBS Lett. 582 (2008), pp. 1203–1209.
16. Y. Choi, D.J. Ballow, Y. Xin, and A. Rajkovic, *Lim homeobox gene, lhx8, is essential for mouse oocyte differentiation and survival*, Biol. Reprod. 79 (2008), pp. 442–449.
17. A.L. Durlinger, J.A. Visser, and A.P. Themmen, *Regulation of ovarian function: The role of anti-Mullerian hormone*, Reproduction 124 (2002), pp. 601–609.
18. O. Oktem and B. Urman, *Understanding follicle growth in vivo*, Hum. Reprod. 25 (2010), pp. 2944–2954.
19. J.J. Eppig, *Oocyte control of ovarian follicular development and function in mammals*, Reproduction 122 (2001), pp. 829–838.
20. K.J. Hutt, E.A. McLaughlin, and M.K. Holland, *Kit ligand and c-Kit have diverse roles during mammalian oogenesis and folliculogenesis*, Mol. Hum. Reprod. 12 (2006), pp. 61–69.
21. M.K. Skinner, *Regulation of primordial follicle assembly and development*, Hum. Reprod. Update 11 (2005), pp. 461–471.

22. E.E. Nilsson and M.K. Skinner, *Bone morphogenetic protein-4 acts as an ovarian follicle survival factor and promotes primordial follicle development*, Biol. Reprod. 69 (2003), pp. 1265–1272.

23. R. Abir, B. Fisch, X.Y. Zhang, C. Felz, G. Kessler-Icekson, H. Krissi, S. Nitke, and A. Ao, *Keratinocyte growth factor and its receptor in human ovaries from fetuses, girls and women*, Mol. Hum. Reprod. 15 (2009), pp. 69–75.

24. K. Reynaud and M.A. Driancourt, *Oocyte attrition*, Mol. Cell. Endocrinol. 163 (2000), pp. 101–108.

25. M. Ito, T. Harada, M. Tanikawa, A. Fujii, G. Shiota, and N. Terakawa, *Hepatocyte growth factor and stem cell factor involvement in paracrine interplays of theca and granulosa cells in the human ovary*, Fertil. Steril. 75 (2001), pp. 973–979.

26. G.M. Kidder and B.C. Vanderhyden, *Bidirectional communication between oocytes and follicle cells: Ensuring oocyte developmental competence*, Can. J. Physiol. Pharmacol. 88 (2010), pp. 399–413.

27. R.B. Gilchrist, M. Lane, and J.G. Thompson, *Oocyte-secreted factors: Regulators of cumulus cell function and oocyte quality*, Hum. Reprod. Update 14 (2008), pp. 159–177.

28. M. Orisaka, K. Tajima, B.K. Tsang, and F. Kotsuji, *Oocyte-granulosa-theca cell interactions during preantral follicular development*, J. Ovarian Res. 2 (2009), p. 9.

29. E.E. Telfer, M. McLaughlin, C. Ding, and K.J. Thong, *A two-step serum-free culture system supports development of human oocytes from primordial follicles in the presence of activin*, Hum. Reprod. 23 (2008), pp. 1151–1158.

30. A.L. Durlinger, M.J. Gruijters, P. Kramer, B. Karels, H.A. Ingraham, M.W. Nachtigal, J.T. Uilenbroek, J.A. Grootegoed, and A.P. Themmen, *Anti-Mullerian hormone inhibits initiation of primordial follicle growth in the mouse ovary*, Endocrinology 143 (2002), pp. 1076–1084.

31. J.L. Juengel and K.P. McNatty, *The role of proteins of the transforming growth factor-beta superfamily in the intraovarian regulation of follicular development*, Hum. Reprod. Update 11 (2005), pp. 143–160.

32. S.G. Hillier, *Current concepts of the roles of follicle stimulating hormone and luteinizing hormone in folliculogenesis*, Hum. Reprod. 9 (1994), pp. 188–191.

33. E.A. McGee and A.J. Hsueh, *Initial and cyclic recruitment of ovarian follicles*, Endocr. Rev. 21 (2000), pp. 200–214.

34. M.R. Hussein, *Apoptosis in the ovary: Molecular mechanisms*, Hum. Reprod. Update 11 (2005), pp. 162–177.

35. R.J. Rodgers and H.F. Irving-Rodgers, *Formation of the ovarian follicular antrum and follicular fluid*, Biol. Reprod. 82 (2010), pp. 1021–1029.

36. S.G. Hillier, *Paracrine support of ovarian stimulation*, Mol. Hum. Reprod. 15 (2009), pp. 843–850.

37. S. Vaccari, K. Horner, L.M. Mehlmann, and M. Conti, *Generation of mouse oocytes defective in cAMP synthesis and degradation: Endogenous cyclic AMP is essential for meiotic arrest*, Dev. Biol. 316 (2008), pp. 124–134.

38. Q.Y. Sun, Y.L. Miao, and H. Schatten, *Towards a new understanding on the regulation of mammalian oocyte meiosis resumption*, Cell Cycle 8 (2009), pp. 2741–2747.

39. R.A. Dragovic, L.J. Ritter, S.J. Schulz, F. Amato, D.T. Armstrong, and R.B. Gilchrist, *Role of oocyte-secreted growth differentiation factor 9 in the regulation of mouse cumulus expansion*, Endocrinology 146 (2005), pp. 2798–2806.

40. S.M. Quirk, R.G. Cowan, R.M. Harman, C.L. Hu, and D.A. Porter, *Ovarian follicular growth and atresia: The relationship between cell proliferation and survival*, J. Anim. Sci. 82 (2004), pp. E40–E52.

41. I. Ben-Ami, S. Freimann, L. Armon, A. Dantes, R. Ron-El, and A. Amsterdam, *Novel function of ovarian growth factors: Combined studies by DNA microarray, biochemical and physiological approaches*, Mol. Hum. Reprod. 12 (2006), pp. 413–419.

42. C. Glister, C.F. Kemp, and P.G. Knight, *Bone morphogenetic protein (BMP) ligands and receptors in bovine ovarian follicle cells: Actions of BMP-4, -6 and -7 on granulosa cells and differential modulation of Smad-1 phosphorylation by follistatin*, Reproduction 127 (2004), pp. 239–254.

43. M. Zhang, H. Ouyang, and G. Xia, *The signal pathway of gonadotrophins-induced mammalian oocyte meiotic resumption*, Mol. Hum. Reprod. 15 (2009), pp. 399–409.

44. D.L. Russell and R.L. Robker, *Molecular mechanisms of ovulation: Co-ordination through the cumulus complex*, Hum. Reprod. Update 13 (2007), pp. 289–312.

45. J.S. Richards, D.L. Russell, S. Ochsner, and L.L. Espey, *Ovulation: New dimensions and new regulators of the inflammatory-like response*, Annu. Rev. Physiol. 64 (2002), pp. 69–92.

46. R.L. Robker, L.K. Akison, and D.L. Russell, *Control of oocyte release by progesterone receptor-regulated gene expression*, Nucl. Recept. Signal. 7 (2009), p. e012.

47. C. Ko, M.C. Gieske, L. Al-Alem, Y. Hahn, W. Su, M.C. Gong, M. Iglarz, and Y. Koo, *Endothelin-2 in ovarian follicle rupture*, Endocrinology 147 (2006), pp. 1770–1779.

48. M. Shimada, Y. Yanai, T. Okazaki, Y. Yamashita, V. Sriraman, M.C. Wilson, and J.S. Richards, *Synaptosomal-associated protein 25 gene expression is hormonally regulated during ovulation and is involved in cytokine/chemokine exocytosis from granulosa cells*, Mol. Endocrinol. 21 (2007), pp. 2487–2502.

49. D.M. Duffy, L.K. McGinnis, C.A. Vandevoort, and L.K. Christenson, *Mammalian oocytes are targets for prostaglandin E2 (PGE2) action*, Reprod. Biol. Endocrinol. 8 (2010), p. 131.

50. L.L. Espey and J.S. Richards, *Temporal and spatial patterns of ovarian gene transcription following an ovulatory dose of gonadotropin in the rat*, Biol. Reprod. 67 (2002), pp. 1662–1670.

51. J. Kim, I.C. Bagchi, and M.K. Bagchi, *Control of ovulation in mice by progesterone receptor-regulated gene networks*, Mol. Hum. Reprod. 15 (2009), pp. 821–828.

52. J.I. Park, W.J. Kim, L. Wang, H.J. Park, J. Lee, J.H. Park, H.B. Kwon, A. Tsafriri, and S.Y. Chun, *Involvement of progesterone in gonadotrophin-induced pituitary adenylate cyclase-activating polypeptide gene expression in pre-ovulatory follicles of rat ovary*, Mol. Hum. Reprod. 6 (2000), pp. 238–245.

53. C. Stocco, C. Telleria, and G. Gibori, *The molecular control of corpus luteum formation, function, and regression*, Endocr. Rev. 28 (2007), pp. 117–149.

54. L. Devoto, A. Fuentes, P. Kohen, P. Cespedes, A. Palomino, R. Pommer, A. Munoz, and J.F. Strauss, 3rd, *The human corpus luteum: Life cycle and function in natural cycles*, Fertil. Steril. 92 (2009), pp. 1067–1079.

55. H.M. Fraser and C. Wulff, *Angiogenesis in the corpus luteum*, Reprod. Biol. Endocrinol. 1 (2003), p. 88.

56. M.M. Kaczmarek, D. Schams, and A.J. Ziecik, *Role of vascular endothelial growth factor in ovarian physiology—An overview*, Reprod. Biol. 5 (2005), pp. 111–136.

57. S.A. Pauli, H. Tang, J. Wang, P. Bohlen, R. Posser, T. Hartman, M.V. Sauer, J. Kitajewski, and R.C. Zimmermann, *The vascular endothelial growth factor (VEGF)/VEGF receptor 2 pathway is critical for blood vessel survival in corpora lutea of pregnancy in the rodent*, Endocrinology 146 (2005), pp. 1301–1311.

58. K. Okuda and R. Sakumoto, *Multiple roles of TNF super family members in corpus luteum function*, Reprod. Biol. Endocrinol. 1 (2003), p. 95.

59. W.C. Duncan, *Paracrine regulation of luteal development and luteolysis in the primate*, Anim. Reprod. 6 (2009), pp. 34–46.

60. G.D. Niswender, J.L. Juengel, P.J. Silva, M.K. Rollyson, and E.W. McIntush, *Mechanisms controlling the function and life span of the corpus luteum*, Physiol. Rev. 80 (2000), pp. 1–29.
61. J.F. Strauss, 3rd, T. Kishida, L.K. Christenson, T. Fujimoto, and H. Hiroi, *START domain proteins and the intracellular trafficking of cholesterol in steroidogenic cells*, Mol. Cell. Endocrinol. 202 (2003), pp. 59–65.
62. L.K. Christenson and L. Devoto, *Cholesterol transport and steroidogenesis by the corpus luteum*, Reprod. Biol. Endocrinol. 1 (2003), p. 90.
63. L.A. Penny, *Monocyte chemoattractant protein 1 in luteolysis*, Rev. Reprod. 5 (2000), pp. 63–66.
64. M.C. Wiltbank and J.S. Ottobre, *Regulation of intraluteal production of prostaglandins*, Reprod. Biol. Endocrinol. 1 (2003), p. 91.
65. M.J. Cannon and J.L. Pate, *The role of major histocompatibility complex molecules in luteal function*, Reprod. Biol. Endocrinol. 1 (2003), p. 93.
66. S. Lee, T.J. Acosta, Y. Nakagawa, and K. Okuda, *Role of nitric oxide in the regulation of superoxide dismutase and prostaglandin F(2alpha) production in bovine luteal endothelial cells*, J. Reprod. Dev. 56 (2010), pp. 454–459.
67. S.F. Carambula, J.K. Pru, M.P. Lynch, T. Matikainen, P.B. Goncalves, R.A. Flavell, J.L. Tilly, and B.R. Rueda, *Prostaglandin F2alpha- and FAS-activating antibody-induced regression of the corpus luteum involves caspase-8 and is defective in caspase-3 deficient mice*, Reprod. Biol. Endocrinol. 1 (2003), p. 15.
68. C. Borgeest, K.P. Miller, D. Tomic, and J.A. Flaws, *Ovarian toxicity caused by pesticides*, in *Ovarian Toxicology*, P. Hoyer, ed., CRC Press, Boca Raton, FL, 2004, pp. 40–60.
69. P. Nicolopoulou-Stamati and M.A. Pitsos, *The impact of endocrine disrupters on the female reproductive system*, Hum. Reprod. Update 7 (2001), pp. 323–330.
70. Z.R. Craig, W. Wang, and J.A. Flaws, *Endocrine disrupting chemicals in ovarian function: Effects on steroidogenesis, metabolism and nuclear receptor signaling*, Reproduction 142 (2011), pp. 633–46.
71. R.J. Hutz, M.J. Carvan, M.G. Baldridge, L.K. Conley, and T.K. Heiden, *Environmental toxicants and effects on female reproductive function*, Tren. Reprod. Bio. 2 (2006), pp. 1–11.

5 Maternal Recognition of Pregnancy

Paulo Marcelo Perin and Mariangela Maluf

CONTENTS

ABSTRACT

The success of implantation process depends on a synchronous and reciprocal complex molecular interaction between the embryo and the hormonally primed maternal endometrium, which occurs approximately 6 or 7 days after fertilization during a restricted window of receptivity lasting 4 or 5 days. This process requires carefully orchestrated interactions between cells and their extracellular matrix environment mediated by numerous factors and the challenge of the maternal immune system by paternal antigens. Various immune cells and molecules present in the endometrium during the window of implantation not only are involved in the control of trophoblast invasion and the maternal response to fetal allograft but also modulate the vascular remodeling during placental development. The maternal immune response is shifted toward humoral immunity and away from cell-mediated immunity that could be harmful to the fetus, and the progesterone produced by the corpus luteum stimulates an anti-inflammatory response, which decreases the secretion of proinflammatory cytokines and suppresses allogeneic response, allowing fetal survival. Human embryo implantation is a three-stage process including apposition, adhesion, and invasion. During apposition, the trophoblast becomes closely apposed to the luminal endometrial epithelium and the pinopodes, and micro protrusions involved in endocytosis and pinocytosis found on the apical surface of endometrial epithelium during the window of receptivity interdigitate with microvilli on the apical syncytiotrophoblast surface of the blastocyst facilitating its adhesion to the luminal epithelium. The adhesion stage is characterized by the attachment of apical plasma membranes of the trophectoderm (*TE*) and endometrial epithelial cells over which a mucin containing glycocalyx is present and represents a barrier to trophoblast invasiveness. During this stage, the blastocyst triggers the local loss of the repelling mucin 1 favoring the communication between *TE* and endometrial epithelial cells, and several chemokines secreted locally either by the embryo or by the endometrium during the implantation window act as signals for receptor polarization and activation of endometrial adhesion molecules. Following adhesion, the blastocyst invades through the luminal epithelium into the stroma up the uterine vessels using different matrix metalloproteinases involved in the proteolysis of the extracellular matrix. Under maternal control, endometrial decidualization is fundamental for placental formation, including the regulation of trophoblast invasion, to modulate local vascular/immune responses and to resist environmental and oxidative stress, all of which contribute to the viability of the pregnancy. The implantation period of development represents a critical time during which the interactions between the embryo and the maternal endometrium are highly susceptible to exogenous insults that can affect future growth and developmental potential, either prenatally or postnatally. Environmental contaminants such as endocrine disruptors may cause an imbalance of ovarian steroid hormone production, which negatively affects uterine receptivity, and exposure to toxicants may result in the disruption of maternal immune response, which compromises the placentation process and pregnancy development. This chapter reviews current knowledge about key molecular, cellular, immunological, and endocrine events involved in maternal recognition of pregnancy. The effects of the exposure to reproductive toxicants on the window of implantation and on the embryo–endometrial communication are briefly discussed.

KEYWORDS

Blastocyst, implantation, maternal immune system, decidualization, embryo–endometrial communication, reproductive toxicants

5.1 INTRODUCTION

Successful implantation involves synchronous and reciprocal complex molecular interaction between the embryo and the hormonally primed maternal endometrium, which occurs approximately 6 or 7 days after fertilization in the midsecretory phase of the menstrual cycle during a restricted window of receptivity (implantation window) that lasts no longer than 4 or 5 days.

This process requires cell-to-cell contact and cell interactions with the extracellular matrix (*ECM*) mediated by hormonal signaling, growth factors and cytokines, cell adhesion molecules, inflammatory factors, *ECM* proteins, extracellular degrading matrix proteins, and transcription factors. In addition, the maternal immune system is challenged by paternal antigens through exposure to trophoblast tissue, which must be tolerated despite being semiallogeneic. Multiple mechanisms originating from both the embryo and the mother contribute to the development and maintenance of tolerance in order to prevent inflammation and ultimately fetal loss. Various immune cells and molecules present in the endometrium during the window of implantation not only are involved in the control of trophoblast invasion and the maternal response to fetal allograft but also modulate the vascular remodeling during placental development. After recognition of paternally derived fetal antigens, maternal immune response is shifted toward humoral immunity and away from cell-mediated immunity that could be harmful to the fetus. Additionally, progesterone (P) stimulates an anti-inflammatory response that decreases the secretion of pro-inflammatory cytokines and suppresses allogeneic response, allowing fetal survival [1–4].

Implantation encompasses three distinct stages: apposition, adhesion, and invasion. After the emergence from the zona pellucida (*ZP*), the blastocyst assumes a particular orientation as it approaches the endometrium and is guided to the site of implantation, usually in the upper posterior (fundal) wall of the uterine cavity, by a chemokine gradient [5]. During apposition, the trophoblast becomes closely apposed to the luminal epithelium, and the initial adhesion of the blastocyst to the endometrial surface is still unstable. At this stage, the pinopodes, microprotrusions involved in endocytosis and pinocytosis found on the apical surface of the endometrial epithelium during the window of receptivity, interdigitate with microvilli on the apical syncytiotrophoblast surface of the blastocyst facilitating its adhesion to the luminal epithelium through mechanisms that involve uptake of macromolecules and withdrawal of uterine fluid [6].

The adhesion stage is characterized by the attachment of apical plasma membranes of the trophectoderm (*TE*) and endometrial epithelial cells over which a mucin containing glycocalyx is present and represents a barrier to trophoblast invasiveness. Blastocyst adhesion to the endometrium occurs in a specific polarized way through the trophoblast adjacent to the embryonic pole. During this stage, the blastocyst triggers the local loss of the repelling mucin 1 (*MUC-1*) favoring the communication between *TE* and endometrial epithelial cells, which is based on specific ligand–receptor interactions through adhesion molecules of the integrin family. Chemokines such as interleukin (*IL*) 8, monocyte chemoattractant protein 1 (*MCP-1*), and chemokine C–C motif ligand 5 (*CCL-5*) secreted locally either by the embryo or by the endometrium during the implantation window act as signals for receptor polarization and activation of endometrial adhesion molecules [5,7,8].

After adhesion, the blastocyst invades through the luminal epithelium into the stroma up the uterine vessels using different matrix metalloproteinases (*MMPs—MMP-2, MMP-9*) involved in the proteolysis of the *ECM*. Although this activity is mainly controlled by trophoblast cells, the decidua also limits the extent of invasion. In humans, the decidual transformation of the endometrium is primarily under maternal control and is initiated in the mid- to late-secretory phase of the menstrual cycle. Endometrial decidualization is fundamental for placental formation (including the regulation of trophoblast invasion), to modulate local vascular/immune responses and to resist environmental and oxidative stress, all of which contribute to the viability of the pregnancy [9].

The implantation period of development represents a critical time during which the interaction between the embryo and the maternal endometrium is highly susceptible to exogenous insults that can affect future growth and developmental potential, either pre- or postnatally [10]. During the periconceptional period, specific estradiol (E2), P, and human chorionic gonadotropin (hCG) concentrations are necessary not only for endometrial maintenance but also for maintenance of hormone production by the corpus luteum (*CL*). Environmental contaminants such as endocrine disruptors (*ED*) may cause an imbalance of ovarian steroid hormone production, which negatively affects uterine receptivity [11]. Both pro- and anti-inflammatory pathways are necessary for the establishment of the window of implantation and for embryo–endometrial communication. Exposure to toxicants may result in the disruption of T-helper (Th) 1/Th2 cytokine balance in immune cells that compromises the placentation process and pregnancy development [12].

This chapter reviews current knowledge about key molecular, cellular, immunological, and endocrine events involved in maternal recognition of pregnancy. The effects of the exposure to reproductive toxicants on the window of implantation and on the embryo–endometrial communication are briefly discussed.

5.2 EMBRYO–MATERNAL SIGNALING IN IMPLANTATION

5.2.1 HORMONAL SIGNALING

A glycoprotein hormone (hCG) and a steroid hormone (P), secreted by syncytiotrophoblast cells, play important roles in the implantation process. hCG is involved in this process through indirect endocrine and direct paracrine effects (Figure 5.1). hCG is

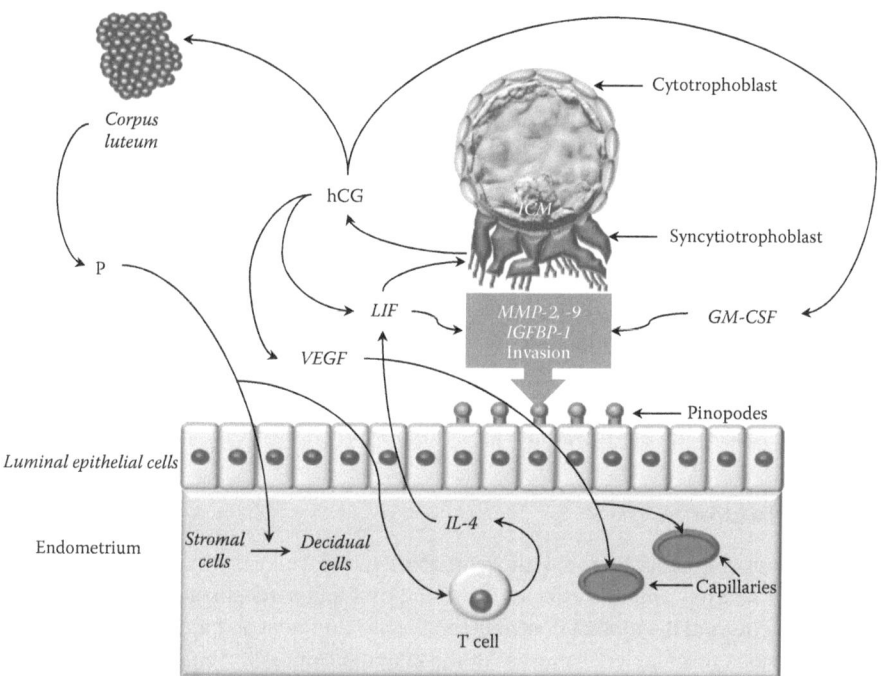

FIGURE 5.1   **(See color insert.)** Schematic representation of the effect of human chorionic gonadotropin on endometrial differentiation, angiogenesis, and tissue remodeling during the implantation process.

continuously produced by trophoblasts and rescues the *CL* from regression via hCG/luteinizing hormone (LH) receptors on ovarian luteal cells that support P production in early pregnancy for approximately 3 to 4 weeks. After this period, syncytiotrophoblast cells in the placenta take over P production from ovarian luteal cells. During implantation, uterine stromal cells differentiate into decidual cells in response to P in a process characterized by morphological changes and the secretion of prolactin (*PRL*) and the endometrium becomes entirely dependent on continuous P stimulation. P is involved in the support of the pregnancy through the stimulation and maintenance of uterine functions necessary for early embryo development, implantation, placentation, and fetal development. Endometrial function is regulated by P through leukemia inhibitory factor (*LIF*) production mediated via *IL-4* originated from T cells. Trophoblast invasiveness is controlled by P through down-regulation of *MMP-9* [13,14].

hCG modulates the production of several endometrial paracrine proteins involved in implantation (*LIF* and granulocyte-macrophage colony-stimulating factor [*GM-CSF*]), angiogenesis (vascular endothelial growth factor [*VEGF*]), endometrial differentiation (insulin-like growth factor binding protein 1 [*IGFBP-1*]), and tissue remodeling (*MMP-2,MMP-9*). *LIF*, a pleiotropic cytokine, is important for both decidualization and implantation. At the time of implantation, it is produced by stromal cells surrounding the blastocyst and is involved in embryonic attachment and intrusion through the epithelium. The development of the placenta is supported by *GM-CSF*, which promotes DNA proliferation, differentiation, and secretory activity of cytotro-phoblast cells. *VEGF* is involved in embryo–endometrium interactions by regulating endometrial vascular permeability and endothelial cell proliferation at the implantation site and by functioning as an endometrial signal for blastocyst development and implantation. The decidualized endometrium expresses *IGFBP-1*, involved in the modulation of the mitogenic and metabolic effects of insulin-like growth factor (*IGF*)-1 and *IGF-2*, which have an important role in growth, apoptosis, metabolism, and development. Cell-to-cell communication between trophoblasts and the decidua involves *IGFBP-1* and *IGF-2* interaction, which regulates invasion. Successful implantation and placentation result from a delicate balance between secretion of *MMPs* from the trophoblast and their inhibition by tissue inhibitors of metalloproteinases (*TIMPs*). *MMP-2* and *-9* are key enzymes in the implantation process enabling the invasion of the trophoblast cells through the decidua and into the maternal vasculature and regulating the bioactivity of growth factors, cytokines, and angiogenic factors [13–16].

5.2.2 Immune System

Throughout pregnancy, the semiallogenic fetus and placenta are in direct contact with decidual maternal immune effector cells, and multiple mechanisms originating from both the fetus and the mother contribute to the development and maintenance of tolerance. Various immune cells present in the endometrium contribute to the successful establishment of pregnancy. The decidua, an active immunological tissue, contains a large population of maternal immune cells involved in the implantation process including uterine natural killer cells (*uNK*), macrophages, T-lymphocyte (T) cells, and dendritic cells (Figure 5.2). *uNK* cells (the most abundant immune cells in the endometrium during the late secretory phase of the menstrual cycle and the window of implantation) not only are involved in the control of the trophoblast invasion and the maternal response to fetal allograft but also modulate the vascular remodeling during placental development. These cells are cluster of differentiation (CD) 56^{bright}/CD16$^-$, coexpressing both inhibitory and activation receptors, differing in function from peripheral natural killer (*NK*) cells, which are CD56^{dim}/CD16$^+$ and express either inhibitory or activation receptors. The absence of CD16 expression in *uNK* cells, which is involved in triggering the lysis of target cells, changes cell function from cytotoxicity to cytokine production. *uNK*-derived interferon γ (*IFN*-γ) facilitates pregnancy-induced artery remodeling and contributes to a normal pregnancy. Tolerance is established through the interaction between *uNK* cells and the trophoblast via human leukocyte antigen G (*HLA-G*). Dendritic cells play a central role for successful implantation, favoring *uNK* cell maturation, tissue remodeling, and angiogenesis. During implantation, macrophages are found in the maternal decidua and in tissues close to trophoblast invasion assisting in tissue remodeling and secreting cytokines and growth factors that control local cellular and tissue interactions. Clearance of apoptotic uterine epithelial cells surrounding the blastocyst by macrophages induces the expression of anti-inflammatory and immunosuppressive cytokines with protective effects on trophoblast survival and immunological tolerance [17–19].

T cells, although less frequent than *uNK* cells in the decidua, play a major role in immune regulation at the fetal–maternal interface. Th cells can be classified into

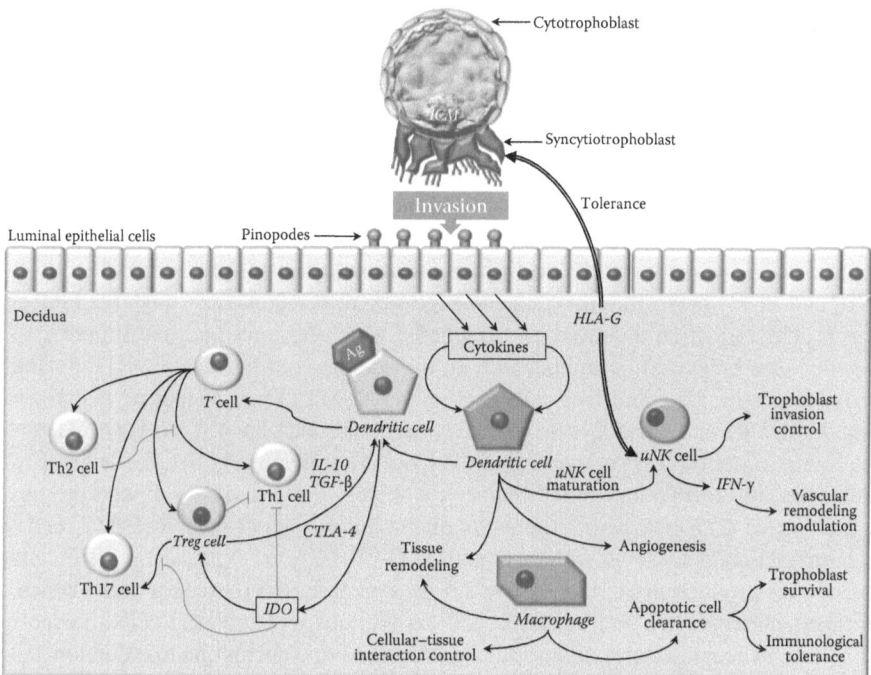

FIGURE 5.2 **(See color insert.)** Effects of decidual maternal immune effector cells present at the feto–maternal interface on the implantation process. During successful pregnancy, the delicate balance between Th1 and Th2 immunity is slightly shifted toward Th2-type immunity at the feto–maternal interface. Cytokines secreted by uterine epithelial cells (*CSF*, *GM-CSF*, *IL-4*, *IL-10*, *PGE$_2$*, and *TGF-β*) modulate the differentiation and functions of dendritic cells.

Th1, Th2, and Th17 cells. Th1 cells, involved in cellular immunity, produce *IL-2*, *IFN-γ*, and tumor necrosis factor-alpha (*TNF-α*), which promote pro-inflammatory immune responses. Th2 cells, involved in humoral immunity, produce *IL-4*, *IL-5*, *IL-6*, *IL-10*, and *IL-13*, which mainly promote anti-inflammatory antibody-dependent immune responses. Th17 cells produce *IL-17*, a pro-inflammatory cytokine, which plays a critical role in the induction of many mediators of inflammation. CD4$^+$CD25$^+$ regulatory T (Treg) cells, defined by transcription factor forkhead box P3 (*FOXP-3*) expression, are important for immunoregulation of the effector T cells (Th1, Th2, and Th17) and for induction and maintenance of peripheral tolerance. Treg cells inhibit proliferation and cytokine production in both CD4$^+$ and CD8$^+$ T cells, inhibit cytotoxic function of *NK* cells, suppress B-lymphocyte (*B*) cell proliferation and immunoglobulin production, and inhibit maturation and function of antigen-presenting cells (*APC*) such as dendritic cells and macrophages [20–22].

Successful implantation and pregnancy development require a delicate balance between Th1 and Th2 immunity that slightly shifted to Th2-type immunity at the feto–maternal interface. After encountering *APC* such as dendritic or B cells, CD4$^+$ T-lymphocytes may differentiate into Th or Treg cells. The expression of nonmaternal inherited proteins on the trophoblast at the feto–maternal interface influences the differentiation of naïve CD4+ T cells into Th1, Th2, Th17, or Treg cells. Th1 activity is decreased in the first trimester of the pregnancy at the feto–maternal interface since the production of *IL-2* and *IFN-γ* by these cells promotes cytotoxicity, which could lead to pregnancy loss. However, *IFN-γ* also plays an important role, ensuring proper placental implantation. Therefore, Th1 activity should be well balanced in order to avoid overstimulation of Th1-type immunity. On the other hand, Th2 activity in the placenta has been shown to dominate during successful pregnancy. In the presence of *IL-4* and *IL-10*, Th2 cells promote antibody responses and produce *IL-4*, *IL-5*, and *IL-10*, protecting the allogeneic fetus against rejection through the inhibition of Th1 induction and activity. Treg cells, which have an immune suppressive function, are increased in number in both the decidua and peripheral blood during early pregnancy. Decidual Treg cells express high levels of human leukocyte antigen DR (*HLA-DR*) and surface cytotoxic T-lymphocyte antigen 4 (*CTLA-4*), indicating higher immune suppressive activity [18,19,22]. Activation of Treg cells leads to the production of two anti-inflammatory cytokines, *IL-10* and *TGF-β*, and the upregulation of *CTLA-4*, which transmits an inhibitory signal to T and *APC* cells. The interaction of *CTLA-4* with its ligands (*CD-80, CD-86*) expressed in *APC* cells induces the production of indoleamine 2,3-dioxygenase (*IDO*) by these cells, which, in target macrophages and dendritic and extravillous trophoblastic cells, prevents immune cell activation through tryptophan deprivation contributing to induction and maintenance of feto–maternal tolerance. Additionally, the presence of *IDO* inhibits the conversion of Treg into Th17 cells that can occur because of their common lineage. The presence of Treg and pro-inflammatory *IL-17*-producing Th17 cells at the feto–maternal interface is inversely related to each other. In normal pregnancy, transforming growth factor β (*TGF-β*) drives the differentiation of Tregs in the absence of *IL-6* and is associated with a decrease in Th17 cells to prevent rejection. In the absence of *TGF-β*, Treg cells act as inducers of Th17 cells and themselves convert to Th17 cells. The mutual antagonism and plasticity between Treg and Th17 cells demonstrates the strict balance between a suppressive or pro-inflammatory immune outcome that is required for successful pregnancy [23,24].

5.2.3   GROWTH FACTORS AND CYTOKINES

5.2.3.1   Leukemia Inhibitory Factor

LIF is a pleiotropic cytokine expressed in the endometrium and decidua that has a variety of autocrine and paracrine functions including stimulation of cell proliferation, differentiation, and survival, all essential for blastocyst development and implantation (Table 5.1). Its actions are mediated by a heterodimeric receptor that consists of two subunits, leukemia inhibitory factor receptor (*LIFR*) and glycoprotein 130 (*gp130*), and activates several signaling pathways including Janus kinase/signal transducer and activator of transcription (*JAK/STAT*), mitogen-activated protein kinase (*MAPK*), and phosphatidylinositol 3-kinase (*PI3K*) pathways. Maximal expression of *LIF* is observed during the implantation window and is regulated by P and local factors such as heparin-binding epidermal growth factor (*EGF*)-like growth factor (*HB-EGF*) and *TGF-β1*. Human decidual leukocytes also express *LIF*,

TABLE 5.1

Overview of the Expression, Production, Action Mode, and Regulation of Growth Factors and Cytokines during the Implantation Process

	Expression	Production	Action	Regulation
LIF	Implantation window	Endometrium, decidua	• Mediates maternal decidual leukocyte and invading cytotrophoblast interactions • Regulates trophoblast differentiation along the invasive pathway • Regulates immune tolerance through *HLA-G*	hCG (blastocyst), P, local factors (*HB-EGF*, *TGF-β1*)
IL-1	Implantation window	Trophoblast, decidua	• Upregulates αvβ3 integrin, *LIF*, and *PGE-2* in endometrial epithelial cells by the blastocyst • Stimulates *MMP-9* expression in trophoblast and endometrial stromal cells	P, *cAMP*
IL-6	Implantation window	Trophoblast, endometrium	• Mediates *IL-1β* actions on the endometrium and trophoblast	hCG, *TGF-β*
IL-11	Decidualization	Trophoblast, epithelial and stromal cells	• Required for decidual-specific maturation of *NK* cells • Regulates endometrial invasion • Promotes *uNK* cell survival and expansion in the human uterus	Steroid hormones, local factors (relaxin, *PGE-2*)
IL-15	Decidualization	Decidua	• Regulates decidualization process	P, *cAMP*, *IFN-γ*, *TGF-β*, *TNF-α*
IGFBP	Decidualization	Decidua	• Regulates trophoblast-derived *IGF* autocrine and paracrine actions at target cells • Increases gelatinolytic activity of trophoblasts and inhibits their migration into decidualized stroma • Limits the effects of *IGF-2* on trophoblast invasion • Alters cellular motility	*IGF-2*

(Continued)

TABLE 5.1 (CONTINUED)
Overview of the Expression, Production, Action Mode, and Regulation of Growth Factors and Cytokines during the Implantation Process

	Expression	Production	Action	Regulation
EGF	Implantation window	Trophoblast, decidua	• Regulates trophoblast invasion, differentiation, and proliferation • Stimulates hCG and *HPL* secretion from trophoblastic cells	Sex steroid hormones
TGF-β	Implantation window	Epithelial and stromal cells	• Increases *ECM* fibronectin • Stimulates trophoblast adhesion to the *ECM* • Inhibits cytotrophoblast proliferation • Promotes cytotrophoblast differentiation into non-invasive syncytiotrophoblast • Modulates maternal immunotolerance • Inhibits trophoblast proliferation and invasion	P
CSF	Midproliferative to midsecretory phase of menstrual cycle	Epithelial cells	• Regulates proliferation, differentiation, and survival of trophoblast cells • Promotes blastocyst attachment • Promotes cytotrophoblast differentiation into non-invasive syncytiotrophoblast	hCG
Glycodelin	Implantation window	Decidua	• Suppresses maternal immune reaction to the fetal allograft • Suppresses trophoblast invasion	P, relaxin
Osteopontin	Implantation window	Cytotrophoblast, endometrial gland, decidua	• Mediates cell–cell attachment and communication of the endometrium and trophoblast	*IL-1, TGF-β, TNF-α*, sex steroid hormones

suggesting that this cytokine may mediate interactions between maternal decidual leukocytes and invading cytotrophoblasts. *LIF* acts on cytotrophoblasts, switching their differentiation to the anchoring phenotype by increasing the synthesis of fibronectin and decreasing the production of hCG, and regulates trophoblast differentiation along the invasive pathway. In addition, the blastocyst also controls endometrial *LIF* expression through hCG secretion in a dose-dependent manner, suggesting that both the preimplantation embryo and the uterus are sites of *LIF* action. Finally, *LIF* is involved in immune tolerance through regulation of *HLA-G*, a class I major histocompatibility complex (*MHC*) molecule specifically expressed by the invasive cytotrophoblast [2,6,14,15].

5.2.3.2   Interleukins *(IL-1, IL-6, IL-11, and IL-15)*

The *IL-1* system includes two ligands (*IL-1α* and *IL-1β*), cell surface receptors (*IL-1* receptor type I [*IL-1R1*], *IL-1* receptor type II [*IL-1R2*]), a nonbinding receptor accessory protein (*IL-1RAcP*), and a naturally occurring receptor antagonist (*IL-1ra*) and is expressed in the human endometrium. Trophoblastic and decidual cells produce *IL-1*, and *IL-1* receptor is present both in endometrial epithelial cells and in trophoblasts. The selective up-regulation of $\alpha v\beta 3$ integrin, *LIF*, and prostaglandin E2 (*PGE-2*) in human endometrial epithelial cells by the blastocyst is mediated in part by the embryonic *IL-1* system and enhances blastocyst implantation. In addition, *IL-1* can stimulate *MMP-9* activity in trophoblasts and expression in endometrial stromal cells, thereby inducing trophoblast invasion. The expression of *IL-1* antagonist is reduced during the implantation window, suggesting the existence of specific mechanisms of regulation that alleviate *IL-1* inhibition, facilitating its preimplantation actions [6,7,14,15].

IL-6 is a pro-inflammatory cytokine that plays important roles in host defense, immune response, acute phase reaction, and hematopoiesis, which has some functional redundancy with *IL-11* and *LIF*. The fact that *IL-6* is maximally expressed during the window of implantation and that its receptor is found on the blastocyst, the endometrium, and the trophoblast suggests a paracrine/autocrine role for *IL-6* in the peri-implantation period. *IL-1β* stimulates endometrial *IL-6* production in a time- and dose-dependent manner, indicating that *IL-6*, in turn, may mediate some actions of *IL-1β* on the endometrium and trophoblast. On the other hand, hCG and *TGF-β* inhibit *IL-6* production [7,18].

IL-11, a multifunctional cytokine with anti-inflammatory activities, is produced by stromal and epithelial cells. Its production is maximal during decidualization and is influenced by steroid hormones and local factors such as relaxin and *PGE-2*. *IL-11* signaling is required for decidual-specific maturation of *NK* cells, suggesting that this cytokine may be important for the establishment of a viable pregnancy. Additionally, the embryo produces *IL-11* during trophoblast invasion, providing a means to exert control over the endometrium [15,18].

IL-15 is a member of the four α-helix bundle cytokine family, which includes *IL-2*, essential for *NK* cell development in bone marrow and stimulates proliferation, cytokine production, and cytotoxicity of activated blood *NK* cells. Opposite to its effects on blood *NK* cells, *IL-15* does not transform *uNK* cells into potent cytolytic cells and plays an important role in promoting their survival and expansion in the human uterus. Decidualization of endometrial stromal cells by P and adenosine 3′,5′-cyclic monophosphate (*cAMP*) enhances *IL-15* expression, which is further enhanced in the presence of *IFN-γ*. Therefore, *uNK* cells may play a role in regulating the process of decidualization itself by secreting *IFN-γ* that will enhance *IL-15* production by adjacent decidualized cells (Table 5.1) [15,25].

5.2.3.3   Insulin-Like Growth Factors

IGF-I and *IGF-II* are low-molecular-weight peptides with antiapoptotic, differentiating, metabolic, and mitogenic functions that are involved in endometrial differentiation and embryo implantation through autocrine and paracrine mechanisms. *IGFBP* are members of a family of proteins (*IGFBP-1* to *-6*) that facilitate the transport and regulate the availability of *IGF* to their specific receptors on target cells. *IGFBP-1*, which is predominantly synthesized by decidual cells, acts by primarily regulating trophoblast-derived *IGF* autocrine and paracrine actions at their target cells. In addition, *IGFBP-1* increases the gelatinolytic activity of trophoblasts and inhibits their migration into decidualized stromal multilayers. During decidualization, *IGF-2* produced by the trophoblast stimulates implantation and invasion while inhibitory to *IGFBP-1* produced by stromal cells. *IGFBP-1* then counteracts this effect by preventing *IGF-2* from binding to the cell surface, thereby limiting the effects of *IGF-2* on trophoblast invasion (Table 5.1). *IGFBP-1* also has *IGF*-independent actions such as binding to cell membranes and altering cellular motility. In this way, the *IGF/ IGFBP* system plays a role in establishing the implantation balance by stimulating and limiting inflammation [14,15,26].

5.2.3.4   Epidermal Growth Factor Family

The *EGF* family of growth factors is composed of *EGF*, *TGF-α*, herapin-binding EGF-like growth factor (HB-EGF), amphiregulin (*AREG*), β-cellulin (*BTC*), epiregulin (*EREG*), and neuroregulins, which interact with receptor subtypes that belong to the erythroblastic leukemia viral oncogene (*ERBb*) gene family of tyrosine kinase receptors (*ERBb-1* [*EGF-R*], *ERBb-2*, *ERBb-3*, and *ERBb-4*) inducing cellular mitosis. *EGF* (a major regulator of the implantation process) is expressed in both decidual and trophoblastic cells and plays an important role in trophoblast invasion, differentiation, and proliferation (Table 5.1). *EGF* induces cell invasion by increasing not only *MMP-2* and *MMP-9* but also urokinase-type plasminogen activator (*UPA*) and plasminogen activator inhibitor 1 (*PAI-1*) activity in trophoblastic cells. Additionally, *EGF* stimulates secretion of hCG and human placental lactogen (*HPL*) from trophoblastic cells [14,15].

Spatiotemporal expression patterns of *EGF* gene family members and *ERBb* in the uterus during the window of implantation suggest compartmentalized functions of *EGF*-like growth factors in implantation [27]. *HB-EGF*, which shares a common receptor with *EGF* and *TGF-α*, is expressed in the stromal and epithelial cells of the uterus and regulates endometrial proliferation, secretion, and decidualization. Its expression reaches a peak level in the midsecretory phase in the uterine luminal epithelium surrounding the blastocyst before the attachment reaction, indicating not only paracrine and juxtacrine interactions with embryonic *ERBb* receptors but also autocrine, paracrine, and juxtacrine interactions with uterine *ERBb* receptors, both important for the attachment reaction. Cells expressing the transmembrane form of

HB-EGF adhere to a blastocyst that displays cell surface *ERBb-4* receptors favoring implantation. On the other hand, *HB-EGF* increases the rate of blastocyst hatching. The expression of *HB-EGF* at the site of an active blastocyst is followed by the expression of *AREG, BTC, EREG,* neuroregulin 1, and cyclooxygenase-2 (*COX-2*) around the time of the attachment reaction, suggesting a compensatory mechanism for rescuing implantation in the absence of one or more members of the *EGF* family [2,14,15].

5.2.3.5 Transforming Growth Factor β Family

The *TGF-β* family is composed of five isoforms (*TGF-β1* to -β5) and has specific roles in proliferation, differentiation, migration, and *ECM* production during the peri-implantation period (Table 5.1). In the human endometrium, *TGF-β1, TGF-β2,* and *TGF-β3* are synthesized by both epithelial and stromal cells. However, only *TGF-β3* secretion varies during the menstrual cycle, being more intense in the glandular epithelium during the late secretory phase. *TGF-β1*, primarily expressed in the endometrial luminal and glandular epithelium during the implantation period, increases *ECM* fibronectin and stimulates trophoblast adhesion to the *ECM*. Additionally, it inhibits the proliferation of cytotrophoblasts and promotes their differentiation into noninvasive syncytiotrophoblasts. During implantation, *TGF-β* modulates maternal immunotolerance and is involved in the regulation of several implantation-related molecules such as *IGFBP-1, LIF, MMP-9,* and *VEGF.* The invading trophoblast secretes a variety of proteases including plasminogen activators and *MMPs. TGF-β1* inhibits trophoblast proliferation and invasion by stimulating *TIMP* secretion and decreasing *MMP* activation through down-regulation of plasminogen activators at the feto–maternal interface [6,15,26].

5.2.3.6 Colony-Stimulating Factor

Colony-stimulating factor (*CSF*) 1, a homodimeric glycoprotein that has a modulatory role in proliferation, differentiation, and survival of trophoblast cells, is preferentially produced by endometrial epithelial glands during the midproliferative up to the midsecretory phase of the menstrual cycle. *CSF-1,* produced by the uterine epithelium, interacts with the *CSF* receptor (encoded by proto-oncogene *c-fms*) on the trophectoderm, promoting blastocyst attachment. During the first trimester of pregnancy, it promotes the differentiation of cytotrophoblast into noninvasive syncytiotrophoblast cells (Table 5.1) [6,28].

5.2.3.7 Glycodelin and Osteopontin

Cytokine glycodelin acts as an immunomodulator and has its expression up-regulated during the implantation window by P and relaxin. Glycodelin has a suppressive effect on the maternal immune reaction to the fetal allograft, establishing tolerance. Glycodelin A is abundantly synthesized by the decidua and suppresses trophoblast invasion by down-regulating proteinase expression and activity. Osteopontin (*OPN*), which is produced by endometrial glands, decidual cells, and cytotrophoblast cells by the time of implantation, mediates cell–cell attachment and communication. Its production is stimulated by *IL-1, TGF-β,* and *TNF-α,* among others, as well as sex steroid hormones (estrogen and P), and it is important for the attachment between the integrins of the human endometrium and the trophoblast (Table 5.1) [18,29].

5.2.4   CELL ADHESION MOLECULES

5.2.4.1   Mucins

Mucins (*MUC*) are highly glycosylated molecules found on the apical side of endometrial epithelial cells functioning mainly as lubricating and protective agents and are involved in embryo attachment. The *MUC-1* barrier protects the embryo from the maternal immune system and prevents its attachment at an improper site. Indeed, the cell surface *MUC-1* is expressed by ciliated endometrial cells and is missing from the surface of nonciliated cells and pinopodes found on the apical surface of the endometrial epithelium during the implantation window. Expression of *MUC-1*, possibly mediated by P, is up-regulated during the implantation window, playing an important role in directing the blastocyst spatially and temporally to the proper implantation site. Paracrine effects from the blastocyst on the endometrium induce a local clearance of *MUC-1* during attachment, allowing embryonic implantation at the specific site. The pro-inflammatory cytokine *TNF-α* (secreted both by the blastocyst and by the endometrium) is involved in the removal of the repelling *MUC-1* at the implantation site through the stimulation of *MUC-1* shedding in the human uterine epithelium [7,18].

5.2.4.2   Integrins and Selectins

Adhesion molecules such as selectins and integrins play important roles in apposition and adhesion processes in human implantation, respectively. L-selectin receptors expressed on the trophoblast cells after hatching interact with selectin oligosaccharide ligands expressed on the maternal receptive endometrium at the time of implantation and are critically involved in the embryonic apposition phase. Integrins, transmembrane glycoproteins, are formed by two different heterodimeric noncovalently bound α and β subunits that contain extracellular, transmembrane, and intracellular domains. The extracellular domain serves as a receptor for *ECM* ligands including collagen, fibronectin, and laminin. Specific recognition and binding of the ligand to integrins activates intracellular transduction pathways and triggers cellular events that have important roles in promoting hormone responsiveness and genomic activation. Although many integrins show constitutive expression in the uterine epithelium and stroma, some such as $\alpha 1\beta 1$, $\alpha 3\beta 1$, $\alpha 6\beta 1$, $\alpha v\beta 3$, and $\alpha v\beta 5$ are cycle dependent, suggesting hormonal regulation. The increased expression of $\alpha v\beta 3$ (during the window of implantation—days 20–24 of the human menstrual cycle) on the endometrial luminal epithelial surface indicates that this integrin may directly participate in implantation as a potential receptor for embryonic attachment. High levels of E2 during the follicular phase of the menstrual cycle inhibit integrin expression. After ovulation, luteal phase P suppresses the inhibitory effect of E2 on integrins. P also acts positively on *OPN* synthesis (an *ECM* component secreted by the uterine epithelium during the implantation window) by increasing paracrine stromal factors (*EGF*, *HB-EGF*) to induce epithelial β3 integrin expression, which is a rate-limiting step in $\alpha v\beta 3$ formation contributing to the establishment of a period of uterine receptivity. The trophectoderm also expresses several integrins ($\alpha 3$, $\alpha 5$, $\beta 1$, $\beta 3$, $\beta 4$, and $\beta 5$) at the time of implantation. Trophoblastic receptors for *ECM* such as α1β1 and α5β1 increase together with the cytotrophoblast cell differentiation to invasive extravillous phenotype, which undergo integrin switching from α6β4 to α1β1 and α5β1, suggesting an active role for the blastocyst in the establishment of a receptive endometrium [2,7,14,18].

5.2.4.3 Cadherins

Cadherins comprise a group of glycoproteins involved in the calcium-dependent cell–cell adhesion mechanism. E-cadherin, which participates in the formation of the epithelial adherens junctions in cooperation with α- and β-catenins, is critical for blastocyst formation and is implicated in uterine–embryo interactions. E-cadherin has a dual function on blastocyst attachment and subsequent invasion. Its expression by both the trophectoderm and the uterine luminal epithelium is required to ensure adhesiveness at the time of the attachment reaction. The regulation of E-cadherin availability at the epithelial cell surface enables cellular adhesion control. Intracellular calcium is essential in E-cadherin regulation. A rise in its concentration, triggered by calcitonin, which is induced by P in the human endometrial epithelium specifically during the midsecretory phase of the menstrual cycle, activates key signaling pathways that mediate cytoskeletal reorganization and disassembly of E-cadherin at cellular contact sites, favoring epithelial cell dissociation and blastocyst invasion [2,7].

5.2.4.4 Immunoglobulins

Members of the immunoglobulin superfamily are involved in the control of cell behavior by acting as cell–cell adhesion or signal-transducing receptors. Cellular adhesion molecules (CAM) play an important role in defining cellular shape and degree of contact with neighboring cells. Intracellular adhesion molecule 1 (ICAM-1), found on the apical surface of the glandular and luminal endometrial epithelial cells at the time of implantation, participates indirectly in the process of blastocyst–endometrium adhesion by interacting with the immune system. The melanoma cell adhesion molecule (MELCAM) is expressed in invasive cytotrophoblasts and is involved in blastocyst attachment and subsequent trophoblastic invasion. Other members of the immunoglobulin superfamily such as cell–cell adhesion molecule (CCAM) and neural cell adhesion molecule (NCAM) are also present on the surface of trophectoderm at the time of implantation [7,30,31].

5.2.4.5 Trophinin

Trophinin is an apical transmembrane glycoprotein that mediates homophilic interactions between two different cell types and is important in trophoblast cell adhesion acting as a molecular switch for trophoblast activation. The cytoplasmic domain of trophinin requires the presence of tastin and bystin to create highly concentrated areas that function as efficient adhesion sites between trophoblastic and endometrial cells. Both trophinin and tastin are uniquely expressed in the human endometrial epithelium and only at the expected time of implantation. The spatially and temporally restricted expression of this cell adhesion complex suggests its role in the process of blastocyst attachment in implantation [6,18,30].

5.2.5 INFLAMMATORY FACTORS

5.2.5.1 Prostaglandins

Prostaglandins (*PGs*) which belong to the eicosanoids family, are involved in distinct steps of the reproductive process including menstruation and ovulation, and are important for successful embryo implantation. They are synthesized from arachidonic acid (*AA*), which is released from plasma membrane phospholipids by the consecutive action of phospholipase A2 (*PLA₂*) and cyclooxygenase (*COX*), two key enzymes in this pathway. Cytosolic *PLA₂* (*cPLA₂*) is specific for arachidonyl-containing phospholipids and is a key intracellular mediator of hormone-stimulated *PG* synthesis. Cyclooxygenase 1 (*COX-1*) is produced by most cells and is primarily responsible for the immediate synthesis of *PGs* in response to agonist stimulation. In contrast, *COX-2* is induced in response to growth factors, cytokines, oncogenes, and inflammatory stimuli, catalyzing *PG* synthesis several hours after the inflammatory insult. *AA* released from membrane phospholipids by *cPLA₂* is oxygenated and reduced by *COX* enzymes to the intermediary *PGH-2*, which will be subsequently metabolized by prostaglandin synthases to form the biologically active end products (*PGD-2*, *PGE-2*, *PGF2-α*, and prostacyclin [*PGI₂*]). After their synthesis, *PGs* are rapidly transported out of the cell by a specific prostaglandin transporter (*PGT*) exerting their autocrine and paracrine effects by binding to cell surface G-protein-coupled receptors to activate intracellular signaling and gene transcription [7,14,32].

In addition to sex steroid hormones, other key molecules are required for successful implantation. During the adhesion phase, which is regulated by steroid hormones, *COX-2* expression is critical for implantation. Its expression is regulated by *EGF*, *IL-1*, and platelet-derived growth factor (*PDGF*). *IL-1* plays a major role in the establishment of endometrial receptivity not only through the up-regulation of αvβ3 integrin expression but also by enhancing *PG* expression via increase of *COX-2*. In addition to its role in embryo implantation, *COX-2* is also involved in decidualization through *PG* synthesis. Initiation of the decidual process is dependent on elevated levels of the second messenger *cAMP*, which is accounted for by increased expression of local factors (relaxin; corticotropin-releasing hormone [*CRH*]; *PGE-2*) that activate adenylate cyclase in stromal cells and the simultaneous down-regulation of phosphodiesterase 4 (*PDE4*), a phosphodiesterase family member that converts *cAMP* to adenosine monophosphate (*AMP*). Endometrial stromal cells become first responsive to, and then dependent on, steroid hormones, foremost on P, continuous *cAMP* stimulation, and protein kinase A (*PKA*) activation. *PGE-2* acting in part via *cAMP* is involved in the regulation of decidualization of human endometrial stromal cells through *IL-11* stimulation and is required for increased vascular permeability and local blood flow at the site of implantation. *PLA* and *PGE-2* receptors are up-regulated in the human endometrium during the window of implantation [7,9,14,26].

COX and *cPLA₂* are expressed in all stages of preimplantation development in human embryos. *COX-1* is mainly expressed during the early stages of embryo development, whereas *COX-2* is predominantly expressed in the later stages (eight-cell, morula, and blastocyst stages). The blastocyst produces *PGs*, which serve as embryonic signals to the uterus and are involved in the modulation of endometrial receptivity. Successful blastocyst implantation requires immune effector mechanisms of both maternal and embryonic origin. *PGE-2* produced by the blastocyst down-regulates *IL-2* receptors on T-lymphocytes, inhibiting the proliferation and cytotoxic activation of these lymphocytes. Through this mechanism, the blastocyst could defend itself from attack by maternal immune cells while also providing an immunosuppressed implantation site [32].

5.2.5.2 Corticotropin-Releasing Hormone

Embryonic trophoblast and maternal decidua produce *CRH* and express *TNF* receptor superfamily member 6 (*FAS*) ligand (*FAS-LG*), a pro-apoptotic cytokine. Epithelial and stromal *CRH* affects the decidualizing effect of P by regulating local modulators including PGE_2 and cytokines (*IL-1* and *IL-6*). In addition, the blastocyst modulates the expression of endometrial *CRH* through *IL-1* or PGE_2 secretion. Subsequently, endometrial *CRH* in association with other local factors participate in the local inflammatory response at the implantation site, rendering the endometrial surface adhesive for blastocyst attachment. The interaction between *FAS* and its ligand (*FAS-LG*) plays an important role in the regulation of immune tolerance by promoting apoptosis of cells carrying *Fas* (such as T- and B-lymphocytes). Locally produced *CRH* promotes implantation and maintenance of early pregnancy primarily by killing activated T cells [14,33].

5.2.5.3 Tumor Necrosis Factor α

A delicate balance between pro- and anti-inflammatory pathways is necessary not only for the establishment of the window of implantation but also for embryo–endometrial communication. Th1 cytokine *TNF-α* and *IL-1* are responsible for the activation of the pro-inflammatory cascade at the feto–maternal interface. These primary pro-inflammatory cytokines activate production of secondary mediators such as cytokines, chemokines, *COX* enzymes, *PGs*, and pentraxin (an effector and modulator of innate resistance); inflammation; and angiogenesis, which play important roles in implantation. In addition, *TNF-α* up-regulates *UPA* secretion from cytotrophoblasts and enhances fibronectin degradation during trophoblast invasion of the endometrial *ECM*. The up-regulation of *UPA* in turn increases activation of *MMP-9* through the plasminogen activator (*PA*)/plasmin system, therefore enhancing trophoblast invasiveness. The pro-inflammatory pathways induced during implantation by *TNF α* and *IL-1* are regulated by *IL-10*, an anti-inflammatory and immune-modulating cytokine. *IL-10* has the ability to reduce inflammation by inhibiting the synthesis of *TNF-α* and *IL-1* [14,34].

5.2.6 EXTRACELLULAR MATRIX PROTEINS

ECM adhesion and proteolysis play important roles during blastocyst implantation. Spatiotemporal regulated changes in the expression of *ECM* molecules at the feto–maternal interface during the window of implantation play a major role in blastocyst attachment, migration, and subsequent invasion of the decidualized endometrium. *ECM* components, including collagens, fibronectin, laminin, type IV collagen (*COL-IV*), trophin, and tastin, influence trophoblastic cell behavior and function by binding to integrins. Both fibronectin and laminin, important players in the invasive process of implantation, are secreted by endometrial stromal cells predominantly in the secretory endometrium and decidua. Fibronectin-mediated trophoblast cell adhesion and migration involve the interaction of cellular receptors of the invading trophoblast with the Arg–Gly–Asp (*RGD*) recognition site of fibronectin, which leads to active focal adhesion kinase and triggers the early signal transduction cascade. Initial recognition of fibronectin by primary trophoblast cells requires apically located integrins of β*1* and β*3* classes. The interaction between cytotrophoblast and fibronectin through α5β*1* integrin restrains invasiveness at the feto–maternal interface. On the other hand, laminin decreases *PRL* and *IGFBP-1* production in endometrial stromal cells facilitating trophoblast invasion. The interaction of laminin and *COL-IV* with the integrin α*1*β*1* receptor accelerates the invasion of trophoblast cells. Trophin and tastin, which are found both in trophoblasts and in endometrial epithelium, are involved in blastocyst attachment through the formation of a cell adhesion molecular complex. Down-regulation of the expression of adhesive molecules and *ECM* proteins in cytotrophoblast cells after their contact with the decidua plays an important role in the control of trophoblast invasion [6,14,28].

5.2.7 EXTRACELLULAR DEGRADING MATRIX PROTEINASES

Trophoblast invasion depends on a delicate balance between activated proteases and protease inhibitors that enable the extracellular degradation and phagocytosis of maternal cells and the *ECM*. This invasive behavior is attributed to the ability of cytotrophoblastic cells to express and secrete three classes of proteases including *MMPs*, serine proteases, and cathepsins, which are implicated in the tissue remodeling and cell invasion processes that take place during implantation and placentation. Regulation of protease activity occurs via differential activation of *MMPs* or by *TIMPs* (*TIMP-1, -2, -3, -4*), which are locally produced and inhibit specifically active forms of *MMP* in extracellular space, limiting cytotrophoblast invasion. Both *MMP* and *TIMP* genes are expressed in the developing embryo preparing for implantation. The *PA*/plasmin system, which includes *UPA*, the tissue-type plasminogen activator (*TPA*), *PA* inhibitors (*PAI—PAI-1, PAI-2*), and cell surface *UPA* receptor, is also involved in tissue remodeling by proteolytic activation of pro-*MMPs* [14,35].

MMPs, a family of endopeptidases capable of degrading all components of the *ECM* (interstitial matrix and basement membrane), are classified into four subgroups according to their substrate specificity and structure: (a) gelatinases (*MMP-2, -9*), (b) collagenases (*MMP-1, -8, -13*), (c) stromelysins (*MMP-3, -7, -10, -11, -12*), and (d) membrane-type *MMPs* (*MMP-14, -15, -16*). Most *MMPs* are secreted as inactive pro-enzymes requiring activation in the extracellular compartment by other *MMPs* or serine proteases (*UPA*, plasminogen, thrombin, and elastase). They regulate cell behavior in various ways including cell–matrix and cell–cell interactions and the release, activation, or inactivation of autocrine or paracrine signaling molecules and cell surface receptors. Multiple subtypes of *MMPs* are expressed not only in invading trophoblast cells but also in the maternal endometrium during the implantation process. Immune cells that infiltrate the uterus during pregnancy also express various proteases including *MMP-2* and *-9*. The expression of *MMPs* is induced by inflammatory cytokines (*IL-1, TNF-α*) but inhibited by P [35,36].

Gelatinases *MMP-2* and *-9*, which have differential expression throughout the first trimester of pregnancy, are produced by the invading cytotrophoblast cells and degrade *COL-IV*, enabling the invasion of these cells through the decidua and into the maternal vasculature. Besides *ECM* degradation, these *MMPs* have distinct actions in the implantation process including the release of *IGF* by degradation of *IGFBP*, the activation of *TGF-β*, and modulation of angiogenic factors such as endothelin-1 (a vasoconstrictor) and angiostatin (an angiogenic inhibitor). The activity of these gelatinases is controlled by specific inhibitors *TIMP-1* (preferentially binding *MMP-9*) and *TIMP-2* (preferentially binding *MMP-2*) or indirectly by cytokines and by interactions with the *ECM*. *TIMPs* function by reversibly binding to the catalytic domain of *MMPs*. Growth factors and cytokines including *TGF-β* and the *EGF* family as well as *LIF* modulate *MMPs* and *TIMPs*. *TGF-β* inhibits trophoblast invasion by up-regulation of *TIMP-1* and *PAI-1* and by down-regulation of *UPA*. *LIF* inhibits gelatinase activity in cytotrophoblasts, thereby affecting their invasiveness. Conversely, *EGF* induces cell invasion by increasing not only *MMP-2* and *-9* but also *UPA* and *PAI-1* activity in trophoblastic cells. The major role for *TIMP-3* (which has its expression in decidualized endometrial stromal cells up-regulated by P) is to limit invasion. On the other hand, *IL-1* inhibits *TIMP-3* expression in these cells, indicating that the trophoblast promotes its own invasiveness by inhibiting the maternal control on invasion. Besides *MMP* inhibition, *TIMPs* have additional roles in increasing cell proliferation and embryo development [6,14].

The *PA*/plasmin system exerts its action on matrix degradation by the conversion of plasminogen into the active serine protease plasmin and indirectly through proteolytic activation of *MMPs*. The activity of this system is balanced by *PAI-1* and *PAI-2*. *PAI-1*, which is the major *PAI* in plasma, has high affinity for tissue-type plasminogen activator (*TPA*) while *PAI-2* inhibits both *UPA* and, less efficiently, *TPA*. *PAI-1* is secreted as an active antiprotease but is rapidly converted to an inactive latent form that can be reactivated by exposure to phospholipids. *ECM* vitronectin stabilizes *PAI-1* in the active conformation and enhances its reactivity toward thrombin. *TGF-β* induces *PAI* and *TIMP-1* secretion and decreases *MMP-2* secretion by cytotrophoblasts and therefore controls trophoblast invasion by regulating proteases and their inhibitors at the feto–maternal interface. Adrenomedullin, a polypeptide belonging to the calcitonin gene-related peptide superfamily, decreases *PAI-1* expression and increases *MMP-2* activity, enhancing trophoblast proliferation and invasion. Trophoblast cells express *UPA* receptors that can bind active *UPA* and localize proteolysis to the leading edge of migrating cells. Proteolysis of protease-activated receptor 1 (*PAR-1*), which is the predominant thrombin receptor on invasive extravillous trophoblast cells, enhances normal cellular invasion, playing an important role in the invasive phase of placentation [14,37,38].

5.2.8 Transcription Factors

5.2.8.1 Homeobox Genes

Homeobox (*HOX*) genes encode transcription factors that regulate differential gene expression within the endometrium, which are essential for endometrial development and for implantation. *HOXA-10* and *HOXA-11* genes have a dynamic temporal pattern of expression throughout the menstrual cycle in the endometrial glands and stroma through autocrine and paracrine mechanisms. Peak expression of *HOXA-10* and *HOXA-11* genes is found during the implantation window. Both genes are up-regulated by estrogen and P and down-regulated by testosterone. In addition to sex steroid regulation, *HOX* genes are also regulated by inflammatory cytokines in the uterus, showing that inflammation plays an important role in regulating genes necessary for human endometrial receptivity. *HOX* genes are involved in the regulation of a number of molecular and morphological markers specific to the implantation window including β*3* integrin, *IGFBP-1*, and pinopodes [39,40].

5.2.8.2   Forkhead Box O Transcription Factor

Forkhead box O (*FOXO*) transcription factors mediate cell fate decisions (e.g., cell cycle arrest, apoptosis) in response to growth factor, hormonal, and environmental signals. These transcription factors are involved in the regulation of decidual marker gene expression such as *PRL* and *IGFBP-1*. In the normal cycling endometrium, *FOXO-1* expression is restricted to the epithelial and stromal compartments and is up-regulated during the luteal phase, which intensifies during pregnancy. *FOXO-1* up-regulates not only IGFBP-1 promoter activity by its association with HOXA-10 but also *PRL* promoter activity cooperatively with *C/EBP*β in endometrial stromal cells. *FOXO-1* interacts with the P receptor, which acquires control of the diverse gene families involved in decidualization, playing an essential role in coordinating different aspects of this process including cell proliferation, differentiation, immune modulation, and resistance to environmental or oxidative stress [15,40–42].

5.2.8.3   JAK/STAT Intracellular Signaling

Migration and invasion of extravillous trophoblast cells are functionally controlled by various cytokines and growth factors (hepatocyte growth factor [*HGF*], *IL-6*, *IL-11*, *LIF*, and *GM-CSF*) acting through *JAK/STAT* signal transduction pathways. Binding of cytokine to its receptor α triggers dimerization with *gp130*, forming a high-affinity receptor leading to activation of the *JAK/STAT* signal transduction pathway. Upon activation, signal transducers and activators of transcription (*STATs*) dissociate from the receptor and form homo- and heterodimers that translocate into the cellular nucleus where they up-regulate the transcription of target proteins. Membrane-bound and soluble forms of the receptor act as inhibitors to cytokine action by competing with the cell surface receptors to limit dimerization with *gp130*. The suppressors of the cytokine signaling (*SOCS*) family of cytoplasmic proteins completes a negative feedback loop to attenuate signal transduction from cytokines that act through the *JAK/STAT* signal transduction pathway [25].

HGF mediates the invasive potential of trophoblast invasion in a dose-dependent manner through the *JAK/STAT* system, playing an important role in the mesenchymal induction of trophoblast growth and differentiation during the development of the placenta. *IL-6*, another cytokine that uses *STATs* to mediate its signal, induces an increased expression of integrins associated with embryo attachment. *IL-11*-induced signaling initiates and maintains the decidualization process through *STAT-3* and *SOCS-3* activation. The signaling via the *JAK/STAT* system is primarily responsible for the onset of receptivity during the implantation window in response to *LIF*. On the other hand, soluble *gp130* (*sgp130*), which is up-regulated during the implantation window in glandular epithelial cells, blocks the biological activity of *LIF* and *IL-6*. *GM-CSF*, another mediator that uses the *JAK/STAT* system, functionally supports the development of the placenta by promoting proliferation, differentiation, and secretory activity of cytotrophoblast cells [43–45].

5.3 CRITICAL WINDOWS OF SUSCEPTIBILITY

Over the past few years, an increasing amount of evidence derived from studies focused on pregnancy outcome showed the negative effects of environmental contaminants on human reproductive health. Important stages of development that define critical windows of susceptibility to reproductive toxicants that can disrupt or interfere with the normal physiology of a cell, tissue, or organ include the periconceptional period (gamete, early embryo development, and implantation), prenatal period (embryo and fetal development), and postnatal period (infancy, childhood, puberty, and adulthood). These stages are characterized by marked cellular proliferation and changing metabolic capabilities in the developing organism. The exposure to environmental contaminants may result in permanent and irreversible adverse effects that can have lifelong and even intergenerational impacts on health if they occur during a critical window, or may still affect development or result in eventual adult disease with a reduced magnitude in comparison to exposure within a critical window if they occur during a sensitive period [10,46–48].

Exchange of hormonal signals and complex molecular interactions between the embryo and the hormonally primed maternal endometrium are required for successful implantation. During the periconceptional period, specific E2, P, and hCG concentrations are necessary not only for endometrial maintenance but also for maintenance of hormone production by the *CL*. Environmental contaminants such as *ED* act through both nuclear receptor-dependent and nonreceptor-dependent mechanisms, mimicking or blocking the action of the steroid hormones, compromising reproductive cyclicity owing to an imbalance in ovarian steroid hormone production and therefore negatively affecting not only implantation but also the early stage of pregnancy (Figure 5.3). *ED* may also act as immune modulators exerting their effects at different levels of the immune regulatory network, including humoral immunity, cell survival, and cytokine synthesis. In addition to the effects on the endocrine and immune systems, *ED* interfere with Ah receptor (*AhR*) and peroxisome proliferator-activated receptor (*PPAR*) signaling pathways, both essential for pre- and peri-implantation development of the embryo and endometrium [49,50]. *HOX* genes are involved in the control of a number of molecular and morphological markers specific to the implantation window. Exposure to xenoestrogens modifies the expression of these genes, compromising decidualization and embryo implantation. Furthermore, xenoestrogens may change a critical set point in the regulation of *HOX* gene expression resulting in epigenetic modifications of autoregulatory loops that become fixed at distinct set points and later determine altered gene expression throughout life [51].

Both pro- and anti-inflammatory pathways are necessary for the establishment of the window of implantation and for embryo–endometrial communication. In general, pregnancy is associated with a reduced Th1 response and increased Th2 cytokine production. Soluble autocrine and paracrine factors including *IFN, IL, CSF,*

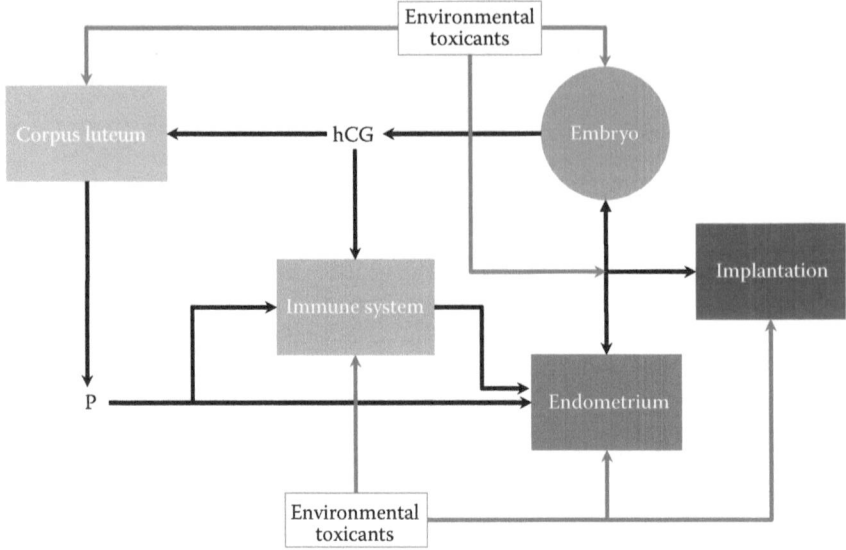

FIGURE 5.3 **(See color insert.)** Critical windows of susceptibility to environmental toxicants during feto–maternal interactions.

TNF, TGF, LIF, and chemokines, which regulate survival, proliferation, migration, and invasion of cytotrophoblast cells, are involved in these processes. The precise balance of these cytokines at the feto–maternal interface is required for proper placental and successful pregnancy development. Exposure to environmental toxicants may result in an imbalance of Th1/Th2 cytokines in immune cells, which compromises the placentation process and pregnancy development. This exposure may also indirectly affect reproductive success through the altered secretion of inflammatory mediators and the dysregulation of tolerance [12].

In addition to endocrine and immune system interactions, implantation and placentation require the presence of several autocrine and paracrine factors including cytokines (*LIF, IL, IGF, EGF* family, *TGF-β* family, *CSF*, glycodelin, and *OPN*), cell adhesion molecules (*MUC*, integrins, selectins, cadherins, immunoglobulin superfamily, and trophinin), and inflammatory factors (*PGs, CRH*, and *TNF-α*) that are essential for reproductive success. The altered balance of these factors caused by the exposure to environmental toxicants may impair trophoblast cell proliferation, differentiation, migration, and invasion, leading to implantation failure, defective placentation, and pregnancy disorders (Figure 5.3) [12,52].

GLOSSARY

AA: arachidonic acid
AMP: adenosine monophosphate
APC: antigen-presenting cells
AREG: amphiregulin

B: B-lymphocyte
BTC: β-cellulin
CAM: cellular adhesion molecules
cAMP: adenosine 3′,5′-cyclic monophosphate
CCAM: cell–cell adhesion molecule
CCL: chemokine C–C motif ligand
CCL-5: chemokine C–C motif ligand 5
CD: cluster of differentiation
CL: corpus luteum
COL-IV: type IV collagen
COX: cyclooxygenase
COX-1: cyclooxygenase-1
COX-2: cyclooxygenase-2
cPLA$_2$: cytosolic phospholipase A2
CRH: corticotropin-releasing hormone
CSF: colony-stimulating factor
CTLA-4: cytotoxic T-lymphocyte antigen 4
E2: estradiol
ECM: extracellular matrix
ED: endocrine disruptors
EGF: epidermal growth factor
ERBb: erythroblastic leukemia viral oncogene
EREG: epiregulin
FAS: Fas (*TNF* receptor superfamily, member 6)
FAS-LG: *FAS* ligand
FOXO: forkhead box O transcription factors
FOXP-3: forkhead box P3
GM-CSF: granulocyte-macrophage colony-stimulating factor
gp130: glycoprotein 130
HB-EGF: heparin-binding EGF-like growth factor
hCG: human chorionic gonadotropin
HGF: hepatocyte growth factor
HLA-DR: human leukocyte antigen DR
HLA-G: human leukocyte antigen G
HOX: homeobox genes
HPL: human placental lactogen
ICAM-1: intracellular adhesion molecule 1
IDO: indoleamine 2,3-dioxygenase
IFN-γ: interferon γ
IGF: insulin-like growth factor
IGFBP-1: insulin-like growth factor binding protein 1
IL: interleukin
IL-1R1: IL-1 receptor type I
IL-1R2: IL-1 receptor type II
IL-1RAcP: IL-1 receptor accessory protein
JAK/STAT: janus kinase/signal transducer and activator of transcription

LH: luteinizing hormone
LIF: leukemia inhibitory factor
LIFR: leukemia inhibitory factor receptor
MAPK: mitogen-activated protein kinase
MCP-1: monocyte chemoattractant protein 1
MELCAM: melanoma cell adhesion molecule
MHC: major histocompatibility complex
MMPs: matrix metalloproteinases
MUC: mucins
MUC-1: mucin 1
NCAM: neural cell adhesion molecule
NK: natural killer
OPN: osteopontin
P: progesterone
PA: plasminogen activator
PAI: plasminogen activator inhibitor
PAI-1: plasminogen activator inhibitor 1
PAR-1: protease-activated receptor 1
PDE4: phosphodiesterase 4
PDGF: platelet-derived growth factor
PGE-2: prostaglandin E2
PGI$_2$: prostacyclin
PGs: prostaglandins
PGT: prostaglandin transporter
PI3K: phosphatidylinositol 3-kinase
PKA: protein kinase A
PLA$_2$: phospholipase A2
PPAR: peroxisome proliferator-activated receptor
PRL: prolactin
RGD: Arg–Gly–Asp fibronectin recognition site
sgp130: soluble glycoprotein 130
SOCS: suppressor of cytokine signaling
STATs: signal transducers and activators of transcription
T: T-lymphocyte
TE: trophectoderm
Th: T-helper cells
TGF-β: transforming growth factor β
TIMPs: tissue inhibitors of metalloproteinases
TNF-α: tumor necrosis factor-alpha
TPA: tissue-type plasminogen activator
Treg: CD4$^+$CD25$^+$ regulatory T cells
uNK: uterine natural killer cells
UPA: urokinase-type plasminogen activator
VEGF: vascular endothelial growth factor
ZP: zona pellucida

REFERENCES

1. J. Szekeres-Bartho, *Immunological relationship between the mother and the fetus*, Int. Rev. Immunol. 21 (2002), pp. 471–495.
2. S.K. Dey, H. Lim, S.K. Das, J. Reese, B.C. Paria, T. Daikoku, and H. Wang, *Molecular cues to implantation*, Endocr. Rev. 25 (2004), pp. 341–373.
3. L. Fraccaroli, J. Alfieri, L. Larocca, M. Calafat, G. Mor, C.P. Leiros, and R. Ramhorst, *A potential tolerogenic immune mechanism in a trophoblast cell line through the activation of chemokine-induced T cell death and regulatory T cell modulation*, Hum. Reprod. 24 (2009), pp. 166–175.
4. A. Munoz-Suano, A.B. Hamilton, and A.G. Betz, *Gimme shelter: The immune system during pregnancy*, Immunol. Rev. 241 (2011), pp. 20–38.
5. F. Dominguez, M. Yanez-Mo, F. Sanchez-Madrid, and C. Simon, *Embryonic implantation and leukocyte transendothelial migration: Different processes with similar players?* FASEB J. 19 (2005), pp. 1056–1060.
6. L.C. Giudice, *Implantation and endometrial function*, in *Molecular Biology in Reproductive Medicine*, B. Fauser, ed., The Parthenon Publishing Group, New York, 1999, pp. 333–352.
7. H. Achache and A. Revel, *Endometrial receptivity markers, the journey to successful embryo implantation*, Hum. Reprod. Update 12 (2006), pp. 731–746.
8. H. Singh, L. Nardo, S.J. Kimber, and J.D. Aplin, *Early stages of implantation as revealed by an in vitro model*, Reproduction 139 (2010), pp. 905–914.
9. G. Teklenburg, M. Salker, C. Heijnen, N.S. Macklon, and J.J. Brosens, *The molecular basis of recurrent pregnancy loss: Impaired natural embryo selection*, Mol. Hum. Reprod. 16 (2010), pp. 886–895.
10. P.M. Perin, M. Maluf, C.E. Czeresnia, D.A. Nicolosi Foltran Januário, and P.H. Nascimento Saldiva, *Effects of exposure to high levels of particulate air pollution during the follicular phase of the conception cycle on pregnancy outcome in couples undergoing in vitro fertilization and embryo transfer*, Fertil. Steril. 93 (2010), pp. 301–303.
11. W.R. Schaefer, L. Fischer, W.R. Deppert, A. Hanjalic-Beck, L. Seebacher, M. Weimer, and H.P. Zahradnik, *In vitro-Ishikawa cell test for assessing tissue-specific chemical effects on human endometrium*, Reprod. Toxicol. 30 (2010), pp. 89–93.
12. N. Bechi, F. Ietta, R. Romagnoli, S. Jantra, M. Cencini, G. Galassi, T. Serchi, I. Corsi, S. Focardi, and L. Paulesu, *Environmental levels of para-nonylphenol are able to affect cytokine secretion in human placenta*, Environ. Health Perspect. 118 (2010), pp. 427–431.
13. A. Herrler, U. von Rango, and H.M. Beier, *Embryo-maternal signalling: How the embryo starts talking to its mother to accomplish implantation*, Reprod. Biomed. Online 6 (2003), pp. 244–256.
14. E. Staun-Ram and E. Shalev, *Human trophoblast function during the implantation process*, Reprod. Biol. Endocrinol. 3 (2005), p. 56.
15. O. Guzeloglu-Kayisli, U.A. Kayisli, and H.S. Taylor, *The role of growth factors and cytokines during implantation: Endocrine and paracrine interactions*, Semin. Reprod. Med. 27 (2009), pp. 62–79.
16. P. Paiva, N.J. Hannan, C. Hincks, K.L. Meehan, E. Pruysers, E. Dimitriadis, and L.A. Salamonsen, *Human chorionic gonadotrophin regulates FGF2 and other cytokines produced by human endometrial epithelial cells, providing a mechanism for enhancing endometrial receptivity*, Hum. Reprod. 26 (2011), pp. 1153–1162.
17. G. Mor and V.M. Abrahams, *Potential role of macrophages as immunoregulators of pregnancy*, Reprod. Biol. Endocrinol. 1 (2003), p. 119.
18. M.S. van Mourik, N.S. Macklon, and C.J. Heijnen, *Embryonic implantation: Cytokines, adhesion molecules, and immune cells in establishing an implantation environment*, J. Leukoc. Biol. 85 (2009), pp. 4–19.

19. J.C. Warning, S.A. McCracken, and J.M. Morris, *A balancing act: Mechanisms by which the fetus avoids rejection by the maternal immune system*, Reproduction 141 (2011), pp. 715–724.

20. Y. Shi, B. Ling, Y. Zhou, T. Gao, D. Feng, M. Xiao, and L. Feng, *Interferon-gamma expression in natural killer cells and natural killer T cells is suppressed in early pregnancy*, Cell. Mol. Immunol. 4 (2007), pp. 389–394.

21. J. Mjosberg, G. Berg, M.C. Jenmalm, and J. Ernerudh, *FOXP3+ regulatory T cells and T helper 1, T helper 2, and T helper 17 cells in human early pregnancy decidua*, Biol. Reprod. 82 (2010), pp. 698–705.

22. S. Saito, A. Nakashima, T. Shima, and M. Ito, *Th1/Th2/Th17 and regulatory T-cell paradigm in pregnancy*, Am. J. Reprod. Immunol. 63 (2010), pp. 601–610.

23. D.A. Somerset, Y. Zheng, M.D. Kilby, D.M. Sansom, and M.T. Drayson, *Normal human pregnancy is associated with an elevation in the immune suppressive CD25+ CD4+ regulatory T-cell subset*, Immunology 112 (2004), pp. 38–43.

24. L.R. Guerin, J.R. Prins, and S.A. Robertson, *Regulatory T-cells and immune tolerance in pregnancy: A new target for infertility treatment?* Hum. Reprod. Update 15 (2009), pp. 517–535.

25. E. Dimitriadis, C.A. White, R.L. Jones, and L.A. Salamonsen, *Cytokines, chemokines and growth factors in endometrium related to implantation*, Hum. Reprod. Update 11 (2005), pp. 613–630.

26. M. Singh, P. Chaudhry, and E. Asselin, *Bridging endometrial receptivity and implantation: Network of hormones, cytokines, and growth factors*, J. Endocrinol. 210 (2011), pp. 5–14.

27. H. Lim, S.K. Das, and S.K. Dey, *erbB genes in the mouse uterus: Cell-specific signaling by epidermal growth factor (EGF) family of growth factors during implantation*, Dev. Biol. 204 (1998), pp. 97–110.

28. T. Strowitzki, A. Germeyer, R. Popovici, and M. von Wolff, *The human endometrium as a fertility-determining factor*, Hum. Reprod. Update 12 (2006), pp. 617–630.

29. K.K. Lam, P.C. Chiu, C.L. Lee, R.T. Pang, C.O. Leung, H. Koistinen, M. Seppala, P.C. Ho, and W.S. Yeung, *Glycodelin-A protein interacts with Siglec-6 protein to suppress trophoblast invasiveness by down-regulating extracellular signal-regulated Kinase (ERK)/c-Jun signaling pathway*, J. Biol. Chem. 286 (2011), pp. 37118–37127.

30. H. Singh and J.D. Aplin, *Adhesion molecules in endometrial epithelium: Tissue integrity and embryo implantation*, J. Anat. 215 (2009), pp. 3–13.

31. D. Haouzi, H. Dechaud, S. Assou, C. Monzo, J. de Vos, and S. Hamamah, *Transcriptome analysis reveals dialogues between human trophectoderm and endometrial cells during the implantation period*, Hum. Reprod. 26 (2011), pp. 1440–1449.

32. H. Wang, Y. Wen, S. Mooney, B. Behr, and M.L. Polan, *Phospholipase A(2) and cyclooxygenase gene expression in human preimplantation embryos*, J. Clin. Endocrinol. Metab. 87 (2002), pp. 2629–2634.

33. N. Vitoratos, D.C. Papatheodorou, S.N. Kalantaridou, and G. Mastorakos, *"Reproductive" corticotropin-releasing hormone*, Ann. N.Y. Acad. Sci. 1092 (2006), pp. 310–318.

34. H.N. Jabbour, K.J. Sales, R.D. Catalano, and J.E. Norman, *Inflammatory pathways in female reproductive health and disease*, Reproduction 138 (2009), pp. 903–919.

35. M. Koizumi, M. Momoeda, H. Hiroi, F. Nakazawa, H. Nakae, T. Ohno, T. Yano, and Y. Taketani, *Inhibition of proteases involved in embryo implantation by cholesterol sulfate*, Hum. Reprod. 25 (2010), pp. 192–197.

36. D.O. Anumba, S. El Gelany, S.L. Elliott, and T.C. Li, *Circulating levels of matrix proteases and their inhibitors in pregnant women with and without a history of recurrent pregnancy loss*, Reprod. Biol. Endocrinol. 8 (2010), p. 62.

37. J.D. Vassalli, A.P. Sappino, and D. Belin, *The plasminogen activator/plasmin system*, J. Clin. Invest. 88 (1991), pp. 1067–1072.

38. R. Soundararajan and A.J. Rao, *Trophoblast "pseudo-tumorigenesis" significance and contributory factors*, Reprod. Biol. Endocrinol. 2 (2004), p. 15.
39. G. Weiss, L.T. Goldsmith, R.N. Taylor, D. Bellet, and H.S. Taylor, *Inflammation in reproductive disorders*, Reprod. Sci. 16 (2009), pp. 216–229.
40. H. Cakmak and H.S. Taylor, *Implantation failure: Molecular mechanisms and clinical treatment*, Hum. Reprod. Update 17 (2011), pp. 242–253.
41. J.J. Kim and A.T. Fazleabas, *Uterine receptivity and implantation: The regulation and action of insulin-like growth factor binding protein-1 (IGFBP-1), HOXA10 and forkhead transcription factor-1 (FOXO-1) in the baboon endometrium*, Reprod. Biol. Endocrinol. 2 (2004), p. 34.
42. G.S. Daftary and H.S. Taylor, *Endocrine regulation of HOX genes*, Endocr. Rev. 27 (2006), pp. 331–355.
43. J.R. Sherwin, S.K. Smith, A. Wilson, and A.M. Sharkey, *Soluble gp130 is up-regulated in the implantation window and shows altered secretion in patients with primary unexplained infertility*, J. Clin. Endocrinol. Metab. 87 (2002), pp. 3953–3960.
44. R.D. Catalano, M.H. Johnson, E.A. Campbell, D.S. Charnock-Jones, S.K. Smith, and A.M. Sharkey, *Inhibition of Stat3 activation in the endometrium prevents implantation: A nonsteroidal approach to contraception*, Proc. Natl. Acad. Sci. U.S.A. 102 (2005), pp. 8585–8590.
45. J.S. Fitzgerald, T.G. Poehlmann, E. Schleussner, and U.R. Markert, *Trophoblast invasion: The role of intracellular cytokine signalling via signal transducer and activator of transcription 3 (STAT3)*, Hum. Reprod. Update 14 (2008), pp. 335–344.
46. G.M. Louis, M.A. Cooney, C.D. Lynch, and A. Handal, *Periconception window: Advising the pregnancy-planning couple*, Fertil. Steril. 89 (2008), pp. e119–e121.
47. I. Silva, A. Lichtenfels, L. Pereira, and P. Saldiva, *Effects of ambient levels of air pollution generated by traffic on birth and placental weights in mice*, Fertil. Steril. 90 (2008), pp. 1921–1924.
48. T.J. Woodruff, A. Carlson, J.M. Schwartz, and L.C. Giudice, *Proceedings of the summit on environmental challenges to reproductive health and fertility: Executive summary*, Fertil. Steril. 89 (2008), pp. 281–300.
49. C.H. Hung, S.N. Yang, P.L. Kuo, Y.T. Chu, H.W. Chang, W.J. Wei, S.K. Huang, and Y.J. Jong, *Modulation of cytokine expression in human myeloid dendritic cells by environmental endocrine-disrupting chemicals involves epigenetic regulation*, Environ. Health Perspect. 118 (2010), pp. 67–72.
50. S.M. Rhind, N.P. Evans, M. Bellingham, R.M. Sharpe, C. Cotinot, B. Mandon-Pepin, B. Loup, K.D. Sinclair, R.G. Lea, P. Pocar, B. Fischer, E. van der Zalm, K. Hart, J.S. Schmidt, M.R. Amezaga, and P.A. Fowler, *Effects of environmental pollutants on the reproduction and welfare of ruminants*, Animal 4 (2010), pp. 1227–1239.
51. C.C. Smith and H.S. Taylor, *Xenoestrogen exposure imprints expression of genes (Hoxa10) required for normal uterine development*, FASEB J. 21 (2007), pp. 239–246.
52. C. Dechanet, T. Anahory, J.C. Mathieu Daude, X. Quantin, L. Reyftmann, S. Hamamah, B. Hedon, and H. Dechaud, *Effects of cigarette smoking on reproduction*, Hum. Reprod. Update 17 (2011), pp. 76–95.

6 Early Embryo Development and Bipotential Gonad Formation

Paulo Marcelo Perin and Mariangela Maluf

CONTENTS

ABSTRACT

The formation of a competent embryo capable to attach to the receptive maternal endometrium and initiate the pregnancy represents a sequence of carefully orchestrated events including fertilization, oocyte activation and pronuclei formation, and embryo growth. The mechanisms involved in the different steps of successful early embryo development include intercellular communication responsible for controlling migration and interaction among distinct cell groups, selective gene expression/suppression responsible for cell differentiation, and strict regulation of cell apoptosis. Following fertilization, the first cleavage divisions are regulated by maternally supplied factors stored within the oocyte. As development proceeds, maternal transcripts are depleted, and the process of early embryogenesis becomes dependent on the genetic information derived from the embryonic genome. Blastocyst formation marks the segregation of the first two cell lineages, the inner cell mass (*ICM*) and the trophectoderm (*TE*), that will eventually become the embryo itself and extraembryonic tissues, respectively. The *ICM* will subsequently differentiate into the primitive endoderm (*PE*) and the pluripotent epiblast (*EPI*). After implantation and further differentiation, the establishment of the germ cell lineage in the embryo involves the segregation of the primordial germ cells (*PGCs*) from the pluripotent cell population of the proximal *EPI* during the pregastrulation period. Pluripotency genes are reactivated in the *PGCs* before migration toward the gonadal ridges. During migration, *PGCs* proliferate actively and express specific genes that regulate their survival, reprogramming, and identity. Once entering the gonadal ridges, *PGCs* lose their motility and form the primary sex cords. The unique microenvironment of this site provides signals that are necessary for proper progression through either male or female gametogenesis. The success or failure of the reproductive process involves a delicate balance of numerous factors including gamete production and reserve, embryo development and quality, and uterine receptivity, all acting in combination. The preimplantation period of development represents a critical time during which the embryo is highly susceptible to exogenous insults that can affect future growth and developmental potential, either prenatally or postnatally. This chapter reviews current knowledge about key molecular and cellular events involved in both ovarian development and differentiation, and early embryo development. Potential targets for environmental contaminants during these developmental stages as well as the potential effects of the exposure to these contaminants on critical or sensitive windows of human development are briefly discussed.

KEYWORDS

Blastocyst, early embryo development, environmental contaminants, implantation, primordial germ cells, ovarian development

6.1 INTRODUCTION

The early embryo development represents a sequence of events including fertilization, oocyte activation and pronuclei formation, and embryo growth, which results in the formation of a competent embryo capable of attaching itself to the maternal endometrium and thus initiating the pregnancy. Fertilization is a dynamic and carefully orchestrated process that occurs in the fallopian tube in which haploid female and male gametes are involved in a timely and orderly sequence of events that culminates in the formation of the zygote and the development of a new independent individual. Preimplantation development, which represents the period between fertilization and implantation into the uterus, is characterized by a sequence of rapid cell divisions controlled by various cellular and molecular mechanisms. These mechanisms include intercellular communication regulated by autocrine, paracrine, and endocrine factors along with cell surface and extracellular matrix (*ECM*) protein expression, which control migration and interaction among distinct cell groups. Selective expression and suppression of genes, responsible for cell differentiation, and a strict balance between proliferation and death of cells (apoptosis) are other important mechanisms and are essential to successful development.

The first cleavage divisions are regulated by maternally derived components stored within the oocyte during oogenesis. As development proceeds, maternal transcripts are depleted and the process of early embryogenesis becomes dependent on the genetic information derived from the embryonic genome. Subsequent cleavage stages lead to the first differentiation process, which results in the formation of the blastocyst, the main achievement of preimplantation development. The first two cell lineages, the inner cell mass (*ICM*) and the trophectoderm (*TE*), which have distinct morphological features, function, gene expression, and developmental fate, give rise to embryonic and extraembryonic tissues, respectively. After the first segregation event, the *ICM* cells differentiate into the primitive endoderm (*PE*) and the pluripotent epiblast (*EPI*). The emergence of the blastocyst from the zona pellucida (*ZP*) represents the final event that occurs during the preimplantation embryo's development [1–4].

After implantation and further differentiation, the establishment of the germ cell lineage in the embryo involves the segregation of the primordial germ cells (*PGCs*) from the pluripotent cell population of the proximal *EPI* during the pregastrulation period [5,6]. Specification of *PGC* precursors from the somatic cells of the embryos is initiated by signals that induce the expression of B-lymphocyte-induced maturation protein 1 (*BLIMP-1*), a transcriptional repressor of the histone methyltransferase (*HMT*) subfamily [7]. Simultaneously, before migration toward the gonadal ridges, pluripotency genes including octamer-binding protein 4 (*OCT-4*), sex-determining region Y chromosome (*SRY*)-box 2 (*SOX-2*), and nanog homeobox (*NANOG*) are reactivated in the *PGCs*. During migration, *PGCs* proliferate actively and express a specific set of genes called stem/germ cell genes that regulate their survival, reprogramming, and identity. Additionally, the surrounding somatic cells also contribute to *PGC* proliferation and survival through the synthesis or secretion of various paracrine factors [8].

Once entering the undifferentiated gonads, *PGCs* lose their motility and begin to coalesce with each other and with the somatic cells to form the primary sex cords while they continue to proliferate for 1 or 2 days. The unique somatic microenvironment of the gonadal ridge provides signals that are necessary to regulate the balance between self-renewal and the differentiation that is needed for proper progression through either male or female gametogenesis [9,10]. Immediately before differentiation, postmitotic/premeiotic germ cells that represent the transition between the end of mitotic and the beginning of meiotic cell cycle are present in both male and female embryonic gonads [11]. The commitment to oogenesis involves premeiotic deoxyribonucleic acid (DNA) replication and entry into and progression through prophase of the first meiotic division during fetal life. On the other hand, the commitment to spermatogenesis involves avoidance of meiotic initiation during fetal life, mitotic arrest, and the expression of certain cell fate markers that lock in the male developmental program [12].

In humans, the success or failure of the reproductive process involves a delicate balance of numerous factors including gamete production and reserve, embryo development and quality, and uterine receptivity among others, all acting in combination. The preimplantation period of development represents a critical time during which the embryo is highly susceptible to exogenous insults that can affect future growth and developmental potential, either pre- or postnatally [13,14]. Abnormalities in developmental potential may arise from the unavoidable maternal exposure to environmental toxicants during the periconceptional period affecting not only reproductive but also pregnancy outcome and postnatal life [15].

This chapter reviews current knowledge about key molecular and cellular events involved in both ovarian development and differentiation and early embryo development. Potential targets for environmental contaminants during these developmental stages as well as the potential effects of the exposure to these contaminants on critical or sensitive windows of human development are briefly discussed.

6.2   EARLY EMBRYO DEVELOPMENT

6.2.1   FERTILIZATION

Fertilization is a dynamic and carefully orchestrated process that occurs in the ampullary-isthmic region of the fallopian tube in which the haploid female and male gametes are involved in a timely and orderly sequence of events that culminates in the formation of the zygote and the development of a new independent individual. After ejaculation, the spermatozoon undergoes capacitation in the female reproductive tract, a process by which it becomes competent to fertilize the oocyte. In this process, cholesterol and other sterols are removed and noncovalently attached glycoproteins acquired in the epididymis are released from the sperm surface, remodeling its head and priming it to bind to the *ZP* and subsequently to undergo the *ZP*-induced acrosome reaction (*AR*), which enables it to penetrate the mature oocyte [16,17]. The sequence of events that occurs during capacitation confers on the spermatozoon the ability to acquire hypermotility, adhere to the *ZP*, respond to physiological inducers of *AR*, and initiate oocyte fusion [18].

Once in the ampullary region of the fallopian tube, cumulus- and follicular fluid–derived chemo-attractants such as progesterone help the sperm to locate and be inwardly guided through the expanded cumulus–oocyte complex (*COC*) [19,20]. The *COC* matrix is composed of high concentrations of *ECM* glycosaminoglycan hyaluronan (*HA*) and *HA*-binding proteins/proteoglycans that are essential not only for ovulation but also for sperm capacitation and in vivo fertilization, and its cumulus cells express various immune cell-related genes including toll-like receptor (*TLR*) family members *TLR-4* and related molecules. The capacitated spermatozoa secrete hyaluronidase, which will determine modification and breakdown of the *COC* matrix generating small *HA* fragments, which, in turn, activate the *TLR* system on cumulus cells. The activated cells then release specific chemokines (interleukin [*IL-6*] and chemokine C–C motif ligand [*CCL-4, CCL-5*]) capable of enhancing sperm capacitation and fertilization [17,21].

Sperm interaction with *ZP3* glycoprotein (a component of the *ZP*) triggers a sequence of events potentiated by progesterone that leads to the *AR* [22]. *ZP3* ligand activates sperm receptors leading to protein phosphorylation and increase in intracellular Ca^{+2}. Three distinct signal transduction pathways including the adenylyl cyclase (*AC*)/adenosine $3',5'$-cyclic monophosphate (*cAMP*)/protein kinase A

(*PKA*), the phospholipase C/diacylglycerol/protein kinase C (*PKC*), and the guanylate cyclase/guanosine 3′,5′-cyclic monophosphate (*cGMP*)/protein G pathways are involved in this process. During the acrosome reaction, a rapid and massive influx of Ca^{+2} ions occurs in response to intracellular alkalinization and membrane depolarization, which, in turn, activate Ca^{+2}-specific ion channels in the plasma membrane. High levels of free intracellular Ca^{+2} induce fusions between the outer acrosomal membrane and the overlying sperm plasma membrane, exposing its contents to the extracellular environment [23]. Acrosome-reacted sperm will penetrate the *ZP* and enter the perivitelline space by a combination of hyperactivated motility and enzymatic digestion from acrosomal and membrane-bound enzymes such as hyaluronidase, acrosin, and collagenase [17].

Fusion is confined to a specific region of each gamete, the microvillar-rich region of the oocyte and the equatorial region of the sperm, which might reflect a unique constitutional or morphological feature of this region. The oocyte plasma membrane fuses with the equatorial segment of the sperm head, which is then incorporated into the ooplasm by a process similar to phagocytosis. Adhesion of gamete membranes is mediated by receptor–ligand pairs located on the oocyte and sperm. The oolemma tetraspanin protein (*CD-9*), which appears around the time the oocyte becomes competent to bind sperm, is critical for oocyte/sperm fusion. Another group of oocyte surface proteins implicated in the gamete fusion process is the lipid-linked glycosylphosphatidylinositol-anchored proteins (*GPI-AP*). Additionally, integrins form multimolecular complexes with *CD-9* and are also involved in sperm binding and fusion through interactions with sperm-associated A disintegrin and metalloprotease (*ADAM*) proteins, including fertilin (α and β—*ADAM1* and *ADAM2*, respectively) and cyritestin (*ADAM3*) [24–26].

6.2.2　Oocyte Activation and Pronuclei Formation

Immediately after fertilization, the oocyte undergoes activation, a process characterized by two main molecular events including modifications of *ZP* to prevent polyspermy and release from metaphase II arrest and completion of the second meiotic division. At this time, a specialized group of secretory vesicles positioned at the cortical area of the oocyte release their content into the perivitelline space through a calcium-dependent exocytotic process (cortical reaction) originated from the site of sperm fusion. As a result of this reaction, the cortical enzymes interact with the molecules present in the inner *ZP*, modifying its structure in a way that prevents further penetration of acrosome-reacted sperm, and the oocyte plasma membrane rapidly becomes refractory to a second fusion event.

Oocyte activation involves repeated oscillations of free cytosolic Ca^{+2} ions (Ca_i^{+2}), a direct result of inositol triphosphate (*IP-3*)-mediated Ca^{+2} release induced by sperm-derived phospholipase C zeta (*PLCζ*), which terminates at the time of pronuclear formation [27,28]. Ca_i^{+2} oscillations are responsible for triggering resumption and completion of meiosis and recruitment of maternal mRNAs necessary for the activation of the embryo genome and influencing further embryonic development. During meiotic maturation, cytoplasmic changes include reorganization of endoplasmic reticulum (*ERE*, main calcium storage in the oocyte), increase in the number and changes in biochemical properties (e.g., sensitivity to *IP-3*) of *IP-3* receptors, increase in Ca_i^{+2} concentration stored in *ERE*, and redistribution of *ERE* Ca^{+2}-binding proteins and are necessary to the oocyte's ability to generate long-lasting Ca_i^{+2} oscillations [29]. Metaphase II arrest is maintained by a c-Mos proto-oncogene product and M-phase promoting factor (*MPF*), an important cell cycle controlling element composed of a complex of cell division control protein kinase 2 (*CDC-2*) and cyclin B. Fertilization results in an immediate increase in intracellular (Ca_i^{+2}), which later activates *PKC*. Activated *PKC* down-regulates mitogen-activated protein kinase (*MAPK*) pathways through dephosphorylation, causing *MPF* inactivation that is accompanied by sperm head decondensation and the formation of a nuclear membrane around the male pronucleus in which the chromatin regains its ability for DNA replication and transcription. Activation of the oocyte results in the extrusion of the second polar body and haploidization of the maternal genome. The haploid set of chromosomes decondenses and a nuclear membrane forms around these chromosomes as the female pronucleus [30].

Upon entry of a sperm into the oocyte, the highly condensed and transcriptionally inert paternal chromatin becomes remodeled into the decondensed and transcriptionally competent chromatin of the male pronucleus. At the same time that the oocyte is progressing through anaphase–telophase II transition, in which the oocyte chromatin decondenses in the female pronucleus, sperm chromatin disperses, transiently recondenses into a small mass, and then extensively decondenses within the male pronucleus, acquiring many of the proteins that are associated with the maternal chromatin. The morphological remodeling of sperm chromatin at fertilization is accompanied by the replacement of protamines by histones supplied by the oocyte, which pool during oogenesis [31]. After binding to maternal histones and other proteins, the paternal DNA is invested with a new nuclear envelope of maternal origin containing nuclear lamins and nuclear pore complexes [32]. At this point of development, low levels of transcription are observed. These levels are slightly higher in the male pronucleus because of the hyperacetylation of the histones and DNA hypomethylation compared with those associated with the female pronucleus. Both epigenetic modifications provide greater access to the genome for activators of transcription in the male pronucleus [33].

6.2.3 EMBRYO DEVELOPMENT

After formation, male and female pronuclei migration toward each other into close apposition and to the center of the oocyte is mediated by the sperm aster, a radially arrayed three-dimensional microtubule structure organized by the sperm centrosome that attracts and binds numerous maternal proteins, including γ-tubulin and nuclear mitotic apparatus protein (*NUMA*). As the pronuclear envelopes surrounding the male and female pronuclei dissolve at first mitosis (syngamy), the paternal centrosome duplicates and separates to form the poles of the first mitotic spindle, and the parental chromosomes intermix as they align at the metaphase plate. The DNA undergoes replication during interphase and a new sequence of cell cycle–mediating events leads to the initiation of the first division and early development [34].

The preimplantation period of development, approximately 5 to 7 days in humans, is characterized by four major events including the transition of maternal-to-zygotic transcripts, compaction, the first cell lineage segregation into *ICM* and *TE*, and zona hatching and implantation to the uterine wall. After syngamy, the zygote undergoes a process of division known as cleavage, in which it divides by mitosis into a number of smaller cells (blastomeres). Since cell proliferation occurs in the absence of significant cell growth, the volume of the preimplantation embryo remains essentially constant between the zygote and the early blastocyst stages. During the early cleavage stages preceding embryonic genome activation (*EGA*), which in humans occurs around the four-cell and eight-cell stages, embryonic functions are largely controlled by maternally inherited mRNAs, proteins, and other macromolecules. These factors are responsible for supporting embryonic metabolism, directing early developmental events (axis formation, cell fate determination), and controlling the process of genome activation by posttranscriptional mechanisms (Figure 6.1) [35].

With the onset of *EGA*, an event that occurs in a stepwise manner for a variety of mammalian species including humans, transcripts that result from activation of the embryonic genome replace depleted maternal transcripts and are involved in early embryogenesis before implantation. The destruction of maternal mRNAs results not only from maternally encoded products but also from proteins and micro RNAs resulting from the transcription of the zygotic genome that provide feedback to enhance the efficiency of maternal mRNA degradation. Additionally, transcriptional activators synthesized de novo in the embryo enhance the efficiency of zygotic transcription. These events result in the transfer of the control of development from the maternal to the zygotic genome [36]. In the absence of appropriate activation and maintenance of embryonic gene expression, the embryo fails to develop beyond early cleavage stages. *EGA* also leads to gene expression reprogramming, which probably accounts for the conversion of the differentiated gamete nuclei into a totipotent zygote nucleus capable of giving rise to a whole, normal, and fertile individual. Blastomeres of the early cleavage-stage embryo equally share this transient totipotent state that lasts only a few cell cycles, disappearing at the blastocyst stage [37,38].

Throughout the earliest stages of embryogenesis, *EGA* involves various essential molecular events that occur simultaneously and establish the subsequent developmental process. Gene reprogramming that is concomitant with *EGA* relies on extensive epigenetic modifications of the genome that coordinate nuclear and cytoplasmic events through bidirectional communication between them and corresponds to both a change in the genetic origin of the transcripts (maternal or embryonic) and a change in the program of gene expression. Embryonic gene expression activation requires chromatin remodeling, which is regulated by two classes of enzymes, adenosine triphosphate (ATP)-dependent nucleosome remodelers (switching defective/sucrose nonfermenting [*SWI/SNF*], imitation switch [*ISWI*], chromodomain [*CHD*], helicase, DNA binding, and inositol requiring 80 [*INO88*]) and histone-modifying enzymes. ATP-dependent remodelers disrupt nucleosome DNA contacts, move nucleosomes along DNA, and remove or exchange nucleosomes, while histone-modifying enzymes determine posttranslational modifications that occur on the N-terminal tails of these proteins to alter the structure of the chromatin and provide binding sites for regulatory proteins [33,39].

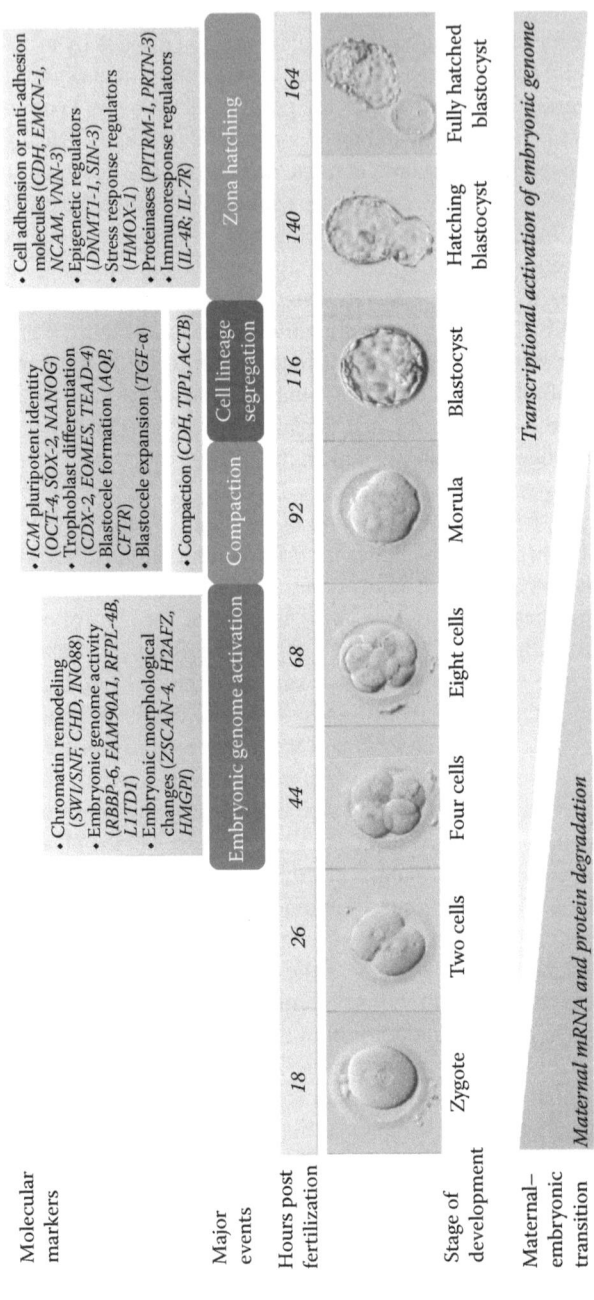

FIGURE 6.1 (See color insert.) Major morphological and molecular markers during the preimplantation period of embryo development.

The processes of translation and degradation of the maternally inherited mRNAs stored in the cytoplasm of the oocyte before ovulation are both concomitant with and required for the successful completion of *EGA*. Two waves of maternal mRNA turnover occur during early embryo development. The first wave (early maternal), which takes place between metaphase II and the two-cell stage, is characterized by over-representation of the nucleic acid and protein catabolism pathways. The second wave (late maternal) of maternal mRNA degradation includes transcripts that decrease gradually over time and is characterized by overrepresentation of genes involved in the metabolism of mRNA and proteins [40]. After remodeling, the embryonic genome becomes accessible to activation by regulatory and transcription factors. Transcriptional activation in the human embryo occurs in three distinct waves at the 2-cell stage, the 4-cell stage, and between the 6- and 8- to 10-cell stages (major activation). Active transcription of genes encoding the DNA-binding proteins retinoblastoma binding protein 6 (*RBBP-6*), which binds to the retinoblastoma (*pRB*) tumor suppressor protein, family with sequence similarity 90, member A1 (*FAM90A1*), a member of the primate-specific family of FAM90A transcripts, RING finger protein 211 (*RFPL-4B*), and LINE-1 type transposase domain containing 1 (*L1TD1*), represents the earliest sign of embryonic genome activity during the first two waves and is required to coordinate the later major activation. During the last transcriptional wave (six- to eight-cell stage), dramatic morphological changes to the embryo including compaction and blastocele formation are observed and genes such as SCAN domain and four zinc finger domains (*ZSCAN-4*), H2A histone family member Z (*H2AFZ*), and high mobility group box protein (*HMGP1*), a novel preimplantation-specific gene that encodes a chromosomal protein containing high mobility group (*HMG*) box domains, are highly expressed [41,42].

Compaction initiated at the 6- to 10-cell stage is signaled by an increase in cell-to-cell contact between blastomeres and does not require a prior round of DNA replication or protein synthesis. The cells flatten against each other and begin to form a variety of junctions between them including the following: (a) gap junctions, which provide channels for the direct passage of small molecules (sugars, amino acids, nucleotides, *cAMP*); (b) adherens junctions, which promote cell-to-cell association via calcium-dependent interactions of the extracellular domains of members of the cadherin family; and (c) tight junctions, which form a permeability seal that prevents the passage of small molecules between cells. Compaction is characterized by the expression of E-cadherin (*CDH*), tight junction protein 1 (*TJP-1*), and actin β (*ACTB*) [43]. After compaction, the boundaries between blastomeres can no longer be distinguished and the cells of the compacted embryo (morula) become highly polarized (surface and cytoplasmic polarity). Surface polarity can be seen by the appearance of dense microvillar (apical) and amicrovillar (basolateral) regions, and cytoplasmic polarity can be seen by the distribution of actin filaments under the apical surface, basolateral location of the cell nucleus, and the presence of endocytotic vesicles located between the apical region and basolateral nucleus [2,44]. The emergence of pluripotent *ICM* lineage from the morula is controlled by metabolic and signaling pathways including wingless-type MMTV integration site (*WNT*), *MAPK*, transforming growth factor β (*TGF-β*), *NOTCH*, integrin-mediated cell adhesion, and apoptosis-signaling pathways. *WNT* and *MAPK* signaling pathways are also active in the differentiation to *TE* cell lineage. Activation of *MAPK* signaling pathway improves proliferative and invasive potential of human trophoblast cells. The activation of the coagulation system, the pathway associated with the role of breast cancer 1 (*BRCA-1*) in DNA damage response, and the leukocyte extravasation signaling pathway play a role in *ECM* and endometrium remodeling and in cytotrophoblast invasion during implantation within the endometrium [3,45].

At the morula stage, the first irreversible segregation of cell commitment becomes apparent. The cells placed in the inside retain pluripotency and express pluripotency-associated genes including *OCT-4*, *SOX-2*, *NANOG*, and teratocarcinoma-derived growth factor 1 (*TDGF-1*), while they exclude markers of trophectodermal lineages such as caudal-related homeobox 2 (*CDX-2*). This group of cells (*ICM*) gives rise to cells that will form the embryo as well as the ectoderm, endoderm, and mesoderm components of the placenta. Transcription factors *OCT-4*, *SOX-2*, and *NANOG* are central to the maintenance of the pluripotent identity of the *ICM* cells, regulating their own and each other's expression in a coordinated manner that involves positive protein–protein and protein–DNA feedback loop interactions. Additionally, these transcription factors promote the transcription of pluripotency-promoting genes such as dosage-sensitive sex reversal, adrenal hypoplasia critical region on chromosome X, gene 1 (*DAX-1*), homeobox expressed in *ESC-1* (*HESX-1*), sal-like 4 (*SALL-4*), repressor element-1 silencing transcription factor (*REST*), reduced expression protein 1 (*REX-1*), signal transducer and activator of transcription 3 (*STAT-3*), T-box family (*TBX-3*), T cell leukemia/lymphoma (*TCL*), transcription factor 3 (*TCF-3*), and zinc finger protein of the cerebellum 3 (*ZIC-3*) and the repression of genes involved in *TE* development (eomesodermin [*EOMES*], heart and neural crest derivatives-expressed 1 [*HAND-1*]) and in lineage commitment to the three germ layers (ectoderm, endoderm, and mesoderm) including homeobox B1 (*HOXB-1*), LIM homeobox 5 (*LHX-5*), and orthodenticle homolog 1 (*OTX-1*), involved in ectoderm development; forkhead box A2 (*FOXA-2*) and GATA-binding protein 6 (*GATA-6*), involved in endoderm development; homeobox gene goosecoid (*GSC*), myogenic factor 5 (*MYF-5*), and brachyury protein T-box transcription factor (*Tb*), involved in mesoderm development [46–48].

By contrast, in the cells positioned on the outside that will develop into the extraembryonic *TE*, which gives rise exclusively to cells that eventually make up the placenta, the activity of key transcription factors such as *CDX-2*, *EOMES*, and TEA domain family member 4 (*TEAD-4*) is up-regulated. Genes with established functions in important processes of trophoblast biology including trophoblast differentiation (GATA-binding protein 2 [*GATA-2*], GATA-binding protein 3 [*GATA-3*], and *HAND-1*), invasion within maternal endometrium (E-cadherin 1 [*CDH-1*], hepatocyte growth factor [*HGF*], insulin-like growth factor 2 [*IGF-2*], and mucin 15 [*MUC-15*]), *ECM* remodeling (collagen alpha-1[IV] [*COL4A-1*], lysyl oxidase-like 4 [*LOXL-4*], microfibrillar-associated protein 5 [*MFAP-5*], and tissue inhibitor of metalloproteinase 2 [*TIMP-2*]), and cell migration and invasion (cathepsin L2 [*CTSL-2*], lumican [*LUM*], and rat sarcoma[*RAS*]-related protein 25 [*RAB-25*]) display significantly increased expression in *TE* cells. Transcription factors estrogen-related receptor β (*ESRRB*), E-twenty-six transcription family 2 (*ETS-2*), and E74-like factor 5 (*ELF-5*) have shown to be essential for subsequent maintenance of trophoblast identity [45].

After compaction, subsequent cleavage divisions allocate cells to the interior of the morula in two different time points: between 8- and 16-cell stages and then again between 16- and 32-cell stages of embryo development. In the first time point, the cleavage plane of the embryo randomly allocates daughter blastomeres to the inside or outside. However, in the second time point, the cleavage plane of the outer cells allocates 80% of the daughter cells to the outside and 20% directly to the inside. This results in approximately 10 to 12 cells comprising the *ICM* of the early blastocyst. At this stage, pluripotency becomes restricted to the cells of the *ICM* and is correlated with developing restrictions in gap junction–mediated intercellular communication. Subsequent divisions of the differentiated *TE* cells always result in *TE* descendants, which communicate among themselves and little with *ICM* cells, forming separate communication compartments that could be involved in the maintenance of the two cell lineages [44]. Once established, cell lineage segregation must be accompanied by epigenetic modifications that ensure the stable inheritance of cell fate. DNA methylation represents one of the earliest global epigenetic marks that distinguish the *ICM* and *TE* lineages and functions as a lineage barrier between embryonic and trophoblast lineage compartments. *ICM* cells become de novo methylated in the early blastocyst while *TE* cells remain hypomethylated, and this global methylation asymmetry observed between embryonic and extraembryonic tissues is maintained throughout development [49].

Blastocyst formation (cavitation), which occurs between the 16- and 32-cell stages, depends on the ion and water transport functions of the differentiated *TE* cells. The formation of a fluid-filled cavity (blastocele) is dependent upon the presence of tight junctions between *TE* cells, which prevent the loss of fluid and ions present in the blastocele, and the polarized distribution of the Na^+/K^+-ATPase confined to the baso-lateral membrane domain of the mural *TE*, which establishes a trans-trophectoderm sodium gradient that facilitates the movement of water across the epithelium through aquaporin water channels (*AQP*) found in the apical and basolateral membranes of the *TE*. The movement of other ions such as chloride and bicarbonate also contrib-utes to blastocele formation and is mediated by cystic fibrosis transmembrane con-ductance regulator (*CFTR*), a *cAMP*-regulated Cl^- channel. The rate of blastocele expansion is stimulated by *cAMP* and growth factors such as transforming growth factor α (*TGF-α*). Molecular association between transmembrane proteins engaged in intercellular adhesion (occludin, claudins, and junction adhesion molecules) with several cytoplasmic plaque proteins (cingulin and zona occludens proteins—*ZO-1*, *ZO-2*, *ZO-3*) and the actin cytoskeleton is necessary for the tight junction assem-bly and function. *CDH*-mediated cell–cell adhesion at embryo compaction is also required for the formation of the *TE* epithelium and blastocele [2,50,51].

Blastocele formation and expansion is critical for subsequent development since it is essential for further *ICM* differentiation. Blastocele fluid contains factors and proteins secreted by both *ICM* and *TE* cells, which are critical for *ICM* cell prolifera-tion. After specification of the *TE*, the *ICM* segregates into two distinct cell lineages, the *PE* and *EPI*. The *PE* forms a monolayer of cells along the surface of the *EPI* that are in contact with the fluid-filled cavity, while the *EPI* remains a mass of cells between the *PE* and the *TE*. Activation of the GATA-binding protein 4 (*GATA-4*) and *GATA-6* transcription factor genes, which antagonize the expression of pluripotency transcription factors such as *NANOG*, is required for *PE* differentiation. The remaining *ICM* cells not involved in this differentiation process express pluripotency genes and become progenitors for all cells of the future organism. Programmed cell death (apoptosis) is essentially restricted to the *ICM* and is critical for later development since it allows for the deletion of a minority of cells with redundant or deleterious potential [52].

The emergence of the blastocyst from the *ZP* (hatching) represents the final event that occurs during the preimplantation embryo development. At day 5 of development, the blastocyst progressively expands and there is a progressive thinning of the *ZP*. As the volume increases, the hydrostatic pressure within the blastocele is increased and the trophoblast epithelium is stretched, forming a continuous robust epithelium with specialized cell junctions. The trophoblast then emerges through a small break in the *ZP*, usually away from the *ICM*. Specialized trophoblast cells (zona-breaker cells), with surface microvilli, bundles of contractile tonofilaments, and lysosomes, secrete clear vesicles that interact with the *ZP* at the hatching point. After leaving the *ZP*, the blastocyst expands instantaneously and an invasive syncytiotrophoblast begins to appear and proliferate at the *ICM* pole. The syncytiotrophoblast is a continuous multinucleated structure with surface microvilli and larger surface protrusions that will aid in the process of implantation in the endometrium [53]. Global gene expression changes observed during the hatching of the blastocyst are essential to the process of implantation. The genes up-regulated at the hatching stage include cell adhesion or antiadhesion molecules (*CDH*, endomucin-1 [*EMCN-1*], neural cell adhesion molecule [*NCAM*], and vanin 3 [*VNN-3*]), epigenetic regulators (DNA methyltransferase 1 [*DNMT-1*] and SIN3 transcription regulator [*SIN-3*]), stress response regulators (heme oxygenase 1 [*HMOX-1*]), proteinases (pitrilysin metallopeptidase 1 [*PITRM-1*] and proteinase 3 [*PRTN-3*]), and immunoresponse regulators (*IL-4R* and *IL-7R*) [3,54].

6.3 OVARIAN DEVELOPMENT AND DIFFERENTIATION

6.3.1 Primordial Germ Cell Specification

Human gamete precursor cells are known as *PGCs*, identified as a small population of specific pluripotent cells located outside the embryo in the developing yolk sac at 2 to 3 weeks of development [55]. At this stage, the cell ultrastructure includes a large round nucleus with dispersed chromatin, centrally located prominent nucleoli and very little cytoplasm containing glycogen particles, mitochondria of vesicular type, bundles of tonofilaments, and occasional lipid droplets [56]. They are characterized by alkaline phosphatase staining and by expression of the transcription factor *OCT-4*, a marker of pluripotent stem cells [57].

Precursors of *PGCs* are located in the *EPI* next to the extraembryonic ectoderm (proximal *EPI*) [58]. Members of the bone morphogenetic protein (*BMP*) family, secreted from the proximal *EPI* (*BMP-4* and *BMP-8b*) and visceral endoderm (*BMP-2*), acting through the *TGF-β* activin receptor-like kinase 2 (*ALK-2*) type I receptor activation and the intracellular mothers against DPP homologs (*SMAD*) signaling molecules phosphorylation (*SMAD-1*, *SMAD-5*, and *SMAD-4*) have critical roles in the early specification of *PGC* precursors from the somatic cells of the embryo [6]. The ability of *PGC* precursors to respond to the *BMP* signaling occurs in a clear temporal fashion for a rather narrow window of time and depends on the expression of wingless-type MMTV integration site family member 3 (*WNT-3*) [59,60].

The expression of Fragilis family genes, which code for interferon-induced transmembrane proteins (*IFITM*) that are associated with the acquisition of germ cell competence and further differentiation, is induced by *BMP-4* in the *EPI* tissue and mediates interactions between germ cells and their surrounding neighbors [6,61]. The subpopulation of *IFITM*-positive *EPI* clustered cells express *BLIMP-1*, a transcriptional repressor of the *HMT* subfamily identified as an important regulator of *PGC* specification and E-cadherin (*CDH*) [5,62]. Active repression of genetic programs that promote somatic differentiation occurs during germ cell specification and is required to maintain germline identity and underlying pluripotency [8,63]. *BLIMP-1* inhibition of somatic transcription factor expression, reactivation of potential pluripotency, induction of germ cell lineage genes, and epigenetic modification mediated by *BLIMP-1*/protein arginine methyltransferase 5 (*PRMT-5*, a methyltransferase with the ability to catalyze the dimethylation of arginine 3 on the N-terminal tails of histones H2A and H4) complex represent the initial events driving germ cell commitment [62,64,65]. The *BLIMP-1/PRMT-5* complex determines a repressive state of the chromatin that precludes the expression of somatic differentiation programs [10]. Shortly after specification, *PGCs* no longer require *CDH*-mediated cell–cell adhesion and start to express the germ cell developmental pluripotency-associated 3 marker (*DPPA-3* also known as *stella*) and pluripotency markers *OCT-4*, *SOX-2*, and *NANOG* [5,64,66–68].

6.3.2 PRIMORDIAL GERM CELL MIGRATION AND PROLIFERATION

Between 4 and 6 weeks of gestation, specified *PGCs* migrate by passive (lateral folding, hindgut expansion) and ameboid (single and grouping) movement through interaction with an *ECM* gradient and chemotactic signaling along the pathway from the base of the yolk sac along the hindgut to the gonadal ridge (a thickened region along the ventral cranial mesonephros that originates the embryonic ovary) [57,69–71]. Kit ligand (*KL*), also known as stem cell factor (*SCF*), a pleiotropic growth factor, can be detected in the somatic cells present along the path of *PGC* migration and its expression progressively increases toward the gonadal ridges. The interaction between *KL* and its cognate tyrosine kinase receptor (*c-KIT*) present on the surface of *PGCs* facilitates their direct migration and promotes general motility. However, *KL* expression does not provide directional information [72,73]. The appearance of the basement membrane molecules such as laminin; fibronectin; types I, III, and IV collagen; perlecan; heparan sulfate; syndecan-4; versican; hyaluronan; and tenascin-C coordinated in time and space in the *PGC* migratory pathway was associated with their acquisition of migratory ability [69]. Recent data suggest that *PGCs* in human embryos migrate along autonomic nerve fibers and Schwann cells from the dorsal mesentery to the developing gonads [74]. The release of lysophosphatidic acid by Schwann cells exerts a chemo-attractant effect on the *PGCs* and acts on cellular adhesion between the two cell types, not only guiding migration but also favoring survival of the *PGCs* [74,75]. Germ cell–germ cell interactions through cell adhesive molecules including cadherins (EP- and N-cadherins), integrins, and the IgG superfamily (platelet/endothelial cell adhesion molecule [*PECAM-1*]) are also involved in their migration control and differential fate [69,76]. The expression of *CDH* and β1-integrin, responsible for cell–cell and cell–*ECM* interactions, respectively, by *PGCs* during migration is necessary for their colonization of the gonadal ridges [12].

During migration, *PGCs* proliferate actively and express a specific set of genes called stem/germ cell genes that regulate their survival, reprogramming, and identity (Figure 6.2). Additionally, the surrounding somatic cells also contribute to *PGC* proliferation and survival through the synthesis or secretion of various paracrine factors [8]. *SCF/c-KIT* signaling is required for proliferation and survival of *PGCs*. Nanos homolog 3 (*NANOS-3*), a member of the *NANOS* family of RNA binding proteins, functions as a translational inhibitor and is involved in the maintenance of *PGC* survival by suppressing both B cell lymphoma/leukemia-2 (*BCL-2*)-associated X protein (*BAX*)-dependent and -independent apoptotic pathways [77]. Fibroblast growth factors (*FGF*), which are actively expressed during the migratory phase by neighboring cells but not *PGCs*, are involved in the initiation of *PGC* proliferation through a paracrine communication pathway [78]. Members of the *BMP* and *IL* (*IL-2*, *IL-4*, *IL-6*, *IL-11*,

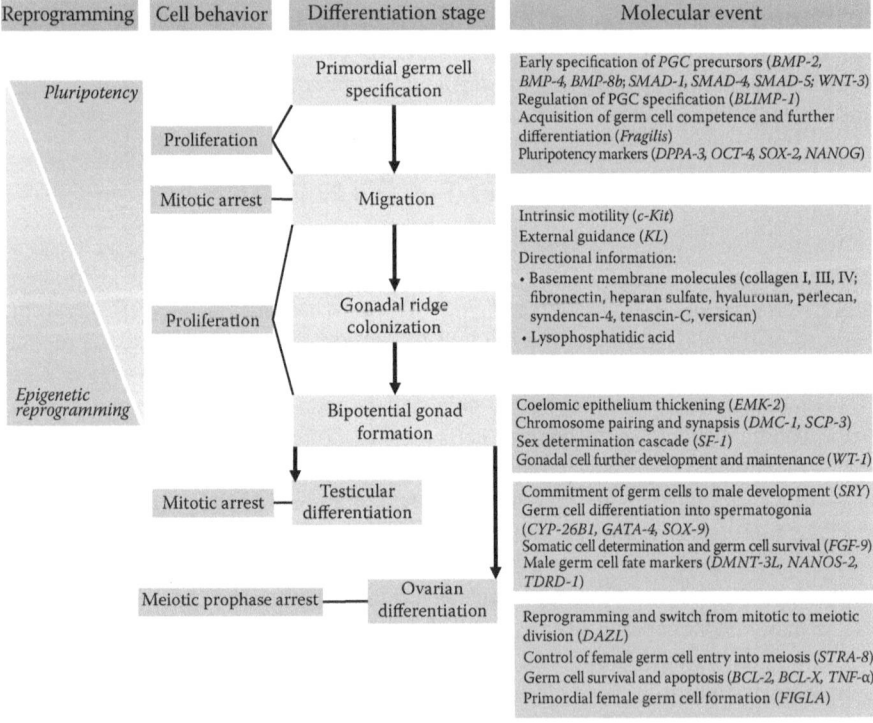

FIGURE 6.2   (See color insert.) Schematic representation of molecular events, differentiation stages, and cell behavior and reprogramming during germline development and gonad formation.

and leukemia inhibitory factor [*LIF*]) families are also required for *PGC* proliferation and survival [79,80]. The up-regulation of proapoptotic genes in *PGCs* and down-regulation of *SCF* (a survival factor) in midline somatic cells during migration suggest that germ cells that do not reach the gonadal ridges are destined to die by apoptosis [12].

6.4   BIPOTENTIAL GONAD FORMATION

The gonadal ridges are formed in humans around week 4 of gestation with the thickening of the mesoderm between the developing mesonephros and the dorsal mesentery root [55]. Several key genes are important for the formation of the bipotential gonad. These include the homeobox gene empty spiracles homolog 2 (*EMX-2*), responsible for the coelomic epithelium thickening, the Wilms tumor 1 homolog (*WT-1*), and LIM homeobox protein 9 (*LHX-9*) involved in the further development and maintenance of gonadal cells. Steroidogenic factor 1 (*SF-1*) is also involved and is expressed in the coelomic epithelium and also in the daughter cells that migrate into the gonadal ridge to become either Sertoli or granulosa cells, playing an important role in the sex determination cascade [6].

The undifferentiated gonads have a cortex and a medulla that give rise to ovary or testis by differential development depending on the genetic makeup of the somatic cells in and surrounding the gonadal ridges rather than on the chromosomal sex of germ cells themselves. Upon arrival at the gonadal ridge, *PGCs* lose their motility and begin to coalesce with each other and with the somatic cells to form the primary sex cords while they continue to proliferate for 1 or 2 days [9]. The unique somatic microenvironment of the gonadal ridge provides signals that are necessary to regulate the balance between self-renewal and the differentiation that is needed for proper progression through either male or female gametogenesis [10]. Immediately before differentiation, postmitotic/premeiotic germ cells that represent the transition between the end of the mitotic and the beginning of the meiotic cell cycle are present in both male and female embryonic gonads [11]. At this point, several meiosis-specific genes including the synaptonemal complex gene (*SCP-3*) and dosage suppressor of *MCK-1* homolog gene (*DMC-1*) are up-regulated to prepare for chromosome pairing and synapsis [81].

6.4.1   TESTICULAR DIFFERENTIATION

The expression of the Y chromosome–linked gene *SRY* in the somatic cells of the bipotential gonad initiates the commitment of germ cells to male development through their differentiation into Sertoli cells that form epithelial aggregates and align into testis cords. The earliest sign of this differentiation is the up-regulation of sex-determining genes such as *SRY*-related HMG-box 9 (*SOX-9*) and *GATA-4* [82,83]. Once specified, Sertoli cells coordinate the events that are involved in the differentiation and topographical organization of the surrounding cells into the distinct cell types of the developing testis and induce mitotic arrest in germ cells, conferring a spermatogenic fate upon them. The migration and proliferation of cells from the adjacent mesonephros occur after the onset of *SRY* expression and determine the increase in gonadal size. The differentiation of these cells into distinct testicular cell populations, in association to *ECM* remodeling, cell associations, and cell movements, leads to the characteristic testis cord structure in which germ cells are surrounded by epithelial Sertoli cells. These Sertoli cells are in turn surrounded by peritubular myoid cells, steroidogenic Leydig cells, and a branching pattern of vasculature found between cords [84,85].

Once the testis cords have formed, germ cells in the developing gonad enter mitotic arrest in the G_0/G_1 phase of the cell cycle as prospermatogonia, remaining in this quiescent state until after birth when they resume proliferation. The location of male germ cells within the testis cords protects them from the effects of retinoic acid (*RA*), an inducer of entry into meiosis through the activation of retinoic acid gene 8 (*STRA-8*). This protection is achieved through the action of a P450 26B1 enzyme (*CYP-26B1*) that degrades *RA* and is expressed by Sertoli cells surrounding the germ cells in the testis cords [81]. Additionally, the progression of germ cells into mitotic arrest in males also involves the regulation of a subset of cell cycle proteins that control G1–S phase transition, including up-regulation of the cyclin-dependent kinase (*CDK*) inhibitors, suppression of cyclin E1 and E2, and regulation of the phosphorylation state of the G1–S phase checkpoint protein *pRB*, which becomes hyperphosphorylated in males [8,86]. The expression of fibroblast growth factor 9 (*FGF-9*), necessary for somatic sex determination and germ cell survival in males, occurs at this point of development and reinforces *SOX-9* expression, which is the hallmark of testis differentiation [87]. Although male germ cells are not dividing, important developmental processes occur during their mitotic arrest including epigenetic programming, regulation of pluripotency, cell signaling and nuclear import/export, sex-specific development, and regulation of gene/protein expression [8]. In addition to the mitotic arrest, the expression of male germ cell fate markers such as nanos homolog 2 (*NANOS-2*), DNA methyltransferase 3L (*DNMT-3L*), and tudor domain containing 1 (*TDRD-1*) and the loss of fucosyltransferase 4 (*FUT-4*), a marker of undifferentiated germ cells, are involved in the commitment to the sex-specific developmental program during fetal life [12].

6.4.2 Ovarian Differentiation

In the absence of *SRY*, ovarian pathways prevail in the bipotential gonad. During migration from the base of the yolk sac to the gonadal ridge, *PGCs* undergo mitotic proliferation with incomplete cytokinesis and germ cells become clusters of oogonia connected by intercellular bridges forming the oogonial nests in which most germ cells divide synchronously [88]. Mitotic activity rapidly increases the number of germ cells from approximately 600,000 oogonia at week 8 of gestation to a maximum of seven to eight million in the two gonads at week 20 of gestation [89]. Through a process called *attrition*, this number decreases to one to two million at birth and constitutes the finite female gamete reserve. In contrast to the continuous proliferation of male germ cells, the proliferation of female germ cells only takes place during embryogenesis and represents the major determinant of the ovarian reserve [90].

A few days after *PGC* arrival at the gonadal ridge (week 6 of gestation) somatic cells forming the secondary sex cords invade the cortex from the medulla and completely surround the oogonial nests. Contact between oogonia and sex cord cells gradually diminishes and each individual oogonium becomes enclosed in a single layer of flattened epithelial cells (pregranulosa cells) surrounded by a basement membrane forming the primordial follicles around week 9 of gestation [91]. At this time, the ovary is reorganized into two compartments: the cortex, where the primordial follicles are located, and the medulla, in which the primary sex cords degenerate into the connective tissue of the ovarian hilus that attaches the ovary to the body wall [85,88].

Oogonia that are not surrounded by an adequate number of somatic cells or are not fit to progress further undergo apoptosis, thus limiting the number of primordial follicles. Germ cell apoptosis is regulated by the balance between the members of the *BCL-2* protein family with opposing functions, *BCL-2* and *BCL-2* like 1 (*BCL-X*) protecting against apoptosis, and *BAX* promoting cell death [6,92]. The apoptosis of random oocytes in the oogonial nests is necessary for the assembly of the primordial follicles, and tumor necrosis factor-alpha (*TNF-α*) is involved in this process [93]. In the human fetal ovary, apoptotic activity is restricted to the germ cells rather than to pregranulosa/granulosa cells and coincides with mitotic proliferation of oogonia and the entrance to meiosis from week 14 to week 20 of gestation [92,94].

During primordial follicle formation, the mitotic activity of the oogonia ceases and the meiotic process is initiated at week 13 of gestation. At this time, the oogonia (defined as primary oocytes) enter prophase of the first meiotic division; progress through leptotene (chromosome condensation), zygotene (homologous chromosome alignment), and pachytene (genetic recombination) stages of the first meiotic prophase; and then arrest at the diplotene stage (dictyate stage arrest). The oocyte arrest persists until just before ovulation when the first meiotic division is completed, the second begins and is followed by another arrest in meiosis II, which is only completed after fertilization [12,95].

The deleted in azoospermia-like gene (*DAZL*) is the first gene necessary to orchestrate these changes and is expressed in *PGCs* upon their arrival at the gonadal ridge. In the presence of *DAZL*, the premeiotic germ cells are responsive to *RA*, which originates from the mesonephros and diffuses into the adjacent gonad where it has a vital role in the onset of germ cell meiosis. In the gonad, *RA* up-regulates the expression of *STRA-8* in *PGCs*, resulting in the entry of female germ cells into meiosis [6,8]. Additionally, *CYP-26B1* is down-regulated in somatic cells of the ovary, allowing the accessibility of female germ cells to *RA* [6]. The germline alpha (*FIGLA*), a basic helix-loop-helix transcription factor expressed exclusively in female germ cells, is required for primordial follicle formation, the establishment of the extracellular *ZP* matrix that surrounds the oocyte to mediate fertilization and to block polyspermy, and has an important role in repressing male germ cell–specific genes in oocytes [96].

6.4.3 Epigenetic Reprogramming

During migration and after gonadal colonization, the genome of *PGCs* undergoes significant reprogramming in which parental imprints are erased and totipotency is restored. In this process, epigenetic marks are largely erased and remodeled without any changes in the sequence of DNA through different mechanisms including DNA methylation, histone modification, and ribonucleic acid (RNA)-mediated silencing [97]. Some of these changes are persistent while others are transient. Chromatin modifications are mediated by a group of highly conserved enzymes such as DNA methyltransferases (*DNMTs*), methyl-CpG-binding proteins (*MeCP*), histone acetyl-transferases (*HAT*), histone deacetylases (*HDAC*), *HMT*, histone demethylases/deiminases (*HMD*), and chromatin remodeling complexes (*ChR*) [98].

In the early female embryo, inactivation of one of the X chromosomes occurs randomly in all cells by DNA methylation. During migration and gonadal ridge colonization, *PGC* imprinted methylation marks are gradually erased and reset to assure that parent-appropriate imprints are transmitted to the next generation, and the inactive X chromosome is reactivated in female germ cells. However, not all epigenetic marks present in imprinted genes may be erased in *PCGs*. While methylation imprints are erased in *PGCs*, other epigenetic modifications may not be removed. Histone modifications may direct DNA methylation by inhibiting interactions between certain *DNMTs* and are important for the oocyte meiotic and developmental competency [64]. Additionally, a specific class of small RNAs, the piwi-interacting RNAs (*piRNAs*), acts as a guide for histone-modifying and chromatin remodeling proteins directing DNA methylation to target genome loci and represents an important pathway for the epigenetic reprogramming of the oocyte [99]. The DNA demethylation process is persistent as sex-specific DNA remethylation does not occur until later during fetal life when male germ cells have committed to the spermatogenic fate or, for each female germ cell, during postnatal life just before ovulation [12].

6.5   CRITICAL WINDOWS OF SUSCEPTIBILITY

Exposure to environmental toxicants during the periconceptional period, a critical window of developmental susceptibility, may alter fertility through direct or indirect effects on different pathways involved in cellular processes including mitotic interference, altered cell signaling, altered energy sources, enzyme inhibition, mutation, alterations in gene expression, alterations in DNA and RNA synthesis and functioning, and programmed cell death or by disrupting the endocrine or immune systems, inducing long-term effects on fetal and offspring health [100–102].

Chemical substances that interfere directly with mitochondrial function may impair oocyte maturation, fertilization, and early embryo development since mitochondria play different roles in the oocyte, zygote, and precompaction cleavage stage embryo processes, including spindle organization and chromosomal segregation, cell cycle timing, and blastocyst compaction, cavitation, and hatching (Figure 6.3) [103]. The diminished ATP-generating capacity necessary to appropriate cell maintenance could inhibit cytokinesis, leading to the arrest of cell division and eventual cell death, resulting in developmentally incompetent embryos [104]. Conversely, other chemicals may act indirectly on a receptor, causing the inappropriate activation (agonist effect) or inhibition (antagonist effect) of the normal ligand-induced signaling. Polycyclic aromatic hydrocarbons (*PAH*), toxic chemicals present in polluted air and cigarette smoke, are released into the environment as by-products of fossil fuel combustion. The activation of the aryl hydrocarbon receptors (*AhR*) by the exposure to *PAH* induces the expression of proapoptotic factor *BAX*, causing oocyte loss and

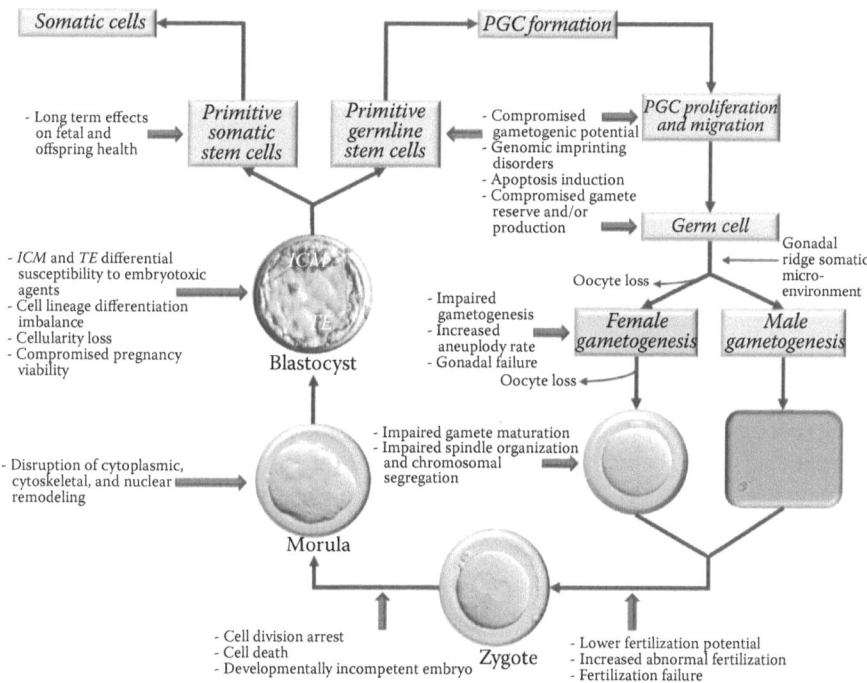

FIGURE 6.3 **(See color insert.)** Potential short- and long-term consequences of the exposure to environmental contaminants during critical steps of embryo and germline development.

diminished ovarian reserve, and could, at least in part, be related to a decline in ovarian response and premature ovarian failure [105–107].

Embryos reaching the compacting morula stage appear to be more susceptible to the effects of environmental contaminants. Major cytoplasmic, nuclear, and cytoskeletal remodeling events occur during compaction, resulting in the first irreversible segregation of cell commitment. The formation of the blastocyst results from the functional differentiation of embryonic cells into two distinct lineages, the *ICM* cells that bring about all embryonic tissues and part of the extraembryonic membranes and the *TE* cells that contribute mainly to the formation of the fetal placenta [108]. Both cell lineages and their ratio have a fundamental role in embryo survival and fetal viability, and the blastocyst has the tools necessary to control the specification of those cell lineages within a relatively narrow range [13]. Because of differences in cell positioning and metabolic requirements, *ICM* and *TE* cell lines have differential susceptibility to embryotoxic agents, and in most instances, *ICM* cells appear less resistant to disruption than *TE* cells [109]. Several transcription factors have been identified as major regulators of the formation and the fate of these two cell lineages. One of the key regulators for *ICM* development is *OCT-4*, a *POU* domain transcription factor (expressed in all blastomeres of the cleavage stage embryo) that becomes restricted to the *ICM* upon blastocyst formation [108]. *CDX-2*, another transcription factor, appears to play a key role in early *TE* specification, and its expression is restricted to the *TE* in peri-implantation embryos [110]. Proper interactions between these and other transcription regulators play decisive roles in the specification and maintenance of these first lineages and are fundamental for the survival of the embryo. *OCT-4* and *CDX-2* negatively regulate the expression of each other to allow proper segregation of *ICM* and *TE* lineages required for normal blastocyst development [108,111]. The imbalance in blastocyst cell lineage differentiation that occurs during early embryonic development may compromise the subsequent postimplantation developmental potential of the embryo [13]. Environmental contaminants that affect the expression of these transcription factors could lead to the loss of the reciprocal inhibition between lineage-specific transcription factors (*OCT-4* and *CDX-2*) involved in the segregation of the first two cell lineages resulting in the loss in cellularity and morphological integrity of the *ICM* and thus negatively influencing pregnancy viability [112–114].

Although protected by surrounding tissues that filter contact with the external environment, both oocytes and embryos can be exposed to various environmental contaminants. The development of ovarian follicles, the individual functional units of the ovary, depends on successful *PGC* migration from the yolk sac to the gonadal ridge, sex differentiation of the gonad, and differentiation into oocytes with associated somatic granulosa and theca cells. Interference with germ cell migration or follicle formation can compromise the gametogenic potential of the gonad with significant reproductive consequences. Sex-specific differentiation of the gonad is based on the expression of specific genes, a process that can be disrupted by mutations in these genes or alterations in factors that control their expression, leading to ovarian dysgenesis. Immediately after sex specification of the gonad, *PGCs* turn into oogonia, which, in turn, transform into oocytes, arrested in the diplotene stage of late prophase until meiotic divisions occur beginning at puberty (meiosis I) and after fertilization (meiosis II). Exposure to chemicals that affect DNA methylation during *PGC* reprogramming may disturb genomic imprinting. Abnormalities in these processes will have a devastating impact on reproductive outcomes including aneuploidy, miscarriage, and premature ovarian failure (Figure 6.3) [115,116].

GLOSSARY

AC: adenylyl cyclase
ACTB: actin β
ADAM: A disintegrin and metalloprotease proteins
AhR: aryl hydrocarbon receptors
ALK-2: activin receptor-like kinase 2
AQP: aquaporin water channels
AR: acrosome reaction
ATP: adenosine triphosphate
BAX: B cell lymphoma/leukemia-2-associated X protein
BCL-2: B cell lymphoma/leukemia-2
BCL-X: BCL-2 like 1
BLIMP-1: B-lymphocyte-induced maturation protein 1
BMP: bone morphogenetic protein

BRCA-1: breast cancer 1
Ca$_i^{+2}$: cytosolic Ca^{+2} ions
cAMP: adenosine 3′,5′-cyclic monophosphate
CCL: chemokine C–C motif ligand
CD-9: tetraspanin protein
CDC-2: cell division control protein kinase 2
CDH: E-cadherin
CDH-1: E-cadherin 1
CDK: cyclin-dependent kinase
CDX-2: caudal-related homeobox 2
CFTF: cystic fibrosis transmembrane conductance regulator
cGMP: guanosine 3′,5′-cyclic monophosphate
CHD: chromodomain
ChR: chromatin remodeling complexes
c-KIT: cognate tyrosine kinase receptor
COC: cumulus–oocyte complex
COL4A-1: collagen alpha-1[IV]
CTSL-2: cathepsin L2
CYP-26B1: P450 26B1 enzyme
DAX-1: dosage-sensitive sex reversal, adrenal hypoplasia critical region on chromosome X, gene 1
DAZL: deleted in azoospermia-like gene
DMC-1: dosage suppressor of *MCK-1* homolog
DNA: deoxyribonucleic acid
DNMTs: DNA methyltransferases
DNMT-1: DNA methyltransferase 1
DNMT-3L: DNA methyltransferase 3L
DPPA-3: developmental pluripotency associated 3
ECM: extracellular matrix
EGA: embryonic genome activation
ELF-5: E74-like factor 5
EMCN-1: cndomucin 1
EMX-2: empty spiracles homolog 2
EOMES: eomesodermin
EPI: epiblast
ERE: endoplasmic reticulum
ESRRB: estrogen-related receptor β
ETS-2: E-twenty-six transcription family 2
FAM90A1: family with sequence similarity 90, member A1
FGF: fibroblast growth factors
FGF-9: fibroblast growth factor 9
FIGLA: factor in the germline alpha
FOXA-2: forkhead box A2
FUT-4: fucosyltransferase 4 (alpha 1,3 fucosyltransferase, myeloid-specific)
GATA-2: GATA-binding protein 2
GATA-3: GATA-binding protein 3

GATA-4: GATA-binding protein 4
GATA-6: GATA-binding protein 6
GPI-AP: glycosylphosphatidylinositol-anchored proteins
GSC: homeobox gene goosecoid
H2AFZ: H2A histone family member Z
HA: hyaluronan
HAND-1: heart and neural crest derivatives-expressed 1
HAT: histone acetyltransferases
HDAC: histone deacetylases
HESX-1: homeobox expressed in ESC-1
HGF: hepatocyte growth factor
HMD: histone demethylases/deiminases
HMG: high mobility group
HMGPI: high mobility group box protein
HMOX-1: heme oxygenase 1
HMT: histone methyltransferase
HOXB-1: homeobox B1
ICM: inner cell mass
IFITM: interferon-induced transmembrane proteins
IGF: insulin-like growth factor
IL: interleukin
INO88: inositol requiring 80
IP-3: inositol triphosphate
ISWI: imitation switch
KL: kit ligand
L1TD1: LINE-1 type transposase domain containing 1
LHX-5: LIM homeobox protein 5
LHX-9: LIM homeobox protein 9
LIF: leukemia inhibitory factor
LOXL-4: lysyl oxidase-like 4
LUM: lumican
MAPK: mitogen-activated protein kinase
MeCP: methyl-CpG-binding proteins
MFAP-5: microfibrillar-associated protein 5
MPF: M-phase promoting factor
MUC-15: mucin 15
MYF-5: myogenic factor 5
NANOG: nanog homeobox
NANOS-2: nanos homolog 2
NANOS-3: nanos homolog 3
NCAM: neural cell adhesion molecule
NUMA: nuclear mitotic apparatus protein
OCT-4: octamer-binding protein 4
OTX-1: orthodenticle homolog 1
PAH: polycyclic aromatic hydrocarbons
PE: primitive endoderm

PECAM-1: platelet/endothelial cell adhesion molecule
PGCs: primordial germ cells
piRNAs: piwi-interacting RNAs
PITRM-1: pitrilysin metallopeptidase 1
PKA: protein kinase A
PKC: protein kinase C
PLCζ: phospholipase C zeta
pRB: retinoblastoma protein
PRMT-5: protein arginine methyltransferase 5
PRTN-3: proteinase 3
RA: retinoic acid
RAB-25: RAS-related protein 25
RAS: rat sarcoma
RBBP-6: retinoblastoma binding protein 6
REST: repressor element-1 silencing transcription factor
REX-1: reduced expression protein 1
RFPL-4B: RING finger protein 211
RNA: ribonucleic acid
SALL-4: sal-like 4
SCF: stem cell factor
SCP-3: synaptonemal complex gene
SF-1: steroidogenic factor 1
SIN-3: SIN3 transcription regulator
SMAD: mothers against DPP homologs
SOX-2: sex-determining region Y (*SRY*)-box 2
SOX-9: *SRY*-related HMG-box 9
SRY: sex-determining region Y chromosome
STAT-3: signal transducer and activator of transcription 3
STRA-8: retinoic acid gene 8
SWI/SNF: switching defective/sucrose nonfermenting
Tb: brachyury protein T-box transcription factor
TBX-3: T-box family
TCF-3: transcription factor 3
TCL: T cell leukemia/lymphoma
TDGF-1: teratocarcinoma-derived growth factor 1
TDRD-1: tudor domain containing 1
TE: trophectoderm
TEAD-4: TEA domain family member 4
TGF-α: transforming growth factor α
TGF-β: transforming growth factor β
TIMP-2: tissue inhibitor of metalloproteinase 2
TJP-1: tight junction protein 1 (zona occludens 1)
TLR: toll-like receptor
TNF-α: tumor necrosis factor-alpha
VNN-3: vanin 3
WNT: wingless-type MMTV integration site

WNT-3: wingless-type MMTV integration site family member 3
WT-1: Wilms tumor 1 homolog
ZIC-3: zinc finger protein of the cerebellum 3
ZO: zona occludens proteins
ZP: zona pellucida
ZSCAN-4: SCAN domain and four zinc finger domains

REFERENCES

1. S. Sudheer and J. Adjaye, *Functional genomics of human pre-implantation develop-ment*, Brief Funct. Genomic Proteomic 6 (2007), pp. 120–132.
2. V. Duranthon, A.J. Watson, and P. Lonergan, *Preimplantation embryo program-ming: Transcription, epigenetics, and culture environment*, Reproduction 135 (2008), pp. 141–150.
3. S. Assou, I. Boumela, D. Haouzi, T. Anahory, H. Dechaud, J. De Vos, and S. Hamamah, *Dynamic changes in gene expression during human early embryo development: From fundamental aspects to clinical applications*, Hum. Reprod. Update 17 (2011), pp. 272–290.
4. M. Gasperowicz and D.R. Natale, *Establishing three blastocyst lineages—Then what?* Biol. Reprod. 84 (2011), pp. 621–630.
5. D. Okamura, T. Kimura, T. Nakano, and Y. Matsui, *Cadherin-mediated cell interaction regulates germ cell determination in mice*, Development 130 (2003), pp. 6423–6430.
6. M.A. Edson, A.K. Nagaraja, and M.M. Matzuk, *The mammalian ovary from genesis to revelation*, Endocr. Rev. 30 (2009), pp. 624–712.
7. M. De Felici, *Germ stem cells in the mammalian adult ovary: Considerations by a fan of the primordial germ cells*, Mol. Hum. Reprod. 16 (2010), pp. 632–636.
8. P. Western, *Foetal germ cells: Striking the balance between pluripotency and differen-tiation*, Int. J. Dev. Biol. 53 (2009), pp. 393–409.
9. M.R. Bendel-Stenzel, M. Gomperts, R. Anderson, J. Heasman, and C. Wylie, *The role of cadherins during primordial germ cell migration and early gonad formation in the mouse*, Mech. Dev. 91 (2000), pp. 143–152.
10. R.M. Cinalli, P. Rangan, and R. Lehmann, *Germ cells are forever*, Cell 132 (2008), pp. 559–562.
11. A. Kocer, J. Reichmann, D. Best, and I.R. Adams, *Germ cell sex determination in mammals*, Mol. Hum. Reprod. 15 (2009), pp. 205–213.
12. J. Bowles and P. Koopman, *Sex determination in mammalian germ cells: Extrinsic versus intrinsic factors*, Reproduction 139 (2010), pp. 943–958.
13. T.P. Fleming, W.Y. Kwong, R. Porter, E. Ursell, I. Fesenko, A. Wilkins, D.J. Miller, A.J. Watkins, and J.J. Eckert, *The embryo and its future*, Biol. Reprod. 71 (2004), pp. 1046–1054.
14. P.M. Perin, M. Maluf, C.E. Czeresnia, D.A. Januario, and P.H. Saldiva, *Impact of short-term preconceptional exposure to particulate air pollution on treatment outcome in couples undergoing in vitro fertilization and embryo transfer (IVF/ET)*, J. Assist. Reprod. Genet. 27 (2010), pp. 371–382.
15. G. Xu, M. Umezawa, and K. Takeda, *Early development origins of adult disease caused by malnutrition and environmental chemical substances*, J. Health Sci. 55 (2009), pp. 11–19.
16. A. Boerke, P.S. Tsai, N. Garcia-Gil, I.A. Brewis, and B.M. Gadella, *Capacitation-dependent reorganization of microdomains in the apical sperm head plasma mem-brane: Functional relationship with zona binding and the zona-induced acrosome reaction*, Theriogenology 70 (2008), pp. 1188–1196.

17. M. Ikawa, N. Inoue, A.M. Benham, and M. Okabe, *Fertilization: A sperm's journey to and interaction with the oocyte*, J. Clin. Invest. 120 (2010), pp. 984–994.

18. A. Barbonetti, M.R. Vassallo, C. Antonangelo, V. Nuccetelli, A. D'Angeli, F. Pelliccione, M. Giorgi, F. Francavilla, and S. Francavilla, *RANTES and human sperm fertilizing ability: Effect on acrosome reaction and sperm/oocyte fusion*, Mol. Hum. Reprod. 14 (2008), pp. 387–391.

19. F. Sun, A. Bahat, A. Gakamsky, E. Girsh, N. Katz, L.C. Giojalas, I. Tur-Kaspa, and M. Eisenbach, *Human sperm chemotaxis: Both the oocyte and its surrounding cumulus cells secrete sperm chemoattractants*, Hum. Reprod. 20 (2005), pp. 761–767.

20. R. Oren-Benaroya, R. Orvieto, A. Gakamsky, M. Pinchasov, and M. Eisenbach, *The sperm chemoattractant secreted from human cumulus cells is progesterone*, Hum. Reprod. 23 (2008), pp. 2339–2345.

21. M. Shimada, Y. Yanai, T. Okazaki, N. Noma, I. Kawashima, T. Mori, and J.S. Richards, *Hyaluronan fragments generated by sperm-secreted hyaluronidase stimulate cytokine/chemokine production via the TLR2 and TLR4 pathway in cumulus cells of ovulated COCs, which may enhance fertilization*, Development 135 (2008), pp. 2001–2011.

22. J. Ramalho-Santos, G. Schatten, and R.D. Moreno, *Control of membrane fusion during spermiogenesis and the acrosome reaction*, Biol. Reprod. 67 (2002), pp. 1043–1051.

23. H.M. Florman, M.K. Jungnickel, and K.A. Sutton, *Regulating the acrosome reaction*, Int. J. Dev. Biol. 52 (2008), pp. 503–510.

24. J.A. Alfieri, A.D. Martin, J. Takeda, G. Kondoh, D.G. Myles, and P. Primakoff, *Infertility in female mice with an oocyte-specific knockout of GPI-anchored proteins*, J. Cell Sci. 116 (2003), pp. 2149–2155.

25. K.K. Stein, P. Primakoff, and D. Myles, *Sperm-egg fusion: Events at the plasma membrane*, J. Cell Sci. 117 (2004), pp. 6269–6274.

26. J.E. Swain and T.B. Pool, *ART failure: Oocyte contributions to unsuccessful fertilization*, Hum. Reprod. Update 14 (2008), pp. 431–446.

27. C.M. Saunders, M.G. Larman, J. Parrington, L.J. Cox, J. Royse, L.M. Blayney, K. Swann, and F.A. Lai, *PLC zeta: A sperm-specific trigger of Ca(2+) oscillations in eggs and embryo development*, Development 129 (2002), pp. 3533–3544.

28. J. Kashir, B. Heindryckx, C. Jones, P. De Sutter, J. Parrington, and K. Coward, *Oocyte activation, phospholipase C zeta and human infertility*, Hum. Reprod. Update 16 (2010), pp. 690–703.

29. A. Ajduk, A. Malagocki, and M. Maleszewski, *Cytoplasmic maturation of mammalian oocytes: Development of a mechanism responsible for sperm-induced Ca2+ oscillations*, Reprod. Biol. 8 (2008), pp. 3–22.

30. Q. Lu, G.D. Smith, D.Y. Chen, Z.M. Han, and Q.Y. Sun, *Activation of protein kinase C induces mitogen-activated protein kinase dephosphorylation and pronucleus formation in rat oocytes*, Biol. Reprod. 67 (2002), pp. 64–69.

31. D.W. McLay and H.J. Clarke, *Remodelling the paternal chromatin at fertilization in mammals*, Reproduction 125 (2003), pp. 625–633.

32. J. Tesarik and V. Kopecny, *Developmental control of the human male pronucleus by ooplasmic factors*, Hum. Reprod. 4 (1989), pp. 962–968.

33. L. Li, P. Zheng, and J. Dean, *Maternal control of early mouse development*, Development 137 (2010), pp. 859–870.

34. K. Chatzimeletiou, E.E. Morrison, N. Prapas, Y. Prapas, and A.H. Handyside, *The centrosome and early embryogenesis: Clinical insights*, Reprod. Biomed. Online 16 (2008), pp. 485–491.

35. P. Braude, V. Bolton, and S. Moore, *Human gene expression first occurs between the four- and eight-cell stages of preimplantation development*, Nature 332 (1988), pp. 459–461.

36. W. Tadros and H.D. Lipshitz, *The maternal-to-zygotic transition: A play in two acts*, Development 136 (2009), pp. 3033–3042.

37. R.M. Schultz, *The molecular foundations of the maternal to zygotic transition in the preimplantation embryo*, Hum. Reprod. Update 8 (2002), pp. 323–331.

38. R.D. Leandri, C. Archilla, L.C. Bui, N. Peynot, Z. Liu, C. Cabau, A. Chastellier, J.P. Renard, and V. Duranthon, *Revealing the dynamics of gene expression during embryonic genome activation and first differentiation in the rabbit embryo with a dedicated array screening*, Physiol. Genomics 36 (2009), pp. 98–113.

39. D.C. Hargreaves and G.R. Crabtree, *ATP-dependent chromatin remodeling: Genetics, genomics and mechanisms*, Cell. Res. 21 (2011), pp. 396–420.

40. A.T. Dobson, R. Raja, M.J. Abeyta, T. Taylor, S. Shen, C. Haqq, and R.A. Pera, *The unique transcriptome through day 3 of human preimplantation development*, Hum. Mol. Genet. 13 (2004), pp. 1461–1470.

41. G. Falco, S.L. Lee, I. Stanghellini, U.C. Bassey, T. Hamatani, and M.S. Ko, *Zscan4: A novel gene expressed exclusively in late 2-cell embryos and embryonic stem cells*, Dev. Biol. 307 (2007), pp. 539–550.

42. M. Yamada, T. Hamatani, H. Akutsu, N. Chikazawa, N. Kuji, Y. Yoshimura, and A. Umezawa, *Involvement of a novel preimplantation-specific gene encoding the high mobility group box protein Hmgpi in early embryonic development*, Hum. Mol. Genet. 19 (2010), pp. 480–493.

43. D.J. Bloor, A.D. Metcalfe, A. Rutherford, D.R. Brison, and S.J. Kimber, *Expression of cell adhesion molecules during human preimplantation embryo development*, Mol. Hum. Reprod. 8 (2002), pp. 237–245.

44. R.M. Schultz, *Preimplantation embryo development*, in *Molecular Biology in Reproductive Medicine*, B. Fauser, ed., The Parthenon Publishing Group, New York, 1999, pp. 313–331.

45. M. Marchand, J.A. Horcajadas, F.J. Esteban, S.L. McElroy, S.J. Fisher, and L.C. Giudice, *Transcriptomic signature of trophoblast differentiation in a human embryonic stem cell model*, Biol. Reprod. 84 (2011), pp. 1258–1271.

46. M. Pesce and H.R. Scholer, *Oct-4: Gatekeeper in the beginnings of mammalian development*, Stem Cells 19 (2001), pp. 271–278.

47. H. Niwa, *How is pluripotency determined and maintained?* Development 134 (2007), pp. 635–646.

48. N.S. Christophersen and K. Helin, *Epigenetic control of embryonic stem cell fate*, J. Exp. Med. 207 (2010), pp. 2287–2295.

49. R.K. Ng, W. Dean, C. Dawson, D. Lucifero, Z. Madeja, W. Reik, and M. Hemberger, *Epigenetic restriction of embryonic cell lineage fate by methylation of Elf5*, Nat. Cell Biol. 10 (2008), pp. 1280–1290.

50. M.R. Ghassemifar, J.J. Eckert, F.D. Houghton, H.M. Picton, H.J. Leese, and T.P. Fleming, *Gene expression regulating epithelial intercellular junction biogenesis during human blastocyst development in vitro*, Mol. Hum. Reprod. 9 (2003), pp. 245–252.

51. A.J. Watson, D.R. Natale, and L.C. Barcroft, *Molecular regulation of blastocyst formation*, Anim. Reprod. Sci. 82–83 (2004), pp. 583–592.

52. K. Cockburn and J. Rossant, *Making the blastocyst: Lessons from the mouse*, J. Clin. Invest. 120 (2010), pp. 995–1003.

53. H. Sathananthan, J. Menezes, and S. Gunasheela, *Mechanics of human blastocyst hatching in vitro*, Reprod. Biomed. Online 7 (2003), pp. 228–234.

54. H.W. Chen, J.J. Chen, S.L. Yu, H.N. Li, P.C. Yang, C.M. Su, H.K. Au, C.W. Chang, L.W. Chien, C.S. Chen, and C.R. Tzeng, *Transcriptome analysis in blastocyst hatching by cDNA microarray*, Hum. Reprod. 20 (2005), pp. 2492–2501.

55. P.M. Motta, S. Makabe, and S.A. Nottola, *The ultrastructure of human reproduction. I. The natural history of the female germ cell: Origin, migration and differentiation inside the developing ovary*, Hum. Reprod. Update 3 (1997), pp. 281–295.

56. T. Funkuda, *Ultrastructure of primordial germ cells in human embryo*, Virchows Arch. B Cell Pathol. 20 (1976), pp. 85–89.

57. T. Goto, J. Adjaye, C.H. Rodeck, and M. Monk, *Identification of genes expressed in human primordial germ cells at the time of entry of the female germ line into meiosis*, Mol. Hum. Reprod. 5 (1999), pp. 851–860.

58. P.P. Tam and S.X. Zhou, *The allocation of epiblast cells to ectodermal and germ-line lineages is influenced by the position of the cells in the gastrulating mouse embryo*, Dev. Biol. 178 (1996), pp. 124–132.

59. J.A. Tucker, K.A. Mintzer, and M.C. Mullins, *The BMP signaling gradient patterns dorsoventral tissues in a temporally progressive manner along the anteroposterior axis*, Dev. Cell 14 (2008), pp. 108–119.

60. Y. Ohinata, H. Ohta, M. Shigeta, K. Yamanaka, T. Wakayama, and M. Saitou, *A signaling principle for the specification of the germ cell lineage in mice*, Cell 137 (2009), pp. 571–584.

61. U.C. Lange, M. Saitou, P.S. Western, S.C. Barton, and M.A. Surani, *The fragilis interferon-inducible gene family of transmembrane proteins is associated with germ cell specification in mice*, BMC Dev. Biol. 3 (2003), p. 1.

62. Y. Ohinata, B. Payer, D. O'Carroll, K. Ancelin, Y. Ono, M. Sano, S.C. Barton, T. Obukhanych, M. Nussenzweig, A. Tarakhovsky, M. Saitou, and M.A. Surani, *Blimp1 is a critical determinant of the germ cell lineage in mice*, Nature 436 (2005), pp. 207–213.

63. S. Strome and R. Lehmann, *Germ versus soma decisions: Lessons from flies and worms*, Science 316 (2007), pp. 392–393.

64. C.R. Nicholas, S.L. Chavez, V.L. Baker, and R.A. Reijo Pera, *Instructing an embryonic stem cell-derived oocyte fate: Lessons from endogenous oogenesis*, Endocr. Rev. 30 (2009), pp. 264–283.

65. M. Saitou, *Germ cell specification in mice*, Curr. Opin. Genet. Dev. 19 (2009), pp. 386–395.

66. M. Pesce and H.R. Scholer, *Oct-4: Control of totipotency and germline determination*, Mol. Reprod. Dev. 55 (2000), pp. 452–457.

67. K. Mitsui, Y. Tokuzawa, H. Itoh, K. Segawa, M. Murakami, K. Takahashi, M. Maruyama, M. Maeda, and S. Yamanaka, *The homeoprotein Nanog is required for maintenance of pluripotency in mouse epiblast and ES cells*, Cell 113 (2003), pp. 631–642.

68. D. Bhartiya, S. Kasiviswanathan, S.K. Unni, P. Pethe, J.V. Dhabalia, S. Patwardhan, and H.B. Tongaonkar, *Newer insights into premeiotic development of germ cells in adult human testis using oct-4 as a stem cell marker*, J. Histochem. Cytochem. 58 (2010), pp. 1093–1106.

69. M. Soto-Suazo and T. Zorn, *Primordial germ cells migration: Morphological and molecular aspects*, Anim. Reprod. 2 (2005), pp. 147–160.

70. Y. Seki, M. Yamaji, Y. Yabuta, M. Sano, M. Shigeta, Y. Matsui, Y. Saga, M. Tachibana, Y. Shinkai, and M. Saitou, *Cellular dynamics associated with the genome-wide epigenetic reprogramming in migrating primordial germ cells in mice*, Development 134 (2007), pp. 2627–2638.

71. M. De Felici, *Primordial germ cell biology at the beginning of the XXI century*, Int. J. Dev. Biol. 53 (2009), pp. 891–894.

72. K.J. Hutt, E.A. McLaughlin, and M.K. Holland, *Kit ligand and c-Kit have diverse roles during mammalian oogenesis and folliculogenesis*, Mol. Hum. Reprod. 12 (2006), pp. 61–69.

73. Y. Gu, C. Runyan, A. Shoemaker, A. Surani, and C. Wylie, *Steel factor controls primordial germ cell survival and motility from the time of their specification in the allantois, and provides a continuous niche throughout their migration*, Development 136 (2009), pp. 1295–1303.

74. K. Mollgard, A. Jespersen, M.C. Lutterodt, C. Yding Andersen, P.E. Hoyer, and A.G. Byskov, *Human primordial germ cells migrate along nerve fibers and Schwann cells from the dorsal hind gut mesentery to the gonadal ridge*, Mol. Hum. Reprod. 16 (2010), pp. 621–631.

75. J.A. Weiner, N. Fukushima, J.J. Contos, S.S. Scherer, and J. Chun, *Regulation of Schwann cell morphology and adhesion by receptor-mediated lysophosphatidic acid signaling*, J. Neurosci. 21 (2001), pp. 7069–7078.

76. M. De Felici, M.L. Scaldaferri, and D. Farini, *Adhesion molecules for mouse primordial germ cells*, Front. Biosci. 10 (2005), pp. 542–551.

77. H. Suzuki, M. Tsuda, M. Kiso, and Y. Saga, *Nanos3 maintains the germ cell lineage in the mouse by suppressing both Bax-dependent and -independent apoptotic pathways*, Dev. Biol. 318 (2008), pp. 133–142.

78. E. Kawase, K. Hashimoto, and R.A. Pedersen, *Autocrine and paracrine mechanisms regulating primordial germ cell proliferation*, Mol. Reprod. Dev. 68 (2004), pp. 5–16.

79. C. Eguizabal, M.D. Boyano, A. Diez-Torre, R. Andrade, N. Andollo, M. De Felici, and J. Arechaga, *Interleukin-2 induces the proliferation of mouse primordial germ cells in vitro*, Int. J. Dev. Biol. 51 (2007), pp. 731–738.

80. A. Ross, S. Munger, and B. Capel, *Bmp7 regulates germ cell proliferation in mouse fetal gonads*, Sex Dev. 1 (2007), pp. 127–137.

81. J. Bowles, D. Knight, C. Smith, D. Wilhelm, J. Richman, S. Mamiya, K. Yashiro, K. Chawengsaksophak, M.J. Wilson, J. Rossant, H. Hamada, and P. Koopman, *Retinoid signaling determines germ cell fate in mice*, Science 312 (2006), pp. 596–600.

82. R. Sekido, I. Bar, V. Narvaez, G. Penny, and R. Lovell-Badge, *SOX9 is up-regulated by the transient expression of SRY specifically in Sertoli cell precursors*, Dev. Biol. 274 (2004), pp. 271–279.

83. M. Bielinska, A. Seehra, J. Toppari, M. Heikinheimo, and D.B. Wilson, *GATA-4 is required for sex steroidogenic cell development in the fetal mouse*, Dev. Dyn. 236 (2007), pp. 203–213.

84. J. Martineau, K. Nordqvist, C. Tilmann, R. Lovell-Badge, and B. Capel, *Male-specific cell migration into the developing gonad*, Curr. Biol. 7 (1997), pp. 958–968.

85. D. Wilhelm, S. Palmer, and P. Koopman, *Sex determination and gonadal development in mammals*, Physiol. Rev. 87 (2007), pp. 1–28.

86. P.S. Western, D.C. Miles, J.A. van den Bergen, M. Burton, and A.H. Sinclair, *Dynamic regulation of mitotic arrest in fetal male germ cells*, Stem Cells 26 (2008), pp. 339–347.

87. L. DiNapoli, J. Batchvarov, and B. Capel, *FGF9 promotes survival of germ cells in the fetal testis*, Development 133 (2006), pp. 1519–1527.

88. D. Suter, *Ovarian physiology*, in *Ovarian Toxicology*, P. Hoyer, ed., CRC Press, Boca Raton, FL, 2004, pp. 1–16.

89. L.A. Kondapalli and T.K. Woodruff, *Development and maturation of the normal female reproductive system*, in *Environmental Impacts on Reproductive Health and Fertility*, T. Woodruff, S. Janssen, L. Guillette Jr., and L. Giudice, eds., Cambridge University Press, Cambridge, 2010, pp. 23–28.

90. T.G. Baker, *A quantitative and cytological study of germ cells in human ovaries*, Proc. R. Soc. Lond. B Biol. Sci. 158 (1963), pp. 417–433.

91. E.A. McLaughlin and S.C. McIver, *Awakening the oocyte: Controlling primordial follicle development*, Reproduction 137 (2009), pp. 1–11.

92. D.N. Modi, S. Sane, and D. Bhartiya, *Accelerated germ cell apoptosis in sex chromosome aneuploid fetal human gonads*, Mol. Hum. Reprod. 9 (2003), pp. 219–225.

93. M.K. Skinner, *Regulation of primordial follicle assembly and development*, Hum. Reprod. Update 11 (2005), pp. 461–471.

94. M.S. Albamonte, M.A. Willis, M.I. Albamonte, F. Jensen, M.B. Espinosa, and A.D. Vitullo, *The developing human ovary: Immunohistochemical analysis of germ-cell-specific VASA protein, BCL-2/BAX expression balance and apoptosis*, Hum. Reprod. 23 (2008), pp. 1895–1901.

95. K.T. Jones, *Meiosis in oocytes: Predisposition to aneuploidy and its increased incidence with age*, Hum. Reprod. Update 14 (2008), pp. 143–158.

96. S. Joshi, H. Davies, L.P. Sims, S.E. Levy, and J. Dean, *Ovarian gene expression in the absence of FIGLA, an oocyte-specific transcription factor*, BMC Dev. Biol. 7 (2007), p. 67.

97. C.B. Schaefer, S.K. Ooi, T.H. Bestor, and D. Bourc'his, *Epigenetic decisions in mammalian germ cells*, Science 316 (2007), pp. 398–399.

98. R.S. Oliveri, M. Kalisz, C.K. Schjerling, C.Y. Andersen, R. Borup, and A.G. Byskov, *Evaluation in mammalian oocytes of gene transcripts linked to epigenetic reprogramming*, Reproduction 134 (2007), pp. 549–558.

99. A.A. Aravin and D. Bourc'his, *Small RNA guides for de novo DNA methylation in mammalian germ cells*, Genes Dev. 22 (2008), pp. 970–975.

100. P.B. Hoyer, *Ovarian toxicity in small pre-antral follicles*, in *Ovarian Toxicology*, P. Hoyer, ed., CRC Press, Boca Raton, FL, 2004, pp. 17–39.

101. E.V. Younglai, A.C. Holloway, and W.G. Foster, *Environmental and occupational factors affecting fertility and IVF success*, Hum. Reprod. Update 11 (2005), pp. 43–57.

102. E.M. Faustman, J.M. Gohlke, R.A. Ponce, T.A. Lewandowski, M.R. Seeley, S.G. Whittaker, and W.C. Griffith, *Experimental approaches to evaluate mechanisms of developmental toxicity*, in *Developmental and Reproductive Toxicology. A Practical Approach*, R. Hood, ed., Taylor & Francis Group, Boca Raton, FL, 2006, pp. 15–60.

103. G.A. Thouas, A.O. Trounson, and G.M. Jones, *Developmental effects of sublethal mitochondrial injury in mouse oocytes*, Biol. Reprod. 74 (2006), pp. 969–977.

104. P.A.A.S. Navarro, L. Liu, and D.L. Keefe, *In vivo effects of arsenite on meiosis, pre-implantation development, and apoptosis in the mouse*, Biol. Reprod. 70 (2004), pp. 980–985.

105. T. Matikainen, G.I. Perez, A. Juriscova, J.K. Pru, J.J. Schlezinger, H.Y. Ryu, J. Laine, T. Sakai, S.J. Korsmeyer, R.F. Casper, D.H. Sherr, and J.L. Tilly, *Aromatic hydrocarbon receptor-driven Bax gene expression is required for premature ovarian failure caused by biohazardous environmental chemicals*, Nat. Genet. 28 (2001), pp. 355–360.

106. M.M. Matzuk, *Eggs in the balance*, Nat. Genet. 28 (2001), pp. 300–301.

107. Y. Takai, J. Canning, G.I. Perez, J.K. Pru, J.J. Schlezinger, D.H. Sherr, R.N. Kolesnick, J. Yuan, R.A. Flavell, S.J. Korsmeyer, and J.L. Tilly, *Bax, caspase-2, and caspase-3 are required for ovarian follicle loss caused by 4-vinylcyclohexene diepoxide exposure of female mice in vivo*, Endocrinology 144 (2003), pp. 69–74.

108. Y. Yamanaka, A. Ralston, R.O. Stephenson, and J. Rossant, *Cell and molecular regulation of the mouse blastocyst*, Dev. Dyn. 235 (2006), pp. 2301–2314.

109. S. Pampfer, *Apoptosis in rodent peri-implantation embryos: Differential susceptibility of inner cell mass and trophectoderm cell lineages—A review*, Placenta 21 (2000), pp. S3–S10.

110. A. Jedrusik, D.E. Parfitt, G. Guo, M. Skamagki, J.B. Grabarek, M.H. Johnson, P. Robson, and M. Zernicka-Goetz, *Role of Cdx2 and cell polarity in cell allocation and specification of trophectoderm and inner cell mass in the mouse embryo*, Genes Dev. 22 (2008), pp. 2692–2706.

111. D. Strumpf, C.A. Mao, Y. Yamanaka, A. Ralston, K. Chawengsaksophak, F. Beck, and J. Rossant, *Cdx2 is required for correct cell fate specification and differentiation of trophectoderm in the mouse blastocyst*, Development 132 (2005), pp. 2093–2102.

112. M. Maluf, P.M. Perin, D.A.N. Foltran Januário, and P.H. Nascimento Saldiva, *In vitro fertilization, embryo development, and cell lineage segregation after pre- and/or postnatal exposure of female mice to ambient fine particulate matter*, Fertil. Steril. 92 (2009), pp. 1725–1735.

113. D.A.N.F. Januário, P.M. Perin, M. Maluf, A.J. Lichtenfels, and P.H. Nascimento Saldiva, *Biological effects and dose-response assessment of diesel exhaust particles on in vitro early embryo development in mice*, Toxicol. Sci. 117 (2010), pp. 200–208.

114. P.M. Perin, M. Maluf, C.E. Czeresnia, D.A. Nicolosi Foltran Januário, and P.H. Nascimento Saldiva, *Effects of exposure to high levels of particulate air pollution during the follicular phase of the conception cycle on pregnancy outcome in couples undergoing in vitro fertilization and embryo transfer*, Fertil. Steril. 93 (2010), pp. 301–303.

115. T.K. Woodruff and C.L. Walker, *Fetal and early postnatal environmental exposures and reproductive health effects in the female*, Fertil. Steril. 89 (2008), pp. e47–e51.

116. E. Diamanti-Kandarakis, J.P. Bourguignon, L.C. Giudice, R. Hauser, G.S. Prins, A.M. Soto, R.T. Zoeller, and A.C. Gore, *Endocrine-disrupting chemicals: An Endocrine Society scientific statement*, Endocr. Rev. 30 (2009), pp. 293–342.

7 Assessing the Reproductive Health of Men with Occupational Exposures*

Steven M. Schrader, Susan Reutman, and Katherine L. Marlow

CONTENTS

ABSTRACT

Male reproductive health is a product of complex synchronies among testicular, accessory sex gland, neuroendocrine, and erectile function. Hypotheses about endogenous or exogenous factors thought to disrupt one or more of these functions are amenable to human studies. The purpose of this chapter is to describe the methods used to design, conduct, and interpret such studies. Common study designs and laboratory analyses applied to study men's

---

* The findings and conclusions in this report are those of the authors and do not necessarily represent the views of the National Institute for Occupational Safety and Health.

reproductive health are presented, together with special research considerations and future assessment methods.

KEYWORDS

Accessory sex glands, accessory sex gland assessment, hormone assessment, male reproductive effects, man, men, occupational exposures, productive epidemiology, reproductive neuroendocrinology, reproductive systematic review, semen analysis, sexual function assessment, sperm, study design, testes, testicular function assessment

7.1   BACKGROUND

The earliest report linking environmental exposure to adverse human male reproductive effects dates back to 1775 when an English physician, Percival Pott, reported a high incidence of scrotal cancer in chimney sweeps. This observation led to safety regulations in the form of bathing requirements for these workers [1]. Brenneke, Hertwig, Muller, and Snell were among the first to formally study effects of exposures on offspring in mice, demonstrating that irradiated males sired smaller litters and linking chromosomal abnormalities in fertilized eggs to sperm irradiation [2]. Similarly, Auerbach and Robson, and Bock and Jackson later used mice to show that chemically exposed males had reduced fertility with induction of chromosomal abnormalities and other male germline mutations [2–5]. That male-mediated reproductive harm may occur in humans as a result of toxicant exposures became firmly established only relatively recently when Lancranjan et al. studied lead-exposed workers in Romania in 1975 [6], and later in 1977, when Whorton et al. examined the effects of dibromochloropropane (DBCP) on male workers in California [7]. Since these discoveries, additional human reproductive toxicants have been identified through the convergence of laboratory and observational findings. It has also been increasingly recognized that men's nonchemical exposures, both exogenous (e.g., physical exposures such as genital hyperthermia, pressure, and radiation therapy) and endogenous (e.g., constitutional factors such as age and genetic variation), may affect men's reproductive health and capacity [8–14]. The purpose of this chapter is to provide an overview of methods used to study the effects of exposures on male reproduction and their reproductive health, with a primary emphasis on the implementation and interpretation of human studies.

7.2   INITIATION OF STUDIES

Most research on human male reproductive health has been stimulated by studies of the effects of exposures on nonhuman animals and their offspring, that is, animal models. Many research gaps remain, as the pool of potential human exposures with undetermined effects on male reproduction is vast. Consider chemical exposures. More than 99 million unique organic and inorganic substances are currently registered in the Chemical Abstract Service database of the American Chemical Society, with approximately 15,000 new substances added per day at this writing [15].

Roughly 84,000 chemicals are in commerce in the United States [16] and more than 100,000 in the European Union [17], but male reproductive toxicity has only been thoroughly investigated in a small fraction of them. Under the 2007 European regulation on Registration, Evaluation, Authorisation and Restriction of Chemicals (REACH), manufacturers and importers are required to identify and share chemical risks that are then added to a European Union registry for subsequent evaluation and public dissemination [18]. European governments have undertaken an ambitious effort to identify reproductive toxicants among a subset of chemicals produced in high volumes [19]. The bulk of this work is currently conducted by testing whole animals, although alternatives are increasingly being developed as cost and animal welfare issues attend mass testing [19–21]. Alternatives to animal testing include in vitro testing (e.g., the mouse embryonic stem cell test for early developmental toxicity or mEST), in silico (i.e., computerized) methods, such as quantitative structure–activity relationship (i.e., QSAR) models, and grouping of related substances. Statistical models, termed QSAR models, are applied to compare and contrast structurally similar chemicals by examining statistical correlations between them on qualitative variables that may affect biological activity, such as polarity, lipophilicity, and molecule size [22]. Results of animal, in vitro, and in silico tests may sometimes be used to group chemicals; the health effects (and appropriate control) for one chemical may sometimes be extrapolated to similar chemicals within the same group. Overall, these nonhuman methods are vital for early identification of potential human reproductive health hazards, but extrapolating results to humans is inherently uncertain; thus, multiple, high-quality human studies are often needed to address these uncertainties and enhance human risk prediction.

Surveillance and anecdotal observations also have led to investigations of male reproductive exposures. Studies of DBCP were initiated after informal discussions of infertility problems among wives attending a softball game [23]. The petroleum refinery industry exemplifies a profession in which the workers themselves had concerns regarding their reproductive health [24]. Work-related accidents such as contamination of a truck driver and rescue workers responding to a truck accident–related bromine spill [25] or the nuclear radiation disaster in Chernobyl [26] also have led to studies. Adverse health effects observed in case studies of high-dose accidental exposures may provide clues to potential health effects that should be studied at lower exposures. Corporations may also initiate occupational research to validate anecdotal claims as with studies on dinitrotoluene and toluenediamine [27].

Relying on anecdotes and surveillance to identify possible male reproductive toxicants, however, is haphazard. In contrast to more overt health hazards, male occupational reproductive hazards can be "silent"; this presents an obstacle to identifying emerging hazards using human populations. To illustrate, suppose hypothetically that an effect such as reversible sterility is, in fact, induced by an unsuspected male reproductive toxicant. Although this is an extremely severe effect, only the subset of nonvasectomized male men trying to achieve pregnancy (or at least having regular sexual intercourse) with reproductive-aged, noncontracepting partners during the exposure period, who underwent a diagnostic workup during the exposure period and then were informed they were sterile, would even be aware a problem exists. As another example, a broader group of workers may be privately aware of an overt outcome such as diminished sexual function but (as with infertility) misattribute it to normal aging, and so on, and have a similar reluctance to disclose it, and thus, even after a reproductive health problem is acknowledged, it may only be known to a man's partner and, perhaps, his private physician. It is probably safe to assert, therefore, that a cluster of male reproductive health problems is far less apt to "sound the alarm" than a cluster of more commonly diagnosed and socially discussed health problems. This underscores the importance of nonhuman hazard screening efforts. Therefore, the toxicologist, the physician, the epidemiologist, the worker himself, the labor union, and the corporation will continue to be "on the lookout" for potential exposures and study populations.

7.3   STUDIES OF HUMAN POPULATIONS— DESIGNS AND CONSIDERATIONS

Animal and human experimentation on the male reproductive system have constraints, both ethical and pragmatic. As methods to reduce animal testing are evolving, so are nonexperimental, observational methods to study factors that may affect men's reproductive health or their offspring. Unlike animal studies, human studies cannot rely on random assignment of subjects to treatment or control groups or controlling all extraneous variables like diet and environment. Consequently, alternative designs and analysis methods for observational studies have been developed with the goal of controlling or minimizing biases introduced by sub- or nonrandomization. Population-based studies, broadly, are one such category of designs. The goal of sample selection in a population-based study is to represent the target human population of interest (e.g., nation, region, demographic group). To the extent sample representativeness is achieved, results of population-based studies may be considered externally valid (i.e., generalizable) for testing associations.

Epidemiological studies of occupational exposures and adverse male reproductive effects may follow several different study epidemiological (e.g., cohort, case–control, cross-sectional) and other (e.g., clinical case studies, etc.) designs. Table 7.1 describes various population-based study designs in terms of subject selection, an important determinant of how population based a study is, together with examples and the potential advantages and disadvantages of each design. These will be discussed at length later in the chapter.

In any study of occupational exposure and adverse reproductive effects, however, there are several challenges that must be considered. Designing reproductive studies of male populations that are externally valid presents some unique challenges. Selection bias may occur when enrollment or study completion varies between men, with versus without known or suspected reproductive health problems. Observational male reproductive studies may involve collection of questionnaires, blood or urine for neuroendocrine and other measurements, semen analysis, and sexual function analysis, either alone or in combination. Questionnaires are the least expensive and least invasive, and so may be more readily accepted. Male reproductive questionnaires, however, typically contain sensitive items (i.e., sexual function and habits, lifestyle, and disease histories may be requested) that may limit their acceptability for some people. Men with reproductive health concerns may be disproportionately

TABLE 7.1
Selected Design Elements for Various Human Male Reproductive Study Designs: Examples, Advantages, and Disadvantages

Study Design	Description	Example	Advantages	Disadvantages
Prospective cohort	*Subject selection*: based on defined group membership (e.g., common industry, birth year, region, etc.) *Aim*: follow entire group forward in time to track emergent exposures and outcomes (e.g., disease) Outcomes are later compared among prior exposure and nonexposure subgroups and with population-based regional or national statistics.	The Japanese atomic bomb survivor cohort has been followed over time for the development of cancers, including cancers of male genitalia, in order to calculate the relative risk (RR) of these cancers associated with radiation dose.	- Typically includes most members of the population of interest (no sampling), so extremely population based. - Permits calculation of incidence and RR.	- Time and $ costs relatively high unless outcomes manifest quickly (e.g., semen alterations). - Inefficient for rare diseases. - Attrition issues. - Special ethical issues pertaining to prospective follow-up.
Retrospective cohort (AKA historical cohort, nonconcurrent prospective)	*Subject selection*: based on defined group membership at some designated earlier point in time. *Aim*: follow entire group over historical time (retrospectively reconstructed) to track emergent exposures and outcomes. Outcomes are then compared among prior exposure and nonexposure subgroups and with concurrent population-based regional or national statistics.	Past paternal occupational group(s) recorded on birth certificates might be linked to registries to calculate the RRs of birth anomalies among offspring associated with occupation type.	- Time and $ costs relatively low. - Permits calculation of incidence and RR if group is population based. - Efficient for outcomes posited to manifest long after exposure (e.g., birth defects, cancer).	- Historical records may not contain variables of interest or may record them in insufficient detail. - Questionnaire responses of cases and controls regarding historical exposures subject to differential recall bias. - Does not permit calculation of incidence unless group is population based.

(Continued)

TABLE 7.1 (CONTINUED)
Selected Design Elements for Various Human Male Reproductive Study Designs: Examples, Advantages, and Disadvantages

Study Design	Description	Example	Advantages	Disadvantages
Case–control (AKA case–referent)	*Subject selection:* based on the outcome's (e.g., disease's) presence (case) or absence (control). When case and control groups are drawn from a cohort study, it is a "nested case control" design (similar to retrospective cohort). *Aim:* using historical exposure and history information, compare prior exposure vs. nonexposure history of cases and controls.	Past workplace exposures of infertile (cases) and fertile (controls) male workers are compared and infertility risk is estimated by calculating the odds ratio associated with exposure.	- Time and $ costs relatively low. - Efficient: requires smaller sample sizes; often used for rare diseases. - Nested cases and controls reduce potential bias.	- Does not permit calculation of incidence. - Generates odds ratios that approximate RRs only for rare outcomes. - Considerable bias potential attributed to use of inappropriate control group (bias addressed by use of nested cases and controls).
Cross-sectional (AKA prevalence)	*Subject selection:* based on defined group membership (e.g., common industry, birth year, region, etc.). *Aim:* using "snapshot" of current outcome and exposures, compare exposure vs. nonexposure histories of those who currently do vs. do not have the outcome.	A group of workers' short-term exposures are assessed using urine samples and current sexual function scores by questionnaire. Prevalence of low scores and association between scores and exposures are described.	- Time and $ costs relatively low. - Permits calculation of prevalence if population based. - Suited to collecting detailed data (surveys, exams, and biomarkers) for outcomes not routinely monitored. - Can be repeated in the future to develop a prospective cohort.	- Noninformative regarding whether exposure preceded disease (or vice versa) unless exposure is acute and outcome is immediate. - Does not permit calculation of incidence. - Hypothesis generating, exploratory design.

(Continued)

TABLE 7.1 (CONTINUED)
Selected Design Elements for Various Human Male Reproductive Study Designs: Examples, Advantages, and Disadvantages

Study Design	Description	Example	Advantages	Disadvantages
Clinical trial (AKA, clinical experiment)	*Subject selection*: based on defined group membership or a convenience sample (e.g., patients). *Aim*: randomly select participants into intervention or nonintervention groups with prospective follow-up to compare outcomes between groups who did and did not receive the intervention.	A group of men are randomized as to whether or not they will consume "Medication X" vs. a placebo to test the effect of the treatment on erectile function.	- Ideal to assess possible cause–effect relationships. - Randomization minimizes confounding bias.	- Ethical and feasibility considerations limit scope of exposure interventions acceptable for human experiments. - Often conducted on small, potentially underpowered sample sizes.

motivated to participate and so less deterred from providing sensitive questionnaire information and samples (e.g., semen, blood) or complying with study requirements (e.g., abstinence before sample collection, attending multiple study appointments). High participation rates are desirable to improve the representativeness of study samples, especially when participants and nonparticipants differ systematically on the reproductive health or exposure factors under study. Such high participation rates are not generally achieved in studies of men's reproductive health. Recruitment is often more successful when populations of interest are defined more narrowly (e.g., clinics, industries), as the pool of participants may be more uniformly motivated and recruitment efforts more targeted. For instance, participation rates for community-based semen quality studies are generally very low [28], but low refusal rates have been achieved among motivated men from narrower source populations, for example, military settings [29] and men whose partners previously had spontaneous abortions [30]. This illustrates a commonly encountered design dilemma when achieving a high participation rate requires a generalizability trade-off in terms of using a more motivated and narrowly defined source population. Even within narrowly defined target populations, individual constituent clinics and companies targeted for recruitment may opt out, or their inclusion may not be feasible; hence, convenience samples are often used but not without potentially incurring more loss of generalizability. Engaging potential participants involves up-front study budget and timeline investments to clearly present the project and "market" the importance of its goals, with emphasis on the vital role of high participation to producing valid results; reasonable financial compensation for time, travel, child care, work hours lost, and so on; and maximizing the convenience and privacy of participation. Mobile vans and shipped semen samples (discussed further in Section 7.10), for example, improve the convenience and privacy of delivering semen samples.

For an observational study to be externally valid, it must also be internally valid or free from internal biases and nonrandom error. Information biases, such as recall, reporting, and misclassification bias, can influence the internal validity of reported and record-based information and cloud interpretation of laboratory-based sample measurements. Recall bias is a memory-related bias that occurs when those who have experienced events (e.g., illnesses) have altered recall (typically keener or even exaggerated exposure recall) when queried in retrospect compared with those who have not. This phenomenon is not unique to reproductive health studies and may happen whenever reported information (e.g., study questionnaires, interviews, and patient-reported medical histories) are used. For example, a person who has experienced an adverse outcome may report an exposure more readily than one with no history of reproductive dysfunction. As Levin [31] describes, a couple that has recently experienced a stillbirth or congenitally malformed child will be more inclined to search for a previous toxic exposure as the source. In an analysis of recall levels in a time-to-pregnancy (TTP) study, most men (74%) and their female partners agreed fairly well on TTP within + 2 menstrual cycles [32]. Men's recall of TTP varied with the number of pregnancies fathered, self-assessed reporting confidence, prenatal diethylstilbestrol cohort membership (positively), planning of said pregnancy, current marital status to said partner, and his education. The authors suggested that recall among men who took part in TTP studies may be heightened in comparison with that of men from the general population. Reporting bias may occur when participants are reluctant to provide information on sensitive topics.

Although blood sampling is a widely accepted medical practice, it is an invasive procedure. In male reproductive studies, hormone levels are often obtained from blood samples, but endocrine profiles may not necessarily reflect the status of the male reproductive system [33]. Semen analysis provides information on spermatogenesis, accessory sex glands, and sperm cell motility [34]. Studies using semen samples might require the participants to have the capacity to produce samples by masturbation, raising concerns regarding participation bias should a toxicant under study concurrently affect the ability to produce ejaculate [35]. Additionally, this procedure employs complex scientific equipment and methodologies [36]. Cohort semen studies that combine all these analysis approaches provide the most complete assessment and thus the greatest likelihood of detecting adverse reproductive effects. However, such studies are expensive, complex, and necessitate a team approach minimally requiring (1) an andrologist, (2) an epidemiologist, (3) an industrial hygienist, (4) a physician, and (5) a statistician [37].

Misclassification is another threat to internal validity. Nondifferential misclassification may occur when groups being compared are equally likely to be misclassified with regard to either their exposure or disease, whereas differential misclassification occurs when the groups being compared are not equally likely to be misclassified. Nondifferential misclassification (e.g., random memory errors) typically biases results toward finding no association, that is, "the null," whereas differential misclassification can potentially bias results toward erroneous findings of either no association or a false association. For example, differential misclassification could result were respondents with sensitive reproductive health–related exposure, or disease histories, systematically skip or inaccurately report their histories on questionnaires or in interviews. Differential misclassification can also magnify the observed association when there is a true (but lesser) association.

Confounding bias can occur when both an outcome and an exposure vary according to yet another factor(s), and adequate adjustment for the confounder is not achieved by the study design and analysis. For example, maternal age could hypothetically confound an unadjusted analysis of paternal age and spontaneous abortion. This is because advanced maternal age is associated with both increased spontaneous abortion risk and older partner (paternal) age.

Use of secondary data sources (e.g., preexisting health history, personnel, surveillance, registry, or company records) originally collected closer in time to health and exposure events can mitigate potential recall bias. For studies that are resource intensive and require large sample sizes, use of secondary data may offer a less costly means to achieve adequate statistical power. One limitation of using secondary data is that information of interest may not have been recorded, and this particularly holds true for "silent" male reproductive conditions among groups outside of clinic settings. Secondary data sources are useful sources of population-based data for studies or surveillance of rare events that require large, otherwise costly, samples, such as birth defects. In the United States, however, medical birth records frequently do not record important information such as parental occupation, whereas this type of data is routinely collected in birth records of European countries. Also, documentation of potential confounders (e.g., alcohol use, drug use, smoking, or job status) may be limited or, when available, categorized too broadly. For example, pesticide exposure of such heterogeneous professions as crop farming and a fishery husbandry may be very different although these two groups may be classified together in existing databases. Alternatively, a researcher may elect to use questionnaires or questionnaire-based interviews or, less frequently, clinical assessments or sample collections (biological or environmental sampling) as the primary source(s) of data. This strategy provides more control over what, when, and how data will be collected, advantages to be weighed against the potential for recall bias and unacceptable nonparticipation levels (and resultant biases and costs) that sometimes hamper primary data collections. Primary data may be collected exclusively or in conjunction with secondary data to augment and validate the primary or secondary data sources. The goal of conducting population-based sampling, data validation, controlling for confounders, and achieving a high participation rate is to produce valid, replicable conclusions about potential human reproductive toxicants via nonexperimental methods.

General strategies to control or reduce bias in studies are available, such as precise definition and ascertainment of larger target populations of interest, random or otherwise representative sampling schemes, achievement of adequate statistical power and a high participation rate, control of confounding, and application of advanced analysis methods to account for residual biases (e.g., sensitivity analysis or Bayesian methods).

7.4 COHORT STUDIES

Reproductive cohort studies evaluate the frequency of adverse outcomes among a group defined by common characteristics (e.g., demography, geography, exposures) by following them over time.

In such a study, baseline data are collected, and individuals are followed longitudinally, either prospectively or retrospectively, for a specific reproductive outcome. In TTP studies, for example, cohorts of couples attempting to become pregnant are followed either prospectively or retrospectively until pregnancy is achieved. Men may be informants for prospective TTP studies, particularly when paternal behaviors or exposures are thought to affect the outcome of interest. Similarly, retrospective TTP cohorts may be constructed on the basis of the male partner's exposures. Results of prospective and retrospective TTP studies may differ, as pregnancy attempts are the usual sampling unit for prospective studies, whereas the pregnancy itself is the usual sampling unit for retrospective studies [38]. In general, less recall bias is anticipated among prospective than retrospective cohorts, and for retrospective cohort studies, less recall bias among shorter-term than longer-term studies. One example of a short-term, prospective cohort study of men involved "summer hire" pesticide applicators [39]. This example illustrates the importance of selecting appropriate variables for a prospective study design. Individuals were evaluated at the beginning of the season before they started working with pesticides and at the conclusion of the spraying season 2 months later. If semen analyses are conducted to predict reproductive outcome, however, correct timing is needed. Since the time for spermatogenesis and delivery of mature sperm to the ejaculate is approximately 72 days, if primary spermatogonia were affected by exposure, this would not be observed in a time frame that covered less than 80 to 90 days. Thus, a study of summer-hire workers could not make valid conclusions regarding the effect on spermatogenesis of a 2-month exposure among pesticide applicators. An example of a longer-term retrospective cohort study was the 1989 Vietnam experience study, in which military veterans were grouped according to whether or not they had served in Vietnam from 1967 to 1972 [40]. This study was able to detect subsequent differences between the groups in semen quality and TTP but revealed little about the reproductive health of the individuals at the time of exposure.

7.5 CASE–CONTROL STUDIES

Case–control or case–referent studies involve comparing the frequency of toxic exposure of men who have experienced reproductive dysfunction to those without such a medical history [31]. Case–control studies provide an efficient design to detect the association of rare outcomes with toxic exposures. For example, Nassar et al. [41] applied the case–control study method to examine a posited association between parental exposure to endocrine-disrupting chemicals (EDCs) and hypospadias among their offspring. Hypospadias is a birth defect that involves the urethra of the penis. Cases were obtained from a state birth defects registry in Australia and controls were a random sample of noncases from birth records from the same state. Maternal and paternal occupations, as well as information on other potential confounders and covariates, were obtained from birth records. Maternal and paternal exposure to EDCs was estimated for the various occupations. Use of this approach permitted separate estimates of the odds of hypospadias given maternal and paternal prenatal EDC exposures. Other similar examples are cited in a previous systematic review and meta-analysis on the same topic [42]. Case–control studies are, however, subject to considerable bias when cases do not arise from the same population as controls. When cases and controls are selected from existing cohorts, such designs are described as "nested" case–control studies; this approach lowers the risk of bias as both cases and controls are drawn from the same population, plus it is cost-effective.

7.6 CROSS-SECTIONAL STUDIES

A cross-sectional study provides a "snapshot" of men's exposures and reproductive outcomes as they exist at a fixed point in time. In contrast with cohort and case–control studies, a purely cross-sectional study does not include either prospective or retrospective exposure or outcome information. Such data may be particularly useful to explore relationships between acute or short-acting exposures and more transient endpoints (e.g., sperm counts). Cross-sectional studies are often less expensive to implement than other study designs and so are often used to examine hypothesized relationships. Even when high-quality, population-based cross-sectional studies may suggest associations, because of the immediate temporal nature of cross-sectional data, these studies are less informative about causality than cohort studies. Cross-sectional data can be used to estimate the prevalence, not incidence, of an outcome.

7.7 CLINIC-BASED STUDIES

Case studies typically involve the report by a physician of clinic or hospital patients exposed to potentially toxic agents. These reports involve the evaluation of individuals, groups of men with the same exposure (e.g., occupation, lifestyle), or clinical treatment after accidental exposure. While such reports rarely provide a definitive relationship between exposure and male reproductive effects, they can serve as sentinel reports that initiate further studies.

Some case studies provide unique information that would not be observed by using other study methods. One such study of a firearms instructor [43] provided possibly the best demonstration of the effect of lead on sperm. The instructor had fathered one son but became infertile as a result of work exposure that elevated his blood lead concentration to 88 µg/dl. During the next 3 years, the exposure was decreased, and the man was placed on chelation therapy. His sperm count increased as his blood levels decreased, and he later fathered another child after his blood level of lead decreased below 30 µg/dl. Similarly, after men exposed to high levels of kepone in the work environment were treated with cholestyramine to offset the toxic action of kepone, their sperm count and sperm motility increased accordingly [44].

Clinical (i.e., hospital and clinic based) studies of treatment outcomes (e.g., cancer, fertility [45]) vastly outnumber studies of men's exposures in the clinical literature on male reproduction. However, the advent of clinical data and specimen biorepositories that include sperm offers potential opportunities to expand research on potential targets and mechanisms of adverse male-mediated reproductive effects in humans. The National Institutes of Health/National Institute of Child Health and Human Development Cooperative Reproductive Medicine Network Biorepository of data (i.e., clinical, demographic, and laboratory data) and samples (i.e., serum, saliva, and sperm) promises to yield "a unique platform to assess developmental outcomes from conception to birth" [46].

7.8 SURVEILLANCE

Surveillance of human male reproductive health encompasses monitoring levels of adverse reproductive health effects in male populations and adverse effects on their offspring. Large-scale surveillance programs are ideally population based and thus describe information useful for tracking rates (e.g., incidence, prevalence) and ratios (e.g., standardized fertility, birth, and sex ratios) over time and comparing rates and ratios within and between populations. Examples of the types of surveillance systems that capture relevant outcomes for men's reproductive health monitoring and studies include sexually transmitted disease, cancer, births, and adverse birth outcome tracking. These systems are primarily registry based and maintained or supported by government agencies. Use of these systems to track male reproductive outcomes among subgroups of exposed men (e.g., occupational groups, etc.) or to study exposure–reproductive disease relationships is limited by the extent such systems fail to capture men's exposures. For instance, Fitzgerald et al. [47] found that "father's usual occupation" is listed on birth certificates by only a third of states in the United

States. Because exposure variables available from surveillance databases are often broad, careful attention to the appropriate use and interpretation of such data is indicated. Brender et al. [48] found agreement on paternal occupation between reported (maternal interview) and recorded (birth certificate) data sources 63% of the time.

Surveillance to monitor male reproductive health among targeted population subgroups (e.g., occupational, clinical) is also conducted. In the United States, a surveillance strategy for evaluating men working with known male reproductive toxicants was proposed and conducted by a team from the University of California [49]. However, this program had many problems and was eventually discontinued [50]. While this first attempt was discouraging, surveillance remains warranted as chemicals such as lead and ethylene glycol ethers remain in the US workplace, posing a potential hazard to the reproductive health of the male worker. Better surveillance is needed to monitor those working with these and other occupational toxicants. Addition of biological markers of reproduction and semen characteristics and evaluation for use of occupational exposure data from existing sources (e.g., birth certificates) are potential activities to enhance human surveillance [51]. Surveillance of reproductive health findings across multiple studies may also be conducted in the form of systematic reviews or meta-analyses, such as multinational efforts to monitor for the existence of declines in men's sperm counts [52].

7.9   SYNTHESIS

Narrative reviews, the traditional means of synthesizing information from multiple studies, continue to dominate the human male reproductive review literature. Increasingly, however, more formal and less ostensibly subjective methods to synthesize studies, such as systematic reviews, meta-analyses, and pooled analyses, are also being published. Narrative reviews provide an overview of past study findings from the perspective of the author(s) with interjection of their opinions on the relevance and quality of individual studies, as well as their interpretation of the body of evidence and its implications. A number of recent narrative reviews have, for example, focused on the relationship between men's body mass index (BMI) and various reproductive health outcomes including infertility [53–58]. Narrative reviews are often subjective in approach. Considered together, though, they represent a cross section of the perspectives of various subject matter experts and offer a platform for dissemination of emerging and novel ideas about potential mechanisms and implications.

Systematic reviews are a more formal approach to synthesizing information from multiple studies, since the content is more explicit and exhaustive. Study search strategies are documented and should be replicable and typically use relevant electronic search engines (e.g., MEDLINE, EMBASE, Biological Abstracts, PsycINFO, and CINAHL). Efforts to include all relevant studies for consideration may involve active discovery of studies in less accessible sources, such as results embedded in unpublished documents and reports. For example, MacDonald et al. [59] published a systematic review of the research literature on the relationship between BMI and men's semen parameters and reproductive hormone levels. They reported adherence to "Quality of Reporting of Meta-analysis of Randomized Controlled Trials and

Observational Studies" standards (QUOROM, MOOSE) [60,61]. The authors concluded that there was strong evidence of an inverse relationship between BMI and testosterone, free testosterone, and sex hormone binding globulin (SHBG). While arguably much less subjective than narrative reviews, the reporting characteristics of systematic reviews have been evaluated and systematic reviews conducted by different authors on the same topic can also be inconsistent [62].

Meta-analysis is a statistical approach for combined analysis of results from multiple distinct but comparable studies. The studies are typically selected for inclusion based on those identified through a systematic review. Meta-analysis is increasingly being applied to observational studies of semen quality and male-mediated birth outcomes. MacDonald et al. [59] coupled their previously described (above) systematic review of BMI, hormones, and semen with a very small (five-study) meta-analysis of semen parameters, finding no evidence of a relationship of semen with BMI. It is noteworthy that meta-analyses conducted with a small number of studies may have very low power, even when the number of subjects across studies is large and the effect size is substantial [63]. Meta-analyses may be undertaken with the primary "analytic" goal of identifying and estimating differences among study-specific effects or, more controversially, with a "synthetic" goal of estimating an average effect across studies [38]. Many texts have been written on procedural and statistical implementation of meta-analyses, and a number of common and specialized software packages are capable of performing it [63]. Most commonly, meta-analysts use results presented by other investigators as their data (i.e., means, standard errors, confidence limits), and thus, meta-analyses may be subject to the same biases as the constituent studies. Differences between methods and participants are generally inherent in human observational studies, and deciding where these study differences fall on a continuum between "fixed" and "random" is important for meta-analysis design and results interpretation. If all studies are so similar as to be "functionally identical" and the goal is to estimate a common effect size for those studies rather than to generalize, then fixed effects may be in order [63]. Alternatively, if the studies were conducted independently by different researchers and hence likely not "functionally identical," then random effects may be more appropriate. Meta-analyses with fewer studies, or where random versus fixed effects is less clear, may opt to present results of both random- and fixed-effects analyses. While large differences between studies may render them noncombinable, it may be argued that, when relatively slight, inherent study diversity may temper biases in the individual studies, that is, if the biases of constituent studies were not overwhelming and did not alter results in the same direction. It is critically important, therefore, that constituent studies be of sufficient quality for inclusion. Also, the level of similarity of constituent study populations and methods must be appropriate, given the primary goal(s) of the meta-analysis. Meta-analyses are subject to an extra layer of potential biases related to study selection. Measures to reduce these biases (or the appearance of bias) include replicable and documented systematic reviews of the literature for relevant studies, plumbing alternative sources for analyses studies, transparent documentation of rationale(s) for study exclusion, identification and screening of constituent studies for inclusion by parties not involved in those studies, appropriate weighting (fixed vs. random) according to study similarity, and adjustment for bias in the analysis. The subjective nature of these activities is a reason that "synthetic" goals may be considered controversial [38]. Thus, meta-analysis "consumers" interested in average effects must be particularly mindful of these potential biases. Meta-regression offers a method of examining associations between variables across studies but has much larger sample size (i.e., number of studies) requirements than meta-analysis.

Pooling of the actual data across studies, as opposed to combining summary statistics as is done for a meta-analysis, is generally the most highly preferred method of data synthesis, when feasible. Continuing with the BMI research example theme, Aggerholm et al. [64] combined data from five population-based environmental studies of the relationship of BMI to male reproductive hormone levels and semen quality into one large database ($N = 2139$). The authors reflected on the degree of homogeneity of the study populations and the comparability of sample collection and laboratory analysis protocols, all key considerations to be weighed before pooling studies.

7.10   ASSESSMENTS OF MALE REPRODUCTIVE HEALTH

Toxicants can attack the male reproductive system at one of several sites, or at multiple sites. These sites and the assays associated with their respective functions are discussed individually. This does not necessarily indicate, however, that there exists an absolute one-to-one relationship between a particular measurement and the associated site of action. These sites include the neuroendocrine system, the testes, accessory sex glands, and sexual function.

The establishment of a male reproductive profile for assessing reproductive potential for both individual and population investigations is essential. The same profile can be used for both types of studies, but there are some basic differences in methodology. The assessment profile illustrated in Table 7.2 is being used by the National Institute for Occupational Safety and Health to assess populations exposed to potential reproductive toxicants. Differences between assessing the individual versus the population will be noted. A summary of assessments and specific methodologies follow. If individual data (vs. population comparisons) are to be used, care should be taken to compare the results with the normal range of results of the laboratory conducting the analysis and not published values. If a population-based study is being conducted, a concurrent comparison cohort must be used and the analyses should be

TABLE 7.2
Endocrine Profile for Assessing Reproductive Toxicant Effects

Hormone	Fluid for Measurement		
	Saliva	Blood	Urine
Luteinizing hormone		X	X
Follicle-stimulating hormone		X	X
Inhibin B		X	
Testosterone—Total		X	X
—Free	X	X	

TABLE 7.3
Semen Characteristics—Reference Limits (5th Centiles)

Semen Characteristics	Lower Reference Limit (95% Confidence Interval)
Semen volume (ml)	1.5 (1.47–1.7)
Total sperm number (10^6 per ejaculate)	39 (33–46)
Sperm concentration (10^6 per ml)	15 (12–16)
Motility (%)	40 (38–42)
Vitality (live spermatozoa, %)	58 (55–63)
Sperm morphology (normal forms, %)	4 (3.5–4.0)

Source: World Health Organization, *WHO Laboratory Manual for the Examination and Processing of Human Semen, 5th edition.* World Press, Geneva, Switzerland, 2010.

blind to exposure status. Table 7.3 provides the World Health Organization (WHO) reference values for various semen parameters [65].

Table 7.4 provides examples of occupational exposures that have been shown to have negative effects on one or more sites of male reproduction. The most effective data collection is achieved by establishing a temporary laboratory near the worksite for blood collection and designed such that the semen samples can be conveniently submitted. Studies with multiple study sites or long recruitment periods may make establishing a temporary laboratory impractical. In this case, blood can be collected by a local nurse or clinic and serum can be shipped to the analytical laboratory. The semen sample can be collected, placed in a cold (not frozen) container, and shipped to the andrology laboratory [66]. When semen is shipped in this manner, sperm motility and viability measures are compromised, but the other semen parameters can be assessed [67].

7.11   NEUROENDOCRINE SYSTEM

The endocrine and nervous systems work in concert to coordinate the function of the various components of the reproductive axis, drawing upon inputs that are external (e.g., sexual cues, temperature) and internal (e.g., checks and balances between endocrine tissue function, metabolic status). The reproductive endocrine status of the male can be assessed by measuring the hormones in the blood, urine, or saliva, depending on the hormone. The principal hormones of interest for assessing the effects of reproductive toxicants in men are luteinizing hormone (LH), follicle-stimulating hormone (FSH), inhibin B, and testosterone.

Since the circulating profile of LH is pulsatile, the status of this hormone for the individual, if measured in blood, is best estimated in serial samples. The pooled results of three samples collected at 20-min intervals will provide the best estimate of mean concentration [68]. Yet, multiple blood draws often result in poor participation rates of workers. If a population is being evaluated, a single blood sample per individual may suffice [69]. Alternatively, an integral of its pulsatile secretion may be obtained by measuring LH in urine [70].

Circulating FSH levels are not as variable as those for LH. This is attributable in part to a longer circulating half-life for FSH compared to LH. Thus, analysis of a

TABLE 7.4

Examples of Workplace Exposures Affecting Reproductive Health

Site of Action	Examples
Neuroendocrine	
Hormone profile	Insecticides [71], lead [72,73]
Testicles	
Sperm concentration	Lead [72,74], diesel exhaust [75], pesticide [76], bisphenol A [77]
Sperm morphology	Insecticides [71], lead [74], carbon disulfide [78], pesticide [76], bisphenol A [77]
Sperm genetics	Phthalate [79], styrene [80], OP pesticides [81], carbyl [82], fenvalerate [83], lead [74]
Accessory Sex Glands	
Toxicant in semen	Lead [73], trichloroethylene [84], boron [85], cadmium [86]
Semen volume	Lead [73], organophosphate [87]
Sperm viability	Carbon disulfide [78], bisphenol A [77], lead [74]
Sperm motility	Insecticides [71], diesel exhaust [75], lead [74], carbon disulfide [78], phthalate [79], pesticide [76], bisphenol A [77], fenvalerate [88]
Sexual Function	
Libido	Carbon disulfide [78], bisphenol A [89]
Erectile function	Bisphenol A [89], bicycle saddles [10]
Penis sensitivity	Bicycle saddles [90]
Ejaculatory function	Bisphenol A [89]

single blood sample for an individual will provide a more reliable estimate of FSH than for LH. FSH can also be measured in urine for the sake of convenience. Neither gonadotropin is exuded into the saliva.

The variability of inhibin B levels secreted by Sertoli cells into the serum is also nominal. Therefore, inhibin B levels can be assessed with a single serum sample. Inhibin B cannot be measured meaningfully in urine or saliva.

Approximately 2% of circulating testosterone is free, with the remainder bound to SHBG, albumin, and other serum proteins. The free circulating testosterone is the active component and therefore provides a more accurate marker of physiologically available testosterone than does total circulating testosterone under conditions when SHBG concentration or binding may be variable [68]. The circhoral fluctuations of circulating testosterone levels, like those for LH, are significant. Estimates of free and total testosterone can be determined in single blood samples but are greatly improved by assaying multiple blood samples and pooling the results.

Serum levels of total and free testosterone can be measured directly. However, serum-free testosterone concentrations are more accurately determined by calculating them from serum concentrations of total testosterone, SHBG, and albumin and association constant [91]. Alternatively, a single measurement in urine of testosterone after sample hydrolysis or of testosterone metabolite (e.g., androsterone, etiocholanolone, or testosterone glucuronide) provides a convenient index of total testosterone [92]. Quantifying testosterone in saliva affords a convenient alternative to blood sampling while providing a measure of the unbound, biologically active component of circulating testosterone [93]. If measuring steroid hormone metabolites in urine, consideration should be given to the potential that the exposure being studied may alter the metabolism of excreted metabolites. This is especially pertinent since most metabolites are formed by the liver, a target of many toxicants. Lead, for example, reduces the amount of sulfated steroids that were excreted into the urine [94]. Precision of urinary measurements is improved by normalizing urinary flow rate (concentration) by adjusting for urinary levels of creatinine or osmolality.

Circulating levels for the reproductive hormones become elevated during night as the male enters puberty. In men, secretion of testosterone and inhibin B maintains this diurnal pattern through adulthood, with peak values in early morning and declining toward late afternoon [95]. This pattern appears to be driven by sleep, not a circadian rhythm [96]. Thus, samples for assessing testosterone and inhibin B should be collected at approximately the same time of day to avoid variations owing to diurnal secretory patterns.

In summary, Table 7.2 lists the primary hormones for assessing reproductive toxicity effects in men. FSH, LH, inhibin B, and testosterone can all be evaluated in a population-based study by assessing the hormone levels in a single blood sample from each man, preferably at about the same time of day. A wide variety of potentially toxic occupational exposures, including DBCP [97], phthalate [98], stilbene [99], trichloroethylene [100], fluoride [101], bisphenol A [102], radiation [103], and sedentariness [104], have been reported to alter serum levels of one or more of these hormones. Recent publications representative of the literature describe the effects of various pesticide and lead exposures on serum levels of all four hormones [105,106]. Alternatively, urine samples typically represent a more convenient way to measure the gonadotropins and testosterone in populations. Few population studies of men have assessed occupational exposures on endocrine effects measured in urine or saliva [107,108]. For the study of an individual, three blood samples collected 20 min apart or urinary assessment will improve the estimate.

7.12   TESTES

Semen analysis provides a useful profile of the function of the male reproductive system. The WHO [65] has published reference ranges for semen parameters and these are provided in Table 7.3 as general information. The various measurements that are routinely used in the assessment of occupational exposure are presented in the list on next page. Specific instructions should be provided to each man to ensure that the semen sample is properly collected by masturbation after a set time of abstinence (usually 2–3 days) and delivered to the laboratory within 1 h from the time of ejaculation. The men should be instructed to maintain the semen at room temperature, avoiding any temperature shock to the sperm cells. At the time of collecting the semen sample, each subject should record the duration of abstinence, time of semen collection, and any information regarding sample collection loss or spillage. Providing a label on the jar facilitates the recording of this information.

Semen analyses can be conducted in two phases. The initial evaluation of the sample should be conducted when the sample arrives at the laboratory (or field site) and should consist of recording the temperature, turbidity, color, liquefaction time, and volume of the semen. Temperature shock to the semen sample can affect many sperm parameters. An inexpensive temperature logging monitor on the collection jar is useful to determine the temperatures to which the semen has been exposed since collection. Motility assessments, viability estimates, sperm counts, the preparation of slides, and preservation of seminal plasma should also be conducted at this time. Sperm motility should be assessed objectively either with computer-assisted sperm analysis (CASA) or by counting nonmotile cells in an aliquot, then counting all cells in a separate aliquot that has been heated to immobilize the sperm. Percent motile is the total in the heated aliquot minus the nonmotile sperm in the first aliquot. CASA can be conducted on-site with the fresh ejaculate or video recorded for future analyses. If CASA is used, several sperm motility variables can be measured (see list below). These variables provide useful information on the progression of sperm cells (curvilinear velocity, straight-line velocity, and linearity) as well as the sperm motility pattern. Sperm motility characteristics should be measured in a chamber at least 10 µm deep in order for the sperm to move freely in all planes. Morphologic and morphometric analyses of sperm on slides may be conducted at a later time.

Semen Profile for Assessing Reproductive Toxicant Effects

Sperm concentration
Sperm viability
   Vital stain
   Hypo-osmotic swelling
Sperm motility
   Percent motile
   Curvilinear velocity
   Straight-line velocity
   Linearity
   Lateral-head amplitude
Sperm size and shape
   Morphology
   Morphometry
Sperm genetics
   DNA stability
Semen parameters
   pH
   Volume
   Marker chemicals from accessory glands
   Toxicant or metabolite concentration

Measurements of sperm motility and velocity should be conducted using a microscope stage warmed to 37°C. An attempt to record 200 motile sperm per sample is desirable if one is interested in the distribution of velocity measurements, but 100 motile sperm will suffice if means are to be compared. When assessing motility, one should avoid "hunting" for motile sperm. All fields examined or searched should be included in the calculations; therefore, assessing a certain number of arbitrary fields is advised. Whole semen should be used for measuring sperm motility. If a CASA system is being used for velocity estimates, the number of sperm per field should be reduced to minimize cell collisions. Using a 10- to 20-μm-deep chamber, the sperm concentration should be less than 40 million/ml. Diluents (including seminal plasma), however, alter sperm velocity up to a dilution of approximately 1:1. The current recommendation for CASA of sperm velocity is to dilute all samples to one part semen in one part iso-osmotic buffer. If this dilution does not reduce the sperm concentration below 40 million/ml, then an additional dilution in the same buffer should be performed on those concentrated samples [36]. Thus, two recordings should be made: whole semen for percent motility and diluted sperm for sperm velocity.

Sperm viability may be determined by vital stain [109] or by hypo-osmotic swelling (HOS) assay [110]. The HOS assay determines the structural and functional integrity of the cell membrane [111].

Sperm morphology should be estimated on fixed, stained semen smears. During the past 30 years, several schemes have been presented for the assessment of normal and abnormal sperm morphologies. Variations in sperm size and shape are not distinct, but rather a continuum. This provides a challenge within and especially among laboratories to establish a repeatable system for morphologic classification [112–115]. Since 1980, the WHO has adopted different sperm morphology classifications several times. Currently, there are two widely accepted classification systems, WHO 3rd Edition [116] (often called traditional morphology) and WHO 5th Edition (often called strict morphology) [65]. The main difference between these classification systems is how they classify a "borderline normal" sperm; normal with traditional scheme, abnormal with strict scheme. With recent advances of computerized image analyses, several methods of sperm morphometry have been introduced [117–123]. These morphometric analysis systems provide objective assessments of individual sperm head size and shape. Sperm morphometry is now routinely used as part of the assessment of reproductive hazards to the male workers [124].

Sperm concentration, sperm morphology, and sperm head morphometry all provide indices of the integrity of spermatogenesis and spermiogenesis. Thus, the number of sperm in the ejaculate is directly correlated with the number of germ cells per gram of testis [125], while abnormal morphology is probably a result of abnormal spermiogenesis. Azoospermia is probably the most severe observation, as it is often an indication that type Λ spermatogonia have been lost and recovery is unlikely.

Genetic damage is difficult to detect in human sperm [126]. Epidemiological studies of large populations have demonstrated increased frequency of adverse pregnancies in women whose husbands were working in various occupations [127]. This is primarily because the chromosomes are in interphase and there is no replication and no production of proteins. Some of the methods being used to detect genetic damage with varying success are fluorescent in situ hybridization [128–130] of certain chromosomes, TUNEL (terminal deoxynucleotidyl transferase dUTP nick end labeling), comet, and the sperm chromatin stability assay [131–134]. DNA adducts may also provide information about spermatogenesis at the genetic level. Several reports have shown that paternal exposure may affect pregnancy or the health of the offspring. These data have stimulated research into the genetic stability of the sperm cell and the cause/effect relationships of damage to sperm.

7.13   ACCESSORY SEX GLANDS

Seminal plasma is not essential for fertilization; thus, the artificial insemination of sperm collected from the epididymis results in conception. On the other hand, seminal plasma contributes importantly to the normal coitus-fertilization scenario. Seminal plasma serves as a vehicle for sperm transport, a buffer from the hostile acidic vaginal environment, and an initial energy source for the sperm. Cervical mucus prevents passage of seminal plasma into the uterus. Some constituents of seminal plasma, however, are carried into the uterus to the site of fertilization by adhering to the sperm membrane.

The viability and motility of spermatozoa in seminal plasma are typically a reflection of seminal plasma quality. Alterations in sperm viability or alterations in sperm motility parameters would suggest an effect on the accessory sex glands.

Biochemical analysis of seminal plasma provides insights into the function of the accessory sex glands. Chemicals that are secreted primarily by each of the glands of this system are typically selected to serve as a marker for each respective gland. For example, the epididymis is represented by glycerylphosphorylcholine; the seminal vesicles, by fructose; and the prostate gland, by zinc. Note that this type of analysis provides only gross information on glandular function and little or no information on the other secretory constituents. Measuring volume provides additional general information on the nature of seminal plasma.

Seminal plasma may be analyzed for the presence of a toxicant or its metabolite. Heavy metals have been detected in seminal plasma using atomic absorption spectrophotometry [135], while halogenated hydrocarbons have been measured in seminal fluid by gas chromatography after extraction [135] or protein-limiting filtration [136].

A toxicant or its metabolite may act directly on accessory sex glands to alter the quality or quantity of their secretions. Alternatively, the toxicant may enter the seminal plasma [137–139] and thereby affect the sperm and the body of the female partner after intercourse or may be carried to the site of fertilization on the sperm membrane and affect the ova or conceptus.

There are few reports of toxicant effects on the accessory sex glands in humans. Ethylene dibromide (EDB) is one example of a toxicant that exerts posttesticular effects. Short-term exposure to the toxicant reduced sperm velocity and semen volume [140]. Chronic exposure decreased sperm motility and viability, decreased seminal fructose levels, and increased semen pH [140]. An EDB metabolite was present within the semen of some exposed workers [136]. Other potential toxicants that have been detected in semen include lead, cadmium, hexachlorobenzene, hexachlorocyclohexane, dieldrin, and polychlorinated biphenyls [135]. Cocaine has been shown to bind to the sperm membrane [141].

Several sperm assessment methods measure the sperm function [142] and may evaluate sperm across more than one of the subjective toxicant site divisions outlined above. The penetration of sperm through cervical mucus (or viscous fluids stimulating cervical mucus) [143–145], the penetration of sperm into a zona-free hamster egg (sperm penetration assay [SPA]) [146], the penetration of sperm through a zona pellucida removed from immature human ova (hemi-zona assay), and the binding to hyaluronic acid [147] have been shown to evaluate different sperm functions [148,149]. With the exception of SPA, these have not been utilized in assessing reproductive toxicants in the field setting. SPA has been used with limited success [146,150].

7.14 SEXUAL FUNCTION

Human sexual function refers to the integrated activities of the testes and secondary sex glands, the endocrine control systems, and the central nervous system–based behavioral and psychological components of reproduction (libido). Erection, ejaculation, and orgasm are three distinct, independent physiological and psychodynamic events that normally occur concurrently in men. If details regarding functions or mechanisms are desired, several reviews and in-depth reports are available [151–153].

Burnett [154] recently published a review on the effects of environmental exposures on erectile function. Assessment of occupational exposure–induced anomalies of sexual function is difficult. The researcher usually must rely on the testimony and recall of the worker regarding his sexual function. This testimony may often be confounded by the bias of the individual to guard his ego or masculine image or to attribute a preexisting libido problem to exposures at work.

Burris et al. [155] reported application of a monitor (Rigiscan®) for assessing erection at home. The assessment of erectile function using the Rigiscan has been used successfully in the occupational setting in studies of the effect of bicycle saddles on bicycle patrol officers [90,156].

The assessment of ejaculate volume may provide information on the integrity of the emission phase of ejaculation. This is, of course, complicated by effects on the accessory sex glands' capacity. Thus, a semen sample of reduced volume but with a normal ratio of constituents (marker chemicals) supports a diagnosis of an emission phase defect.

The numbness or loss of feeling of the penis can be objectively measured using a biothesiometer. The equipment can easily be set up in a private room (i.e., a restroom) and the computer operator can be in an adjoining room. The study subject places his penis in a plastic trough and the computer operator sends signals to the apparatus to increase or decrease vibration to detect the level of vibration that can be sensed by the penis [90].

7.15 FUTURE ASSESSMENT METHODS

There are several new methods that may play a key role in future studies of toxicant exposures and male reproduction [157,158], especially those that detect genomic damage [159]. As new methods are added to the reproductive health profile, there are some potential limitations that need to be considered: Are methods practical in an environmental or occupational field setting (or easily preserved for later assessment)? Is there adequate statistical power with typical field study sample sizes assessing accuracy and precision? Is there enough semen available in most specimens to analyze all of the measures (a prioritization scheme may be needed)?

REFERENCES

1. H.W. Herr, *Percivall Pott, the environment and cancer*, Br. J. Urology 108 (2011), pp. 479–481.
2. D.E. Sawyer and R.J. Aitken, *Male-mediated developmental defects and childhood disease*, Reprod. Health 8 (2000), pp. 107–126.
3. C. Auerbach, J.M. Robson, and J.G. Carr, *The chemical production of mutations*, Science 105 (1947), pp. 243–247.
4. C. Auerbach and J.M. Robson, *Chemical production of mutations*, Nature 157 (1946), pp. 302–305.
5. M. Bock and H. Jackson, *The action of triethylenemelamine on the fertility of male rats*, Brit. J. Pharmacol. 12 (1957), pp. 1–7.
6. I. Lancranjan, H.I. Popescu, O. Gavanescu, I. Klepsch, and M. Serbanescu, *Reproductive ability of workmen occupationally exposed to lead*, Arch. Environ. Health 30 (1975), pp. 396–401.
7. D. Whorton, R.M. Krauss, S. Marshall, and T.H. Milby, *Infertility in male pesticide workers*, Lancet 2 (1977), pp. 1259–1260.
8. A. Jung and H.C. Schuppe, *Influence of genital heat stress on semen quality in humans*, Andrologia 39 (2007), pp. 203–215.
9. P. Thonneau, L. Bujan, L. Multigner, and R. Mieusset, *Occupational heat exposure and male fertility: A review*, Hum. Reprod. 13 (1998), pp. 2122–2125. Review.
10. S.M. Schrader, M.J. Breitenstein, J.C. Clark, B.D. Lowe, and T.W. Turner, *Nocturnal penile tumescence and rigidity testing in bicycling patrol officers*, J. Androl. 23 (2002), pp. 927–934.
11. G.R. Dohle, *Male infertility in cancer patients: Review of the literature*, Int. J. Urol. 17 (2010), pp. 327–331. Epub February 22, 2010.
12. A.F. Stewart and E.D. Kim, *Fertility concerns for the aging male*, Urology 78 (2011), pp. 496–499.
13. S. Hinchliff and M. Gott, *Seeking medical help for sexual concerns in mid- and later life: A review of the literature*, J. Sex. Res. 48 (2011), pp. 106–117.
14. J. Axelsson, J.P. Bonde, Y.L. Giwercman, L. Rylander, and A. Giwercman, *Gene-environment interaction and male reproductive function*, Asian J. Androl. 12 (2010), pp. 298–307. Epub March 29, 2010.
15. American Chemical Society, CAS Abstract Service Division [2011]. Chemical Abstract Service (CAS) registry and CAS registry numbers. [http://www.cas.org/expertise/cas content/registry/regsys.html]. Date accessed: June 23, 2015, Copyrighted 2015.
16. US EPA [http://www.epa.gov/opptintr/existingchemicals/pubs/tscainventory/basic.html]. Accessed 10/7/2011, last updated 8/15/11.
17. Environmental Directorate General, European Commission [2007]. REACH in brief. [http://ec.europa.eu/environment/chemicals/reach/pdf/publications/2007_02_reach _in_brief.pdf]. Date accessed: June 23, 2015.
18. European Commission. [http://ec.europa.eu/environment/chemicals/reach/reach_intro .htm]. Accessed 10/11/2011, last updated 9/28/2011.
19. R. Vogel, T. Seidle, and S. Horst, *A modular one-generation reproduction study as a flexible testing system for regulatory safety assessment*, Reprod. Toxicol. 29 (2010), pp. 242–245.
20. H. Speilman, *The way forward in reproductive/developmental toxicity testing*, ATLA 37 (2009), pp. 641–656.

21. A.R. Scialli, *The challenge of reproductive and developmental toxicology under REACH*, Regul. Toxicol. Pharmacol. 51 (2008), pp. 244–250.

22. C. Gianinis, A. Zira, S. Theocharis, and A. Tsantili-Kakoulidou, *Application of quantitative structure–activity relationships for modeling drug and chemical transport across the human placenta barrier: A multivariate data analysis approach*, J. Appl. Toxicol. 29 (2009), pp. 724–733.

23. D. Whorton and D. Foliart, *DBCP: Eleven years later. Symposium on the assessment of reproductive hazards in the workplace*, Presented Cincinnati, OH, June 16, 1988. Reprod. Tox. 2 (1988), pp. 155–161.

24. M.J. Rosenberg, A.J. Wyrobek, J. Ratcliffe, L.A. Gordon, G. Watchmaker, S.H. Fox, D.H. Moore II, and R.W. Hornung, *Sperm as an indicator of reproductive risk among petroleum refinery workers*, Br. J. Ind. Med. 42 (1985), pp. 123–127.

25. G. Potashnik, R. Carel, I. Belmaker, and M. Levine, *Spermatogenesis and reproductive performance following human accidental exposure to bromine vapor*, Reprod. Toxicol. 6 (1992), pp. 171–174.

26. A. Birioukov, M. Meurer, R.U. Peter, O. Braun-Falco, and G. Plewig, *Male reproductive system in patients exposed to ionizing irradiation in the Chernobyl accident*, Arch. Androl. 3 (1993), pp. 99–104.

27. P.V.V. Hamill, E. Steinberger, R.J. Levine, L.J. Rodriguez-Rigau, S. Lemeshow, and J.S. Avrunin, *The epidemiologic assessment of male reproductive hazard from occupational exposure to TDA and DNT*, J. Occup. Med. 24 (1982), pp. 985–993.

28. T.M. Stewart, D.Y. Liu, C. Garrett, E.H. Brown, and H.W. Baker, *Recruitment bias in studies of semen and other factors affecting pregnancy rates in fertile men*, Human Reprod. 24 (2009), pp. 2401–2408.

29. G.K. Lemasters, D.M. Olsen, J.H. Yiin, J.E. Lockey, R. Shukla, S.G. Selevan, S.M. Schrader, G.P. Toth, D.P. Evenson, and G.B. Huszar, *Male reproductive effects on solvent and fuel exposure during aircraft maintenance*, Reprod. Toxicol. 13 (1999), pp. 155–166.

30. A.F. Olshan, S.D. Perreault, L. Bradley, R.M. Buus, L.F. Strader, S.C. Jeffay, L. Lansdell, D.A. Savitz, and A. Herring, *The healthy men study: Design and recruitment considerations for environmental epidemiologic studies in male reproductive health*, Fertil. Steril. 87 (2007), pp. 554–564.

31. S.M. Levin, *Problems and pitfalls in conducting epidemiological research in the area of reproductive toxicology*, Am. J. Ind. Med. 4 (1983), pp. 349–364.

32. R.H.N. Nguyen and D.D. Baird, *Accuracy of men's recall of their partner's time to pregnancy*, Epidemiol. 16 (2005), pp. 694–698.

33. G. Assennato, C. Paci, M.E. Baser, R. Molinini, G. Candela, B.M. Altamura, and R. Giorgino, *Sperm count suppression without endocrine dysfunction in lead-exposed men*, Arch. Environ. Health 41 (1986), pp. 387–390.

34. S.M. Schrader and J.S. Kesner, *Mechanisms of male reproductive toxicology*, in *Occupational and Environmental Reproductive Hazards: A Guide for Clinicians*, M. Paul, ed., Williams and Wilkins Publishers, Baltimore, 1992, pp. 3–17.

35. R.J. Levine, P.B. Blunden, D. Dalcorso, T.B. Starr, and C.E. Ross, *Superiority of reproductive histories to sperm count in detecting infertility at dibromochloropropane manufacturing plant*, J. Occup. Med. 25 (1983), pp. 591–597.

36. S.M. Schrader, R.E. Chapin, E.D. Clegg, R.O. Davis, J.L. Fourcroy, D.F. Katz, S.A. Rothmann, G. Toth, T.W. Turner, and M. Zinaman, *Laboratory methods for assessing human semen in epidemiologic studies: A consensus report*, Reprod. Toxicol. 6 (1992), pp. 275–279.

37. S.M. Schrader, *General techniques for assessing male reproductive potential in human field studies*, in *Methods in Reproductive Toxicology*, Vol. 3A, R. Chapin and J. Heindel, eds., Academic Press, San Diego, 1993, pp. 362–371.

38. K.J. Rothman, S. Greenland, and L.T. Lash, *Chapters 31 and 33*, in *Modern Epidemiology, 3rd ed.* S. Seigafuse and L. Bierig, eds., Lippincott Williams & Wilkins, Philadelphia, 2008, pp. 620–640, 652–680.

39. S.M. Schrader, T.W. Turner, and J.M. Ratcliffe, *The effects of ethylene dibromide on semen quality: A comparison of short term and chronic exposure*, Reprod. Toxicol. 2 (1988), pp. 191–198.

40. F. Destefano, J.L. Annest, M.J. Kresnow, S.M. Schrader, and D.F. Katz, *Semen characteristics of Vietnam veterans*, Reprod. Toxicol. 3 (1989), pp. 165–173.

41. N. Nassar, P. Abeywardana, A. Barker, and C. Bower, *Parental occupational exposure to potential endocrine disrupting chemicals and risk of hypospadias in infants*, Occup. Environ. Med. 67 (2010), pp. 585–589.

42. C.M. Rocheleau, P.A. Romitti, and L.K. Dennis, *Pesticides and hypospadias: A meta-analysis*, Pediatric. Urol. 5 (2009), pp. 17–24.

43. J. Fisher-fischbein, A. Fischbein, H.D. Melnick, and C.W. Bardin, *Correlation between biochemical indicators of lead exposure and semen quality in a lead-poisoned firearms instructor*, JAMA 257 (1987), pp. 803–805.

44. P.S. Guzelian, *Therapeutic approaches for chlordecone poisoning in humans*, J. Toxicol. Environ. Health 8 (1981), pp. 757–766.

45. R. Soules, *Assisted reproductive technology has been detrimental to academic reproductive endocrinology and infertility*, Fertil. Steril. 84 (2005), pp. 570–572.

46. S.A. Krawetz, P.R. Casson, M.P. Diamond, H. Zhang, R.S. Legro, W.D. Schlaff, C. Coutifaris, R.G. Brzyski, G.M. Christman, N. Santoro, and E. Eisenberg, *Reproductive medicine network (RMN), establishing a biological specimens repository for reproductive clinical trials: Technical aspects*, Systems Bio. Reprod. Med. 57 (2011), pp. 222–227.

47. E. Fitzgerald, D. Wartenberg, W.D. Thompson, and A. Houston, *Birth and fetal death records and environmental exposures: Promising data elements for environmental public health tracking of reproductive outcomes*, Public Health Reports 124 (2009), pp. 825–830.

48. J.D. Brender, L. Suarez, and P. Langlois, *Validity of parental work information on the birth certificate*, BMC Public Health 8 (2008), p. 95.

49. M.B. Schenker, S.J. Samules, C. Perkins, E.L. Lewis, D.F. Katz, and J.W. Overstreet, *Prospective surveillance of semen quality in the workplace*, J. Occup. Med. 30 (1988), pp. 336–344.

50. S.J. Samuels, *Lessons from a surveillance program of semen quality*, Reprod. Toxicol. 2 (1988), pp. 229–231.

51. C.C. Lawson, T.M. Schnorr, G.P. Daston, B. Grajewski, M. Marcus, M. McDiarmid, E. Murono, S.D. Perreault, S.M. Schrader, and M. Shelby, *An occupational reproductive research agenda for the third millennium*, Environ. Health Perspect. 111 (2003), pp. 584–592.

52. S.H. Swan, E.P. Elkin, and L. Fenster, *The question of declining sperm density revisited: An analysis of 101 studies published 1934–1996*, Environ. Health Perspect. 108 (2000), pp. 961–966.

53. A.O. Hammoud, M. Gibson, C.M. Peterson, A.W. Meikle, and D.T. Carrell, *Impact of male obesity on infertility: A critical review of the current literature*, Fertil. Steril. 90 (2008), pp. 897–904.

54. J.R. Loret de Mola, *Obesity and its relationship to infertility in men and women*, Obstet. Gynecol. Clin. N. Am. 36 (2009), pp. 333–346.

55. S. Cabler, A. Agarwal, M. Flint, and S.S. Du Plessis, *Obesity: Modern man's fertility nemesis*, Asian J. Androl. 12 (2010), pp. 480–489.

56. R. Pasquali, *Obesity and androgens: Facts and perspectives*, Fertil. Steril. 85 (2006), pp. 1319–1340.

57. S.S. Kasturi, J. Tannir, and R.E. Brannigan, *The metabolic syndrome and male infertility*, J. Androl. 29 (2008), pp. 251–259.

58. S.S. Du Plessis, S. Cabler, D.A. McAlister, E. Sabanegh, and A. Agarwal, *The effect of obesity on sperm disorders and male infertility*, Nat. Rev. Urol. 7 (2010), pp. 153–161.

59. A.A. MacDonald, G.P. Herbison, M. Showell, and C.M. Farquhar, *The impact of body mass index on semen parameters and reproductive hormones in human males: A systematic review with meta-analysis*, Human Reprod. Update 16 (2010), pp. 293–311.

60. D. Moher, D.J. Cook, S. Eastwood, I. Olkin, D. Rennie, and D.F. Stroup, *Improving the quality of reports of meta-analyses of randomized controlled trials: The QUOROM statement*, Lancet 354 (1999), pp. 1896–1900.

61. D.F. Stroup, J.A. Berlin, S.C. Morton, I. Olkin, G.D. Williamson, D. Rennie, D. Moher, B.J. Becker, T.A. Sipe, and S.B. Thacker, *Meta-analysis of observational studies in epidemiology*, JAMA 283 (2000), pp. 2008–2012.

62. D. Moher, J. Tetzlaff, A.C. Tricco, M. Sampson, and D.G. Altman, *and reporting characteristics of systematic reviews*, PLoS Med. 4 (2007), p. e78.

63. M. Borenstein, L.V. Hedges, J.P.T. Higgins, and H.R. Rothstein, *Introduction to meta-analysis*, John Wiley & Sons, Ltd, West Sussex, UK, 2009.

64. A.S. Aggerholm, A.M. Thulstrup, G. Toft, C.H. Ramlau-Hansen, and J.P. Bonde, *Is overweight a risk factor for reduced semen quality and altered serum sex hormone profile?*, Fertil. Steril. 90 (2008), pp. 619–626.

65. World Health Organization, *WHO laboratory manual for the examination and processing of human semen, 5th edition*, World Press. Geneva, Switzerland, 2010.

66. G.M. Buck Louis, E.F. Schisterman, A.M. Sweeney, T.C. Wilcosky, R.E. Gore-Langton, C.D. Lynch, D. Boyd Barr, S.M. Schrader, S. Kim, Z. Chen, and R. Sundaram, *Designing prospective cohort studies for assessing reproductive and developmental toxicity during sensitive windows of human reproduction and development—The LIFE Study*, Pediatr. Perinat. Epidemiol. 25 (2011), pp. 413–424.

67. T.W. Turner and S.M. Schrader, *Sperm migration assay as measure of recently ejaculated sperm motility in specimens shipped overnight*, J. Andrology 27 (2006), p. 58.

68. R.Z. Sokol, *Endocrine evaluation in the assessment of male reproductive hazards*, Reprod. Toxicol. 2 (1988), pp. 217–222.

69. S.M. Schrader, T.W. Turner, M.J. Breitenstein, and S.D. Simon, *Male reproductive hormones for occupational field studies*, J. Occup. Med. 35 (1993), pp. 574–576.

70. H.E. Kulin, O.M. Bell, R.J. Santen, and A.J. Ferber, *Integration of pulsatile gonadotropin secretion by timed urinary measurements: An accurate and sensitive 3-hour test*, J. Clin. Endocrinol. Metab. 40 (1975), pp. 783–789.

71. M. Kamijima, H. Hibi, M. Gotoh, K. Taki, I. Saito, H. Wang, S. Itohara, T. Yamada, G. Ichihara, E. Shibata, T. Nakajima, and Y. Takeuchi, *A survey of semen indices in insecticide sprayers*, J. Occup. Health 46 (2004), pp. 109–118.

72. A. Mahmoud, P. Kiss, M. Vanhoorne, D. De Bacquer, and F. Comhaire, *Is inhibin B involved in the toxic effect of lead on male reproduction?*, Int. J. Androl. 38 (2005), pp. 150–155.

73. B.H. Alexander, H. Checkoway, E.M. Faustman, C. van Netter, C.H. Muller, and T.G. Ewers, *Contrasting associations of blood and semen lead concentrations with semen quality among lead smelter workers*, Am. J. Ind. Med. 34 (1998), pp. 464–469.

74. N. Naha and A.R. Chowdhury, *Inorganic lead exposure in battery and paint factory: Effect on human sperm structure and functional activity*, J. UOEH 28 (2006), pp. 157–171.

75. A. Guven, A. Kayikci, K. Cam, P. Arbak, O. Balbay, and M. Cam, *Alterations in semen parameters of toll collectors working at motorways: Does diesel exposure induce detrimental effects on semen?*, Andrologia 40 (2008), pp. 346–351.

76. F. Hossain, O. Ali, U.J.A. D'Souza, and D.K.S. Naing, *Effects of pesticide use on semen quality among farmers in rural areas of Sabah, Malaysia*, J. Occup. Health 52 (2010), pp. 353–360.

77. D.K. Li, Z. Zhou, M. Miao, Y. He, J. Wang, J. Ferber, L.J. Herrinton, E. Goa, and W. Yuan, *Urine bisphenol-A (BPA) level in relation to semen quality*, Fertil. Steril. 95 (2011), pp. 625–630.

78. J.Y. Ma, J.J. Ji, Q. Ding, W.D. Liu, S.Q. Wang, Ning Wang, and G.Y. Chen, *The effects of carbon disulfide on male sexual function and semen quality*, Toxicol. Ind. Health 26 (2010), pp. 375–382.

79. L.P. Huang, C.C. Lee, P.C. Hsu, and T.S. Shih, *The association between semen quality in workers and the concentration of di(2-ethylhexyl)phthalate in polyvinyl chloride pellet plant air*, Fertil. Steril. 96 (2011), pp. 90–94.

80. L. Migilore, A. Naccarati, A. Zanello, R. Scarpato, L. Bramanti, and M. Mariani, *Assessment of sperm DNA integrity in workers exposed to styrene*, Human Repro. 17 (2002), pp. 2912–2918.

81. L.C. Sánchez-Peña, B.E. Reyes, L. López-Carrillo, R. Recio, J. Morán-Martínez, M.E. Cebrián, and B. Quintanilla-Vega, *Organophosphorous pesticide exposure alters sperm chromatin structure in Mexican agricultural workers*, Toxicol. Appl. Pharmacol. 196 (2004), pp. 108–113.

82. Y. Xia, S. Cheng, Q. Bian, L. Xu, M.D. Collins, H.C. Chang, L. Song, J. Liu, S. Wang, and X. Wang, *Genotix effects of spermatozoa of carbaryl-exposed workers*, Toxicol. Sci. 85 (2005), pp. 615–623.

83. Q. Bian, L.C. Xu, S.L. Wang, Y.K. Xia, L.F. Tan, J.F. Chen, L. Song, H.C. Chang, and X.R. Wang, *Study on the relation between occupational fenvalerate exposure and spermatozoa DNA damage of pesticide factory workers*, Occup. Environ. Med. 61 (2004), pp. 999–1005.

84. P.-G. Forkert, L. Lash, R. Tardif, N. Tanphaichitr, C. Candevoort, and M. Moussa, *Identification of trichloroethylene and its metabolites in human seminal fluid of workers exposed to trichloroethylene*, Drug Metab. Disp. 31 (2003), pp. 306–311.

85. Y. Duydu, N. Başaran, A. Üstündağ, S. Aydin, U. Ündeğer, O.Y. Ataman, K. Aydos, Y. Döker, K. Ickstadt, B.S. Waltrup, K. Golka, and H.M. Bolt, *Reproductive toxicity parameters and biological monitoring in occupationally and environmentally boron-exposed persons in Bandirma, Turkey*, Arch. Toxicol. 85 (2011), pp. 589–600.

86. O. Akinloye, A.O. Arowojolu, O.B. Shittu, and J.L. Anetor, *Cadmium toxicity: A possible cause of male infertility in Nigeria*, Reprod. Biol. 6 (2006), pp. 17–30.

87. S. Yucra, M. Gasco, J. Rubio, and G.F. Gonxales, *Semen quality in Peruvian pesticide applicators: Association between urinary organophosphate metabolites and semen parameters*, Environ. Health 7 (2008), p. 59.

88. T. Lifeng, W. Shoulin, J. Junmin, S. Xuezhao, L. Yannan, W. Qianli, and C. Longsheng, *Effects of fenvalerate exposure on semen quality among occupational workers*, Contraception 73 (2006), pp. 92–96.

89. D.-K. Li, Z. Zhou, M. Miao, Y. He, D. Qing, T. Wu, J. Wang, X. Weng, J. Ferber, L.J. Herrington, Q. Zhu, E. Gao, and W. Yuan, *Relationship between urine bisphenol A level and declining male sexual function*, J. Androl. 31 (2010), pp. 500–506.

90. S.M. Schrader, M.J. Breitenstein, and B.D. Lowe, *Cutting off the nose to save the penis*, J. Sex. Med. 5 (2008), pp. 1932–1940.

91. A. Vermeulen, L. Verdonck, and J.M. Kaufman, *A critical evaluation of simple methods for the estimation of free testosterone in serum*, J. Clin. Endocrinol. Metab. 84 (1999) pp. 3666–3672.

92. C.W. Bardin, *Pituitary-testicular axis, reproductive endocrinology*, S.S.C. Yen and R.B. Jaffe, eds., WB Saunders, Philadelphia, 1986, pp. 177–199.

93. D. Riad-Fahmy, G.F. Read, R.F. Walker, and K. Griffiths, *Steroids in saliva for assessing endocrine function*, Endocr. Rev. 3 (1982), pp. 367–395.

94. P. Apostoli, L. Romeo, E. Peroni, A. Ferioli, S. Ferrari, F. Pasini, and F. Aprili, *Steroid hormone sulphation in lead workers*, Br. J. Ind. Med. 46 (1989), pp. 204–208.

95. E. Carlsen, C. Olsson, J.H. Petersen, A. Anderssen, and N.E. Skakkebaek, *Diurnal rhythm in serum levels of inhibin B in normal men: Relation to testicular steroids and gonadotropins*, J. Clin. Endocrinol. Metab. 84 (1999), pp. 1664–1669.

96. J. Axelsson, M. Ingre, T. Akerstedt, and U. Holmbäck, *Effects of acutely displaced sleep on testosterone*, J. Clin. Endocrinol. Metab. 90 (2005), pp. 4530–4535.

97. D.G. Egnatz, M.G. Ott, J.C. Townsend, R.D. Olson, and D.B. Johns, *DBCP and testicular effects in chemical workers: An epidemiological survey in Midland, Michigan*, J. Occup. Med. 22 (1980), pp. 727–732.

98. G. Pan, T. Hanaoka, M. Yoshimura, S. Zhang, P. Wang, H. Tsukino, K. Inoue, H. Nakazawa, S. Tsugane, and K. Takahashi, *Decreased serum free testosterone in workers exposed to high levels of di-n-butyl phthalate (DBP) and di-2-ethylhexyl phthalate (DEHP): A cross-sectional study in China*, Environ. Health Perspect. 114 (2006), pp. 1643–1648.

99. B. Grajewski, E.A. Whelan, T.M. Schnorr, R. Mouradian, R. Alderfer, and D.K. Wild, *Evaluation of reproductive function among men occupationally exposed to a stilbene derivative: I. Hormonal and physical status*, Am. J. Ind. Med. 29 (1996), pp. 49–57.

100. S.E. Chia, V.H. Goh, and C.N. Ong, *Endocrine profiles of male workers with exposure to trichloroethylene*, Am. J. Ind. Med. 32 (1997), pp. 217–222.

101. D. Ortiz-Pérez, M. Rodríguez-Martínez, F. Martínez, V.H. Borja-Aburto, J. Castelo, J.I. Grimaldo, E. de la Cruz, L. Carrizales, and F. Díaz-Barriga, *Fluoride-induced disruption of reproductive hormones in men*, Environ. Res. 93 (2003), pp. 20–30.

102. T. Hanaoka, N. Kawamura, K. Hara, and S. Tsugane, *Urinary bisphenol A and plasma hormone concentrations in male workers exposed to bisphenol A diglycidyl ether and mixed organic solvents*, Occup. Environ. Med. 59 (2002), pp. 625–628.

103. B. Grajewski, C. Cox, S.M. Schrader, W.E. Murray, R.M. Edwards, T.W. Turner, J.M. Smith, S.S. Shekar, D.P. Evenson, S.D. Simon, and D.L. Conover, *Semen quality and hormone levels among radiofrequency heater operators*, J. Occup. Environ. Med. 42 (2000), pp. 993–1005.

104. N.H. Hjollund, L. Storgaard, E. Ernst, J.P. Bonde, and J. Olsen, *Impact of diurnal scrotal temperature on semen quality*, Reprod. Toxicol. 16 (2002), pp. 215–221.

105. J. Blanco-Muñoz, M.M. Morales, M. Lacasaña, C. Aguilar-Garduño, S. Bassol, and M.E. Cebrián, *Exposure to organophosphate pesticides and male hormone profile in floriculturist of the state of Morelos, Mexico*, Hum. Reprod. 25 (2010), pp. 1787–1795.

106. S.J. Hsieh, Y.W. Chiu, W.F. Li, C.H. Wu, H.I. Chen, and H.Y. Chuang, *Increased concentrations of serum inhibin B among male workers with long-term moderate lead exposure*, Sci. Total Environ. 407 (2009), pp. 2603–2607.

107. C. Padungtod, B.L. Lasley, D.C. Christiani, L.M. Ryan, and X. Xu, *Reproductive hormone profile among pesticide factory workers*, J. Occup. Environ. Med. 40 (1998), pp. 1038–1047.

108. J.M. Dabbs Jr, D. de La Rue, and P.M. Williams, *Testosterone and occupational choice: Actors, ministers, and other men*, J. Pers. Soc. Psychol. 59 (1990), pp. 1261–1265.

109. R. Eliasson and L. Treichl, *Supravital staining of human spermatozoa*, Fertil. Steril. 22 (1971), pp. 134–137.

110. R.S. Jeyendran, H.H. Van den Ven, M. Perez-Palaez, B.G. Crabo, and L.J.D. Zaneveld, *Development of an assay to assess the functional integrity of the human sperm membrane and its relationship to other semen characteristics*, J. Reprod. Fertil. 70 (1984), pp. 219–228.

111. S.M. Schrader, S.F. Platek, L.J.D. Zaneveld, M. Perez-Palaez, and R.S. Jeyendran, *Sperm viability: A comparison of analytical methods*, Andrologia 18 (1986), pp. 530–538.

112. J. Macleod, *Semen quality in 1000 men of known fertility and in 800 cases of infertile marriage*, Fertil. Steril. 2 (1951), pp. 115–139.

113. M. Freund, *Standards for the rating of human sperm morphology*, Int. J. Fertil. 11 (1966), pp. 97–180.

114. B. Fredricson, *Morphologic evaluation of spermatozoa in different laboratories*, Andrologia 11 (1979), pp. 57–61.

115. L.J. Hanke, Comparison of laboratories conducting sperm morphology, Report TA78-28, National Institute for Occupational Safety and Health, Cincinnati, OH, 1981.

116. World Health Organization, *Laboratory manual for the examination of human semen and sperm-cervical mucus interaction, 3rd Edition.* Cambridge University Press, Cambridge, England, 1992.

117. A. Schmassmann, G. Mikuz, G. Bartsch, and H. Rohr, *Quantification of human sperm morphology and motility by means of semi-automatic image analysis systems*, Microscopica Acta 82 (1979), pp. 163–178.

118. D.F. Katz, J.W. Overstreet, and R.J. Pelprey, *Integrated assessment of the motility, morphology and morphometry of human spermatozoa*, INSERM 103 (1981), pp. 97–100.

119. S.M. Schrader, T.W. Turner, B.D. Hardin, R.W. Niemeier, and J.R. Burg, *Morphometric analysis of human spermatozoa*, J. Androl. 5 (1984), p. 22.

120. J.R. Jagoe, N.P. Washbrook, and E.A. Hudson, *Morphometry of spermatozoa using semiautomatic image analysis*, J. Clin. Pathol. 139 (1986), pp. 1347–1352.

121. F. DeStefano, J.L. Annest, M.J. Kresnow, M.L. Flock, and S.M. Schrader, *Automated semen analysis in large epidemiologic studies*, J. Androl. 8 (1987), p. 24.

122. T.W. Turner, S.M. Schrader, and S.D. Simon, *Sperm head morphometry as measured by three different computer systems*, J. Androl. 9 (1988), p. 45.

123. J.F. Moruzzim, A.J. Wyrobek, B.H. Mayall, and B.L. Gledhill, *Classification of human sperm morphology by CASA*, Fertil. Steril. 50 (1988), pp. 142–152.

124. S.M. Schrader, J.M. Ratcliffe, T.W. Turner, and R.W. Hornung, *Use of new field methods of semen analysis in the study of occupational hazards to reproduction: The example of ethylene dibromide*, J. Occup. Med. 29 (1987), pp. 963–966.

125. Z. Zukerman, L.J. Rodriguez-Rigau, D.B. Weiss, A.K. Chowdhury, K.D. Smith, and E. Steinberger, *Analysis of the seminiferous epithelium in human testicular biopsies, and the relation of spermatogenesis to sperm density*, Fertil. Steril. 30 (1978), pp. 448–455.

126. D. Delbès, B.F. Hales, and B. Robaire, *Toxicants and human sperm chromatin*, Mol. Hum. Reprod. 16 (2010), pp. 14–22.

127. S.M. Schrader and M.H. Kanitz, *Occupational hazards to male reproduction*, in *State of the Art Reviews in Occupational Medicine: Reproductive Hazards*, E. Gold, M. Schenker, and B. Lasley, eds., Hanley and Belfus, Inc., Philadelphia, 1994, pp. 405–414.

128. J.M. Holmes and R.H. Martin, *Aneuploidy detection in human sperm nuclei using fluorescence in situ hybridization*, Human. Genet. 91 (1993), pp. 20–24.

129. A.J. Wyrobek, W.A. Robbins, Y. Mehraein, D. Pinkel, and H.U. Weier, *Detection of sex chromosomal aneuploidies X-X, Y-Y, and X-Y in human sperm using two-chromosome fluorescence in situ hybridization*, Am. J. Med. Genet. 53 (1994), pp. 1–7.

130. F.Z. Bischoff, D.D. Nguyen, K.J. Burt, and L.G. Shaffer, *Estimates of aneuploidy using multicolor fluorescence in situ hybridization on human sperm*, Cytogenet. Cell Genet. 66 (1994), pp. 237–243.

131. D.P. Evenson, *Flow cytometry of acridine orange-stained sperm is a rapid and practical method for monitoring occupational exposure to genotoxicants*, in *Monitoring Occupational Genotoxicity*, M. Sorsa and H. Norppa, eds., Alan R. Liss, New York, 1986, pp. 121–132.

132. D.P. Evenson, L.K. Jost, R.K. Baer, T.W. Turner, and S.M. Schrader, *Individuality of DNA denaturation patterns in human sperm as measured by the sperm chromatin structure assay*, Reprod. Toxicol. 5 (1991), pp. 115–125.

133. M. Spano and D.P. Evenson, *Flow cytometric studies in reproductive toxicology*, in *New Horizons in Biological Dosimetry*, B.L. Gledhill and F. Mauro, eds., Wiley-Liss, New York, 1991, pp. 497–511.

134. D. Evenson and R. Wixon, *Environmental toxicants cause sperm DNA fragmentation as detected by sperm chromatin structure assay (SCSA)*, Toxicol. Appl. Pharmacol. 207 (2005), pp. 532–537, Review.

135. B. Stachel, R.C. Dougherty, U. Lahl, M. Schlosser, and B. Zeschmar, *Toxic environmental chemicals in human semen: Analytical method and case studies*. Andrologia 21 (1989), pp. 282–291.

136. A. Zikarge, *Cross-sectional study of ethylene dibromide-induced alterations of seminal plasma biochemistry as a function of post-testicular toxicity with relationships to some indices of semen analysis and endocrine profile*, Dissertation to the University of Texas Health Science Center, Houston, TX, 1986.

137. T. Mann and C. Lutwak-Mann, *Passage of chemicals into human and animal semen: Mechanisms and significance*, CRC Crit. Rev. Toxicol. 11 (1982), pp. 1–14.

138. S. Kumar and V.V. Mishra, *Review: Toxicants in reproductive fluid and in vitro fertilization (IVF) outcome*, Toxicol. Ind. Hlth. 26 (2010), pp. 505–511.

139. T.E. Arbuckle, S.M. Schrader, D. Cole, J.C. Hall, C.M. Bancej, L.A. Turner, and P. Claman, *2,4 D residues in semen of Ontario farmers*, Repro. Tox. 13 (1999), pp. 421–429.

140. S.M. Schrader, T.W. Turner, and J.M. Ratcliffe, *The effects of ethylene dibromide on semen quality: A comparison of short term and chronic exposure*, Reprod. Toxicol. 2 (1988), pp. 191–198.

141. R.A. Yazigi, R.R. Odem, and K.L. Polakoski, *Demonstration of specific binding of cocaine to human spermatozoa*, JAMA 266 (1991), pp. 1956–1959.

142. R.J. Aitkin, *Development of in vitro tests of human sperm function: A diagnostic toll and model system for toxicological analyses*, Toxic In Vitro 4 (1990), pp. 560–569.

143. D.F. Katz, J.W. Overstreet, and F.W. Hanson, *A new quantitative test for sperm penetration into cervical mucus*, Fertil. Steril. 33 (1980), pp. 179–186.

144. C.S. Niederberger, D.J. Lamb, M. Glinz, L.I. Lipshultz, and N.F. Scully, *Tests of sperm function for evaluation of the male-Penetrak-Asterisk and Tru-Trax*, Fertil. Steril. 60 (1993), pp. 319–323.

145. M.M. Biljan, C.T. Taylor, P.R. Manasse, E.C. Joughin, C.R. Kingsland, and D.I. Lewisjones, *Evaluation of different sperm function tests as screening methods for male fertilization potential-the value of the sperm migration test*, Fertil. Steril. 62 (1994), pp. 591–598.

146. B.J. Rogers, *Use of SPA in assessing toxic effects on male fertilizing potential*, Reprod. Toxicol. 2 (1988), pp. 233–240.

147. P. Prinosilova, T. Kruger, L. Sati, S. Ozkavukcu, L. Vigue, E. Kovanci, and G. Huszar, *Selectivity of hyaluronic acid binding for spermatozoa with normal Tygerberg strict morphology*, Reprod. Biomed. Online 18 (2009), pp. 177–183.

148. C.C. Coddington, D.R. Franken, L.J. Burkman, W.T. Oosthuizen, T. Kruger, and G.D. Hodgen, *Functional aspects of human sperm binding to the zona-pellucida using the hemizona assay*, J. Andrology 12 (1991), pp. 1–8.

149. D.R. Franken, A.A. Acosta, T.F. Kruger, C.J. Lombard, S. Ochninger, and G.D. Hodgen, *The hemizona assay—Its role in identifying male factor infertility in assisted reproduction*, Fertil. Steril. 59 (1993), pp. 1075–1080.

150. S.M. Schrader, T.W. Turner, and S.D. Simon, *Sources of variation of the sperm penetration assay under field study conditions*, Assist. Reprod. Technol. Androl. 2 (1991), pp. 63–74.

151. W.C. deGroat and A.M. Booth, *Physiology of male sexual function*, Am. Intern. Med. 92 (1980), pp. 329–331.

152. A.J. Thomas Jr, *Ejaculatory dysfunction*, Fertil. Steril. 39 (1983), pp. 445–454.

153. R.J. Krane, I. Goldstein, and I.S. de Tejada, *Impotence*, N. Engl. J. Med. 321 (1989), pp. 1648–1659.

154. A.L. Burnett, *Environmental erectile dysfunction: Can the environment really be hazardous to your erectile health?*, J. Androl. 29 (2008), pp. 229–236.

155. A.S. Burris, S.M. Banks, and R.J. Sherins, *Quantitative assessment of nocturnal penile tumescence and rigidity in normal men using a home monitor*, J. Androl. 10 (1989), pp. 492–497.

156. S.M. Schrader, M.J. Breitenstein, J.C. Clark, B.D. Lowe, and T.W. Turner, *Nocturnal Penile Tumescence and Rigidity Testing of Bicycling Patrol Officers*, J. Androl. 23 (2002), pp. 927–934.

157. A. Natali and P.J. Turek, *An assessment of new sperm tests for male infertility*, Urology 77 (2011), pp. 1027–1034.

158. M.A. Baker, *The 'omics revolution and our understanding of sperm cell biology*, Asian J. Androl. 13 (2011), pp. 6–10.

159. C. DeJung, *Semen analysis: Looking for an upgrade in class*, Fertil. Steril. 97 (2012), pp. 260–266.

Section III

Reproduction and Development
Biological and Computational Methods

8 Animal Study Protocols and Alternative Assays for the Assessment of Reproductive and Developmental Toxicity

Aldert H. Piersma

CONTENTS

ABSTRACT

This chapter reviews existing and emerging test systems in regulatory reproductive toxicity assessment. Current OECD (Organisation for Economic Co-operation and Development) globally harmonized practice, which stems from the early 1980s, has shown its strengths and weaknesses, and the endocrine disrupter issue as well as the EU REACH (European Union Registration, Evaluation and Authorisation of CHemicals) legislation for chemical safety and animal welfare issues have stimulated innovation in the area. Standardized test protocols are being amended and new protocols are being defined, including animal and nonanimal tests. Testing strategies, which combine individual tests into tiered approaches, are likewise under discussion. Finally, integration of testing strategies within the wider realm of toxicity testing is under extensive study. This array of developments is ongoing at the national, EU, OECD, and United Nations levels. They support improved efficiency of safety testing of chemicals with reduced experimental animal use, while increasing the quality of basic toxicological information used for risk assessment and classification and labeling of chemicals.

KEYWORDS

Reproductive toxicology, test guidelines, alternatives, integrated testing strategy, REACH, risk assessment, hazard identification, fertility, development

8.1   INTRODUCTION

Globally harmonized reproductive toxicity testing protocols have been in use since the 1980s. In recent years, these tests have become the subject of renewed discussion. This has led to changes in existing protocols as well as to the development of new study designs. As a consequence and in parallel, testing strategies for regulatory safety assessment of chemicals are also being reformulated. This chapter gives an overview of existing test systems and current activities worldwide, with an outlook toward developments in the future. This section will give a bird's-eye view of current developments, whereas details of guidelines and test systems are discussed in Section 8.2. Conclusions follow in Section 8.3.

8.1.1 Reproductive Toxicology

Reproductive toxicology is concerned with all possible adverse effects of chemical exposures on any aspect of the reproductive cycle (Figure 8.1). Classically, fertility and prenatal development have been the two areas of main concern. They constitute a wide variety of mechanisms at the molecular, cellular, tissue, and organism level, with different windows of sensitivity in time. Classically, morphological effect assessment and functional integrity of the reproductive system have been used as endpoints. Novel functional endpoints of toxicity have received increasing interest, such as developmental neurotoxicity and behavior, and developmental immune toxicity assessed through immune function tests in offspring at adulthood.

8.1.2 Current Regulatory Reproductive Toxicity Tests

Standardized regulatory reproductive toxicity testing dates back to the early 80s of the 20th century, when OECD (Organisation for Economic Co-operation and Development) protocols were published for the prenatal developmental toxicity study (OECD TG 414), the one-generation reproductive toxicity study (OECD TG 415), and the two-generation reproductive toxicity study (OECD TG 416). Protocols can be retrieved from the OECD website [1]. The underlying justification for the design of this set of three tests was as follows. The OECD TG 414 prescribes prenatal exposure starting after implantation and necropsy 1 day before expected birth. This design avoids interference of effects on implantation and also precludes the possibility of

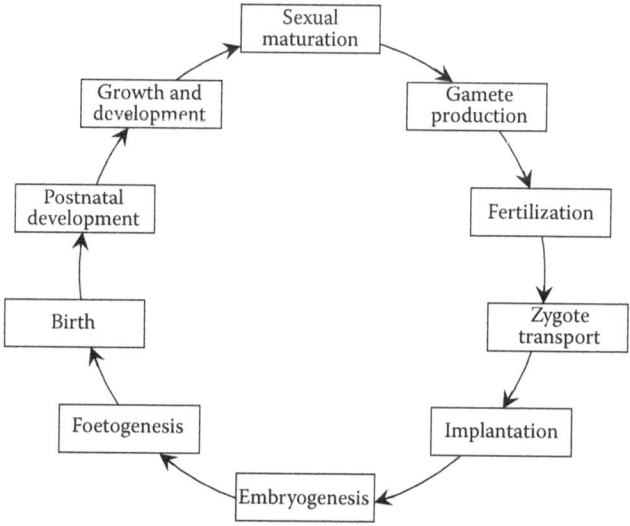

FIGURE 8.1 Schematic representation of the reproductive cycle.

maternal cannibalism of malformed pups. Thus, developmental effects on implanted embryo-fetuses can be studied after explantation from the uterus shortly before birth. The OECD TG 415 one-generation study is primarily a study to detect effects on fertility, prescribing parental exposure of both sexes and of dams throughout pregnancy and until weaning of the pups. The OECD TG 416 two-generation study is designed to allow fertility assessment of animals prenatally exposed, and therefore only this study design in principle covers the entire reproductive cycle. In addition to these definitive tests, a relatively quick screening method that can give initial clues about possible fertility and developmental effects of chemicals is the OECD TG 421 screening study adopted in 1995. In this study, exposure of parental animals is for 2 weeks premating and until postnatal day 6 in dams, which is the day of necropsy. Males are dosed to a total duration of 4 weeks and necropsied. This protocol prescribes only eight pregnant dams per dose group. As noted above, this method is designed as a screening protocol and is not adequate for reliable conclusions on the reproductive toxicity of the test compound in case of absence of observed toxicity. However, if toxicity is found, this may trigger further studies or hazard- and risk-related measures.

8.1.3 CURRENT REPRODUCTIVE TOXICITY TESTING STRATEGY

The toxicity testing strategy in the European Union (EU) under the REACH (Registration, Evaluation and Authorisation of CHemicals) regulation for chemical safety is primarily based on production tonnage levels (Figure 8.2) [2]. Reproductive toxicity testing commences with an OECD TG 421 screening study at tonnage level 1 to gather some but not comprehensive information both on development and fertility.

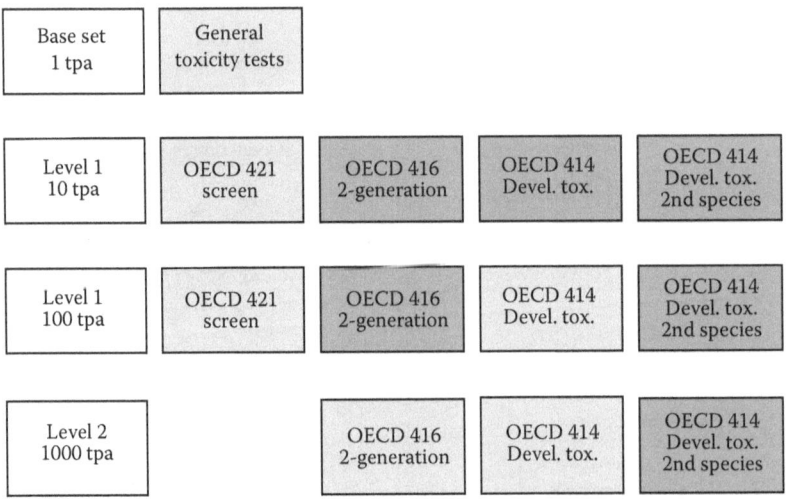

FIGURE 8.2 Schematic representation of the current EU REACH reproductive toxicity testing strategy. Tests marked in light gray are basic requirements. Tests marked in dark gray are only required in case of concern. tpa, ton(s) per annum.

Dependent on the outcome of this study and on tonnage level, a developmental toxicity study (in case of concern for developmental effects) or a two-generation study (in case of concern for fertility effects) may follow. A developmental toxicity study in a second species (usually the rabbit) can be warranted on the basis of equivocal findings in the first developmental toxicity study (usually the rat). The justification for a second species goes back to the thalidomide episode that occurred around 1960 [3,4]. Thalidomide, used as a sedative, caused severe limb reduction defects in children of mothers taking the drug in pregnancy. These defects could not be reproduced in the rat, in which only general fetotoxicity was observed. In the rabbit, limited limb reduction defects were observed, although more sporadically and at higher doses. This prompted regulators in some regulatory frameworks to request developmental toxicity testing in two species in order to prevent similar devastating consequences of other chemicals in the future.

8.1.4   ENDOCRINE DISRUPTION

The issue of endocrine disruption has received considerable attention since the early 1990s, when observations on wildlife and human fertility were considered in association with environmental chemical exposures [5,6]. The question was raised whether current test protocols were sufficiently sensitive to detect endocrine-disrupting effects. The OECD initiated the Endocrine Disrupter Testing and Assessment task force, which later listed a wealth of animal and nonanimal tests proposed to detect endocrine activities of chemicals [7]. A series of specific in vitro receptor binding and activation assays was listed. In addition, two short-term in vivo assays received renewed attention. The uterotrophic assay for (anti-)estrogenicity and the Hershberger assay for (anti-)androgenicity came into consideration for endocrine disruption assessment. In addition, enhancements of existing reproductive toxicity guidelines (OECD TG 414/5/6) were proposed, specifically with reference to hormone-level assessments and histopathology of reproductive organs and thyroid and pituitary glands. Finally, such enhancements were also implemented for the 28-day subchronic toxicity study (OECD TG 407).

8.1.5   EU REACH

The European chemicals regulation REACH requires specific scrutiny for reproductive toxicants together with carcinogenic and mutagenic compounds [2]. As a consequence, such compounds may need additional testing if the current toxicity profile is incomplete. It has been estimated that 30,000 chemicals will enter the REACH legislation, of which around 5000 chemicals reach tonnage level 1 and another 5000 chemicals reach higher levels of testing, both requiring reproductive toxicity testing. Testing will require millions of experimental animals, of which no less than an estimated 65% will be needed in reproductive toxicity testing [8]. The very high percentage is attributed to the fact that, in reproductive toxicity testing, always more than one generation is included in each study and that fetuses from gestation day 18 onward (rat) and pups count as experimental animals under animal ethics legislations. This realization has greatly accelerated the attention for simplifying existing animal test protocols in reproductive toxicology and has further stimulated the process of development, validation, and implementation of animal-free alternatives. In addition, the extended one-generation reproduction study protocol under global discussion in OECD as a possible replacement for the current two-generation study is a promising new development in this respect. Nontesting in silico methods such as read-across, grouping approach, and (quantitative) structure–activity relationships ([Q]SARs) are increasingly being employed to optimize hazard assessment.

8.1.6 INNOVATING TESTS AND TESTING STRATEGIES FOR HAZARD IDENTIFICATION

After around 25 years of experience with the existing system of OECD reproductive toxicity test protocols, and given the high animal use and high cost and time consumed by these tests, efforts are now underway to review the efficiency and necessity of test protocols and current testing strategies in actual practice. In addition, novel endpoints for endocrine effects, developmental effects on immunity, and neurobehavioral effects have received interest as they may represent important parameters that may be affected after prenatal and juvenile exposure, which are not yet measured to the extent necessary in current tests and testing strategies. Current major activities in this area are briefly listed here and described in more detail in Section 8.2. The enhancement of the OECD TG 407 subchronic toxicity study in adult rats with endocrine parameters has resulted in an updated guideline in 2008. The need for addressing developmental neurotoxicity has resulted in the OECD TG 426 Guideline in 2007. An important development is the design of an extended one-generation study, in which the offspring is raised to adulthood for the assessment of various additional parameters and increasing the number of observations within generations, but omitting the production of a second generation as in the OECD TG 416 two-generation study. The increased concern for possible specific sensitivity of children to (endocrine acting) chemicals has been translated into experimental protocols using direct exposure of pups before weaning to achieve relevant exposure levels that may not be achieved after lactational exposure via exposed dams. In reproductive toxicity testing, there are several issues related to strategically employing existing tests more efficiently and without redundancy. One issue relates to the added value of the developmental toxicity study in a second species, for which a review of past experience has been published to underpin decisions on possible eventual alterations in testing strategies [9,10]. Furthermore, the question has been put forward to what extent reproductive toxicity endpoints have an added value in the determination of the overall no-observed-adverse-effect level (NOAEL). A relevant comparison for addressing this question is that between generation study outcome and subchronic adult toxicity outcome, which has been addressed recently [11]. In 2008, OECD finalized Guidance Document 43 (GD 43), which gives guidance on strategic aspects of mammalian reproductive toxicity testing and assessment. In 2005, OECD Guidance Document GD 34 was adopted, which addresses the validation and international acceptance of new or updated test methods for hazard assessment. On a higher level of integration, the question can be put forward at what levels and in what order the different animal tests should be performed. In relation to that, the role and place of in silico and in vitro alternatives as well as possible situations in which tests can be waived dependent on outcomes in previous studies should be considered. Within the REACH Implementation Plans, such a strategy has been proposed for reproductive toxicity testing. The same type of question is being addressed for toxicity testing in general, where reproductive toxicity testing is one of a series of classes of endpoints that have to be assessed in an integrated testing strategy. As an activity under the auspices of the United Nations, a globally harmonized system for classification and labeling has been developed, including criteria for substances toxic to reproduction, which has recently become effective [12].

8.2 TEST GUIDELINES

This section reviews existing and emerging guidelines and discusses their possible application and use in regulatory reproductive toxicity testing.

8.2.1 OECD TG 440 Uterotrophic Assay

8.2.1.1 Principle and Status

The uterotrophic assay protocol was originally developed in the 1930s. It is based on the increase in uterine weight or uterotrophic response. It evaluates the ability of a chemical to elicit biological activities consistent with agonists or antagonists of natural estrogens (e.g., 17β-estradiol); however, its use for antagonist detection is much less common than that for agonists. The uterus responds to estrogens in two ways. An initial response is an increase in weight because of water imbibition. This response is followed by a weight gain owing to tissue growth. The uterus responses in rats and mice are qualitatively comparable. The uterotrophic assay is intended to be included in a battery of in vitro and in vivo tests to identify substances with a potential to interact with the endocrine system, ultimately leading to risk assessments for human health or the environment. The recent OECD validation program used both strong and weak estrogen agonists to evaluate the performance of the assay to identify estrogenic compounds [13]. Thereby, the sensitivity of the test procedure for estrogen agonists was well demonstrated besides a good intra- and inter-laboratory reproducibility. The specificity of the test was not addressed in detail in the validation study and is still an issue for discussion. The OECD adopted the draft guideline (OECD TG 440) in 2007.

8.2.1.2 Test Protocol

Juvenile or ovariectomized young adult rats are exposed with three daily doses of the test compound, either orally or subcutaneously (Figure 8.3). Graduated test substance doses are administered to a minimum of two treatment groups of experimental animals using one dose level per group and a minimum administration period of three consecutive days. The animals are necropsied approximately 24 h after the last dose. For estrogen agonists, the mean uterine weight of the treated animal groups relative to the vehicle group is assessed for a statistically significant increase. A statistically significant increase in the mean uterine weight of a test group indicates a positive response in this bioassay.

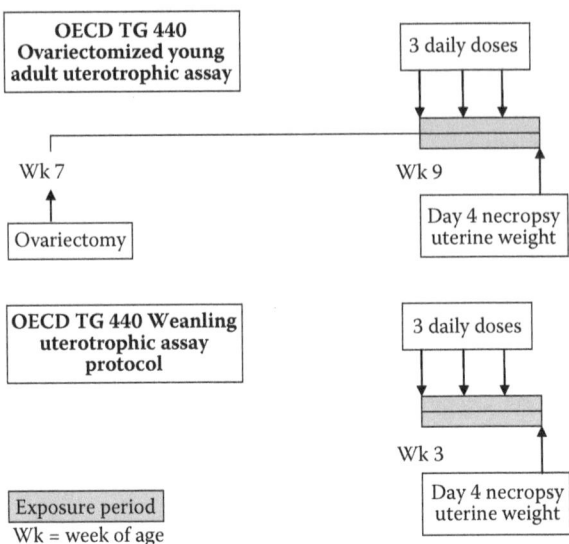

FIGURE 8.3   Schematic representation of the uterotrophic assay.

8.2.1.3   Application in the Testing Strategy

The uterotrophic assay was revived for use in endocrine disrupter screening. It can detect estrogen receptor alpha agonists, and if coadministered with an agonist, compounds acting as antagonists can also be detected. The discussion on the application of this test focuses on the necessity of using animal experimentation for this specific goal, whereas a wealth of in vitro receptor binding and activation assays are available. The counterargument mostly used is that, in an in vivo assay, the kinetics of compounds also comes into play, which would make this in vivo test more relevant for the human situation. In response to that, first, the subcutaneous route is less relevant in view of human exposure, and second, the short exposure duration of three daily doses (with usually relatively high doses) is hard to extrapolate to long-term low-dose exposures. It is anticipated that at least in the EU, the uterotrophic assay will not become a first-choice method for (anti-)estrogenicity testing. Rather, the ongoing development of sophisticated in vitro receptor binding and activation assays will probably reduce the need for the uterotrophic assay. In conjunction with in vivo kinetic studies, in vitro assay results will likely give sufficient relevant information to decide on the (anti-)estrogenic potential of test compounds.

8.2.2   OECD TG 441 Hershberger Assay

8.2.2.1   Principle and Status

The basis of the Hershberger assay is the absolute requirement for testosterone (produced in the testis) or dihydrotestosterone (converted from testosterone by 5α-reductase in the testis and other end target organs) for the rapid growth and maturation of the accessory sex organs during puberty in intact males, for their maintenance postpuberty, for their rapid regression and involution after castration (removal of the source of testosterone), and for their rapid regrowth in the castrate administered an exogenous androgen (typically testosterone propionate). The accessory sex organs of interest are predominantly the epididymides (if they are not removed at castration), the prostate (the ventral lobe or ventral plus dorsolateral lobes), and the levator ani plus bulbocavernosus complex muscle, as well as the other accessory sex organs (seminal vesicles with coagulating glands, Cowper's glands, and preputial glands). The Hershberger assay has been proposed by both the Endocrine Disruptor Screening and Testing Advisory Committee (1998) and OECD (1998) to be validated for use in a comprehensive screen to detect potential endocrine disruptors. An OECD Test Guideline (OECD TG 441) for this assay was adopted in 2009.

8.2.2.2 Test Protocol

The Hershberger assay has been used in various versions (Figure 8.4). The most important versions include the prepubertal intact male assay, the castrated adult male assay, and the peripubertal castrated male assay. The protocols have different advantages. The intact assay contains the complete intact hypothalamic–pituitary–gonadal axis, which makes this an apical assay, and it uses an age window with

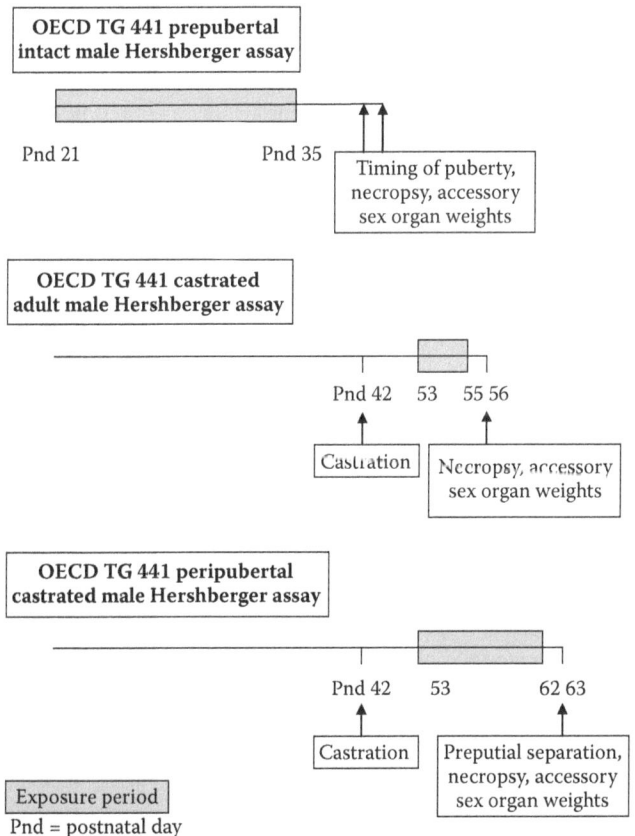

FIGURE 8.4   Schematic representation of typical versions of the Hershberger assay.

relatively high sensitivity to androgens. The concomitant disadvantages include the lack of mechanistic information and the relatively small time window of opportunity, and in this respect, the castrated adult male assay is superior. The peripubertal castrated male assay has been advocated for the high sensitivity of accessory sex organs around puberty, in combination with the possibility for mechanistic information in this castrated model [14].

8.2.2.3 Application in the Testing Strategy

The Hershberger assay has been revived and further developed in its several forms for endocrine disrupter detection, much in a similar sequence as for the uterotrophic assay. The argumentation relevant for the application of the Hershberger assay in a testing strategy is also very similar to the uterotrophic assay. The question of justification of the use of experimental animals for what is basically a receptor-activation assay that could be mimicked to a great extent by in vitro assays is also pertinent here. The exposure routes can be oral or subcutaneous; the latter should be considered less relevant for actual human exposures, although this route may be used for achieving higher internal exposures necessary for the detection of weak (ant-)agonists. It is anticipated that at least in the EU, the Hershberger assay will not become a first-choice method for (anti-)androgenicity testing. Rather, the ongoing development of sophisticated in vitro receptor binding and activation assays will probably reduce the need for the Hershberger assay. In conjunction with in vivo kinetic studies, in vitro assay results will likely give sufficient relevant information in the long run to decide on the (anti-)estrogenic potential of test compounds.

8.2.3 OECD TG 407 Updated Subchronic Toxicity Study

8.2.3.1 Principle and Status

The OECD TG 407 guideline is basically not a specific reproductive toxicity test; however, at an early stage of toxicity testing, it may give clues of possible effects of compound exposure on adult reproductive organs. In view of the endocrine disrupter issue, an initiative was taken to update the existing OECD TG 407 guideline with additional parameters suitable to detect endocrine activity of test substances. This procedure underwent an extensive international program to test for the relevance and practicability of the additional parameters; the performance of these parameters for chemicals with (anti-)estrogenic, (anti-)androgenic, and (anti-)thyroid activity; the intra- and interlaboratory reproducibility; and the interference of the new parameters with those required by the prior TG 407. A detailed description of the procedure and outcome of the project has been published [15]. The updated TG 407 allows the evaluation of endocrine-mediated effects within the context of other toxicological effects. These activities lead to an updated OECD guideline in 2008.

8.2.3.2 Test Protocol

Male and female rats, five per sex per dose group, are exposed for 28 days starting at age 7–9 weeks (Figure 8.5). At the end of the exposure period, extensive necropsy is performed, which is enhanced for endocrine parameters as compared to the existing OECD TG 407 guideline. Enhancements are listed in the next page. The additional

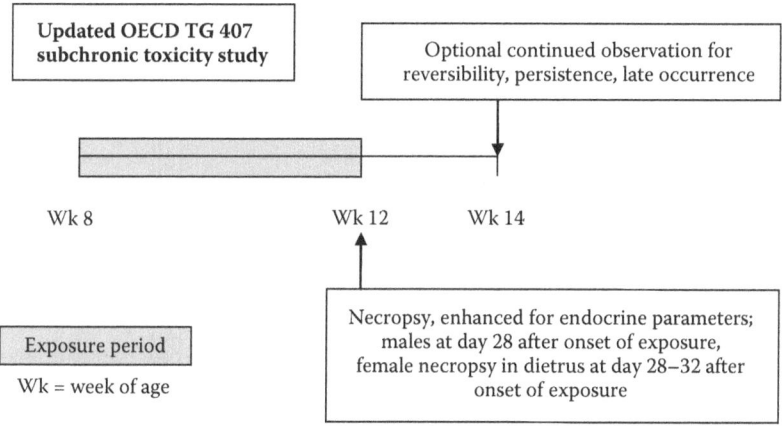

FIGURE 8.5   Schematic representation of the updated OECD TG 407 subchronic toxicity test protocol.

parameters were selected from a larger series that was included in the validation studies. From those studies, it was concluded that most hormonal parameters such as sex steroids, luteinizing hormone, and follicle-stimulating hormone were too variable to be usefully included. Only thyroid hormone levels were considered informative in the protocol. In addition, it was concluded that the protocol is able to detect high- and medium-potency endocrine compounds, whereas low-potency compounds are unlikely to be detected in the system. An optional extended postdosing observational period of 2 weeks is suggested for observation of reversibility, persistence, and late occurrence of effects, specifically for the control and high-dose groups, which may increase the sensitivity of the test. Furthermore, it is stated that the protocol is not performed in a life stage that is most sensitive to endocrine disruption. Extension of the exposure period from 28 up to 90 days was considered unlikely to improve the chance of detection of endocrine activity.

Proposed enhancement endpoints to the OECD407 subchronic toxicity study.

- Organ/Tissue Weights
  - Testes (each weighed separately)
  - Seminal vesicles + coagulating glands
  - Prostate (possible dissection and separate weights for ventral and dorsolateral prostate), ovaries
  - Thyroid
  - Uterus
- Histopathology
  - Pituitary
  - Vagina
  - Epididymides, seminal vesicles + coagulation glands
  - Mammary gland

- Thyroid Hormones
  - Circulating levels of T_3 and T_4
  - Circulating levels of TSH
- Spermatology
  - Epididymal sperm number
  - Sperm morphology
- Estrous Cycle
  - Daily vaginal smears to assess estrous cycling via epithelial cytology for at least 5 days to ensure necropsy during diestrus

8.2.3.3   Application in the Testing Strategy

The updated OECD TG 407 guideline is useful at the base set tonnage level in the EU system for toxicological hazard identification of REACH (see below). The enhancements proposed do not cost extra animals, except for the optional extension of observation postdosing in the control and high-dose group. In the absence of dedicated reproductive toxicity testing at the base set level, the information from this protocol may yield unique and important information that may direct further dedicated testing for reproductive toxicity. On the other hand, it should be realized that, as the draft guideline rightly states, the age of the animals in this test, young adult, is likely not the most sensitive for detection of adverse effects on the endocrine system. Therefore, the absence of findings on the endocrine system cannot be taken as proof of absence of endocrine activity of the compound tested. The usefulness of the updated OECD TG 407 should be weighed against that of in vitro and in vivo screening assays for sex and thyroid hormone (ant)agonistic properties. In vitro hormone receptor binding and activation assays are animal free and more sensitive but lack in vivo kinetics, and in vivo screens such as uterotrophic and Hershberger assays are more sensitive but have limited kinetic relevance and do use additional animals for a single endpoint test.

8.2.4   OECD TG 426 Developmental Neurotoxicity Study

8.2.4.1   Principle and Status

Neurodevelopmental toxicity became of interest in view of increasing awareness of behavioral abnormalities in children such as attention deficit hyperactivity disorder and findings in experimental studies that several compounds affected behavior after prenatal exposure in otherwise unaffected animals [16]. It was felt that important developmental parameters of brain development and functionality were not adequately addressed in existing safety testing of chemicals, which warranted this additional protocol. Developmental neurotoxicity studies are designed to provide data, including dose–response characterizations, on the potential functional and morphological effects on the developing nervous system of the offspring that may arise from exposure in utero and during early life. A developmental neurotoxicity study can be either conducted as a separate study or incorporated into a reproductive toxicity or adult neurotoxicity study (e.g., OECD TG 415, 416) or added onto a prenatal developmental toxicity study (e.g., OECD TG 414). The OECD TG 426 protocol was adopted in 2007.

8.2.4.2 Test Protocol

The test substance is administered to animals during gestation and lactation (Figure 8.6). Dams are tested to assess effects in pregnant and lactating females and to provide comparative information (dams vs. offspring). Offspring are randomly selected from within litters for neurotoxicity evaluation. The evaluation consists of observations to detect gross neurologic and behavioral abnormalities, including the assessment of physical development, behavioral ontogeny, motor activity, motor and sensory function, and learning and memory, and the evaluation of brain weights and neuropathology during postnatal development and adulthood.

8.2.4.3 Application in the Testing Strategy

At this moment in time, there is no absolute requirement for this test at any tonnage level in the EU. However, the use of the test has been extensively discussed in the expert group that developed a guidance document for reproductive toxicity testing under REACH. No mandatory requirement could be decided upon. Discussions in the group also considered the relatively high animal use and the practicality of the labor-intensive protocol, in addition to the limited past experience with the test system. The majority feeling was, however, that the test could probably be optimally used as an adjunct to the two-generation study (OECD TG 416) at the time when the latter would be required in cases triggered by findings in earlier studies pointing to a possible effect on the central nervous system. The subsequent development of the extended one-generation study (see below) in which developmental neurotoxicity testing is incorporated provided a new approach for testing this class of endpoints.

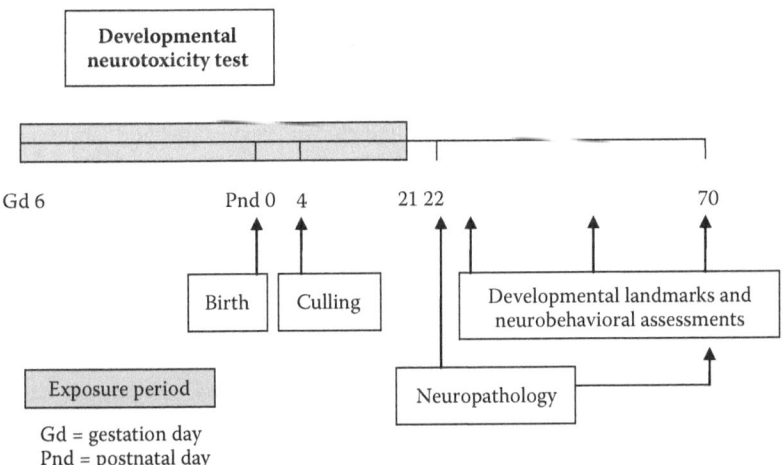

FIGURE 8.6 Schematic representation of a typical stand-alone version of the developmental neurotoxicity test.

8.2.5   EXTENDED ONE-GENERATION STUDY

8.2.5.1   Principle and Status

Various initiatives worldwide have explored the possibility of replacing the existing OECD TG 416 two-generation reproductive toxicity study with an extended one-generation study. Proposals for protocols have been generated independently by US and Japanese groups, and in Germany and in the Netherlands, activities along the same lines are ongoing [17]. The rationale for these activities is manifold. The usefulness in terms of informative yield for risk assessment and classification and labeling of effects observed in the second generation of the two-generation study has become an issue of detailed study. The possible reduction of animal use and the efficiency gain are important additional aspects. The critical issue is whether testing functional fertility in a prenatally exposed generation is essential for reproductive hazard and risk assessment. This aspect is included in the two-generation study but not in the extended one-generation study.

The Netherlands' National Institute for Public Health and the Environment (RIVM) recently performed an analysis of 176 multigeneration studies to assess potential differences between the first and the second generation, both in terms of the types of effects observed and in terms of the effective doses [18]. All substances classified as reproductive toxicants by the Directive 92/32/EEC or considered as toxic to fertility by the California Environmental Protection Agency for which a multigeneration study was found were included ($n = 58$ studies). The rest of the studies ($n = 118$) related to substances that are not classified as reproductive toxicants by these institutions. The second generation in the two-generation studies considered affected neither the overall NOAEL nor the critical effect type. Therefore, it had no impact on the ensuing risk assessment or on classification and labeling. These results clearly support the proposal of replacing the current two-generation study by a one-generation study with a more extensive assessment of parameters at F1 adulthood.

This analysis was followed by a global collection of available studies, resulting in a database of 498 multigeneration studies, brought together in a collaboration between RIVM, the US Environmental Protection Agency, Health Canada, and German BfR [19]. The analysis of this database confirmed earlier conclusions and facilitated the acceptance of the protocol as TG 443 by the OECD Council in July 2011.

8.2.5.2   Test Protocol

Cooper et al. [17] were the first to publish a proposal for an extended one-generation study, which is given here as an example (Figure 8.7). The male and female P-generation is exposed for 4 and 2 weeks, respectively, before mating. Males are exposed up to 6 weeks postmating (total male exposure at least 10 weeks), and females are exposed throughout pregnancy and weaning. F1 animals are exposed up to postnatal day 70 and developmental landmarks are monitored. At postnatal day 70, three sets of assessments are listed: (1) clinical pathology and developmental neurotoxicity, (2) immunotoxicity, and (3) estrus cyclicity and reproductive toxicity parameters. Necropsy of F1 generation animals at adulthood allows detailed pathological assessment of organs relevant for sex and thyroid hormone homeostasis,

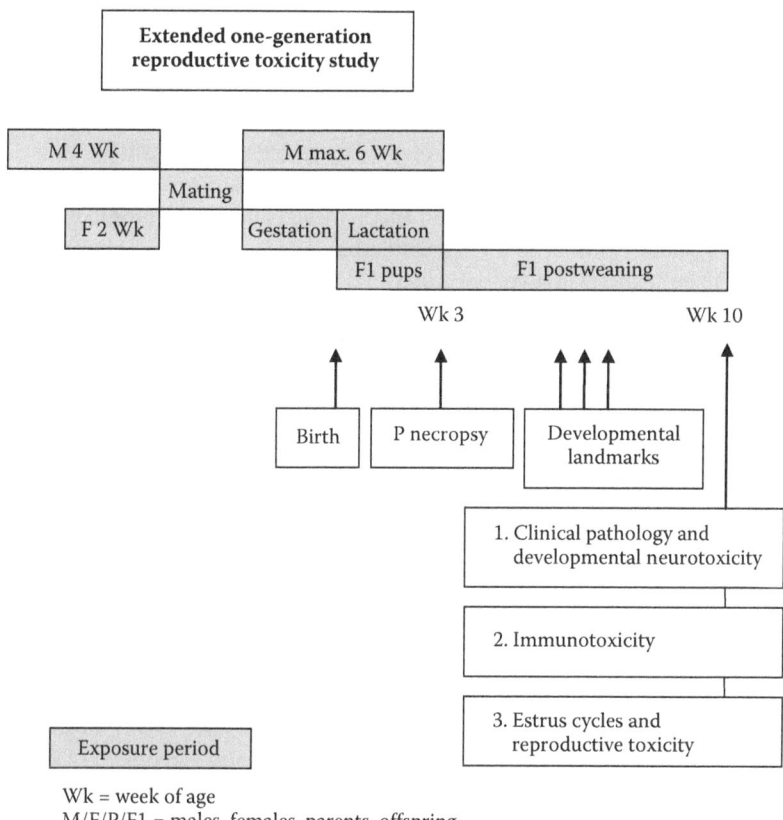

FIGURE 8.7 Schematic representation of an extended one-generation reproduction study.

which, together with the data on parental reproduction, is suggested to give sufficient insight into the possible reproductive toxicity of the test compound.

8.2.5.3   Application in the Testing Strategy

The extended one-generation study may ultimately replace the OECD TG 416 two-generation study as the definitive study on fertility and reproduction effects on which hazard and risk assessment for these endpoints is based. At present, this development appears as a promising one, both on the basis of the retrospective studies performed and in view of foreseen enhanced power and reduction of animal use.

8.2.6   Direct Pup Exposure for Juvenile Toxicity Testing

The need for juvenile toxicity testing in rats via direct oral exposure of suckling pups is a subject of discussion primarily in the world of pharmaceuticals testing. Although it has not been a prime issue in the chemical domain so far, it may be of relevance there as well, and the subject is therefore briefly touched upon. The idea is that assessment of the safety of drugs for pediatric use meets with considerable difficulty when based on adult toxicity studies only. In addition, perinatal toxicity studies that use maternal exposures only (gavage or feed exposure) are dependent to a large extent on lactational transfer of compounds for pup exposure during the first 2 to 3 weeks of life. To fill this exposure gap, it may, in some cases, be warranted to use direct pup exposure of sucklings in order to achieve doses relevant for risk assessment. In addition, exposure may be continued throughout growth and development until adulthood. In 2005, the European Medicines Agency's Committee for Human Medicinal Products produced a draft guideline for testing human pharmaceuticals for pediatric indications in juvenile animals [20]. Whereas most of this design is covered in the chemicals domain by the OECD TG 416 study and the OECD extended one-generation study (see Section 8.2.5), the direct exposure of pups is an aspect that is not covered. Direct pup exposure may, however, be warranted for chemicals as well in view of developing parameters related to, for example, sexual maturation, brain function, and immune competence.

8.2.7   ALTERNATIVE TEST SYSTEMS FOR REPRODUCTIVE HAZARD IDENTIFICATION

A wealth of alternatives have been developed in the area of reproductive and developmental toxicology, boosted by the relatively high animal use in reproductive toxicology (Table 8.1). Overviews of assays available have been published regularly [21–25]. Assays vary widely in the biological domain, ranging from whole embryo cultures (WECs) to hormone receptor activation assays. By their nature, these tests represent a reductionistic approach to reproduction and development, each containing only a limited part of the reproductive cycle or of the (pregnant) individual. This raises important issues about their applicability and predictability. Rat WEC, the zebrafish embryotoxicity test, and embryonic stem cell (ESC) differentiation are among the more promising systems for developmental toxicity screening that are being extensively studied [26–28]. The validity and applicability domain of these test systems for chemical screening are currently issues for research and debate. Although some have been formally validated, alternatives still have not reached a stage where they can replace animal testing in regulatory toxicology. Issues of concern are, for example, the limited kinetic information and the virtual lack of metabolism in in vitro tests. However, in a prescreening situation for prioritizing in vivo testing, these tests prove helpful. Those assays with clearly described mechanisms such as the reproductive hormone receptor activation assays can be considered to provide relatively transparent data. The outcomes of more complex assays such as cell differentiation and embryogenesis assays are more difficult to interpret. Implementation in a regulatory context awaits further definition of their applicability domain (in terms of both chemical domain and biological processes covered) and their predictive capacity. Three of these developmental toxicity assays are briefly described below.

8.2.7.1   Rodent Postimplantation Embryo Culture

Rodent postimplantation embryo culture was first described by New [49,50]. After removal of decidual tissue and parietal yolk sac, embryos can be cultured in their intact visceral yolk sac, from the early somite stage onward until for nutrition the embryo becomes dependent on the placenta. Optimal culture medium is 100% rat

TABLE 8.1

Overview of Selected Alternative Tests in Developmental Toxicology

Test	Material Used	Endpoint	Reference
		Embryos	
Hydra	*Hydra attenuata* single-cell suspended polyps	Reaggregation into polyp	[29]
Drosophila	*Drosophila melanogaster* larvae	Morphological development	[30]
FETAX	*Xenopus laevis* frog eggs	Morphological development	[31]
CHEST	Chick embryos	Morphological development	[32]
WEC	Rat or mouse embryos	Morphological development	[33]
Zebrafish	*Danio rerio* zebrafish eggs	Morphological development	[34]
		Primary Cells	
Drosophila	Neuroblasts and myoblasts	Differentiation	[35]
Chick retina	Chick embryo retinal cell culture	Growth, adhesion, and differentiation	[36]
Chick MM	Chick embryo neural crest and limb bud cells	Differentiation	[37]
Rat MM	Midbrain and limb bud cells	Neural and cartilage differentiation	[38]
		Cell Lines	
HEPM	Human embryonic palatal mesenchyme cells	Proliferation	[39]
MOT	Mouse ovarian tumor cells	Adhesion	[40]
V79	Chinese hamster lung fibroblasts	Metabolic cooperation	[41]
N115	Neuroblastoma cells	Differentiation	[42]
EC	Mouse embryocarcinoma cells	Differentiation	[43]
EST	Mouse embryonic stem cells	Differentiation	[44]
T47D	Human breast cancer cell line	Estrogen receptor activation	[45]
U2-OS	Human osteosarcoma cell line	Androgen receptor activation	[46]
CHO	Chinese hamster ovary cells	Progestagen receptor activation	[47]
H295R	Human adrenal tumor cell line	Steroidogenesis	[48]

serum, cultures have to be oxygenated with increasing oxygen levels during culture, and continuous rotation of cultures is needed to facilitate gas exchange with the culture medium. During the 48-h culture period, roughly between gestation days 10 and 12 in the rat, the embryo keeps the same developmental speed as the in utero situation. A standardized morphological scoring system for rat embryogenesis in culture was developed by Brown and Fabro [51], which was adapted for the mouse by Van Maele-Fabry et al. [52]. In this system, the development of each organ anlagen receives a score, and scores are summed up to arrive at a total morphological score. Endpoints that are used to monitor embryo growth during culture include total protein and DNA content, head length, crown–rump length, and yolk sac diameter. Since the early 1980s, WEC has been used extensively for the study of the embryotoxicity of chemicals [53,54]. In the largest validation study of alternatives in developmental toxicology to date, 20 chemicals were tested in WEC in four independent laboratories [26,55]. Although the results were favorable, continued studies showed that predictability was variable dependent on chemicals tested, which was probably partly attributed to the insufficiency of prediction models used [56]. More recent approaches have used potency correlations to compare in vitro and in vivo results [57] and have applied gene expression modulation as alternative endpoints in order to explore new ways of improving predictability of the test system [58].

8.2.7.2   Embryonic Stem Cell Test

Murine pluripotent ESCs derived from the blastocyst can be induced to differentiate in culture into a wide variety of cell types, including cardiomyocytes. This differentiation pathway is the basis of the embryonic stem cell test (EST) [59]. Cardiomyocytes have the advantage of their very easy identification as contracting cells in differentiated ESC culture. Using the hanging drop culture technique [60], ESCs form multicellular aggregates, named embryoid bodies (EBs). Further differentiation into cardiomyocytes can be induced by suspension culture for 2 days followed by plating onto tissue culture plates. This specific culture method results in EBs with foci of contracting cardiomyocytes. The effect of compound exposure on differentiation of ESC into cardiomyocytes is regarded as a measure of the embryotoxicity of chemicals. In an interlaboratory validation study, 20 test chemicals were tested under blind conditions in four different laboratories [27,55]. Although the EST performed promisingly in this study, the validity of the EST for prediction of embryotoxicity of previously untested compounds later proved less certain. The predictivity of the EST was studied for in-house and marketed pharmaceuticals [61,62]. Although receptor-mediated chemicals were excluded from the validation set, the overall accuracy for marketed pharmaceuticals showed to be 85%. However, for in-house compounds, the overall accuracy was only 53%. Marx-Stoelting et al. [63] reviewed the EST performance and concluded that prediction models had not been optimal. To improve the predictability of the EST, other differentiation pathways may be studied in the test. Differentiation toward cell types of major target tissues such as the nervous and skeletal system can provide additional information about the teratogenic potential of compounds [64,65]. By implementation of transcriptomics techniques into the differentiation assay of the EST, pathways can be revealed that are activated during different phases of differentiation [66,67]. Gene expression changes may be useful in determining the effects of chemicals on embryonic cell differentiation at the molecular level and may improve predictability of the assay [68,69].

8.2.7.3   Zebrafish Embryotoxicity Assay

The zebrafish (*Danio rerio*) has been a major model in biomedical research for several decades, resulting in an immense body of information on zebrafish development and (molecular) biology [70–72]. The morphology of embryonic development of zebrafish, which has been described in great detail [73], can be followed in real time because of the transparency of the eggs and embryos. Further advantages of the zebrafish embryo model are the short embryonic period (organogenesis stage at 48 h compared with gestational day 5–6 in rats and gestational day 21–56 in humans) and the minimal space and test volume requirements. General concepts apply to early development and the signaling repertoire between species in general [74] and between zebrafish and mammals specifically [75]. Braunbeck and Lammer recently reviewed the existing database on fish embryo toxicity tests (FETs) as a basis for discussions on the potential of FET as an alternative for acute toxicity testing in adult fish [76]. A variety of developmental and teratogenic endpoints can be scored in a standardized way [77,78]. Although no formal validation of zebrafish embryo toxicity test (ZFET) for toxic effects in mammals, such as developed by the European Centre for the Validation of Alternative Methods [79], has yet been performed, several reports showed the predictive potential of zebrafish embryo tests for mammals [80,81]. Results may even improve when molecular tools are applied [82]. Applying the ZFET for more detailed developmental toxicity testing may require the inclusion of further morphological endpoints and application of molecular tools. Since the zebrafish genome is well defined, there is good availability of a wide variety of molecular tools [83].

8.2.8   IN SILICO NONTESTING METHODS FOR HAZARD IDENTIFICATION

Besides in vivo and in vitro test methods, there is increasing use worldwide of (Q)SAR, grouping, and read-across techniques in regulatory testing programs for chemical safety, concurrent with the development and validation of such methods. These methods are based on chemical and physical properties of compounds, which are compared to predict toxicity across defined classes of compounds. In brief, the toxicity of an untested compound is predicted on the basis of knowledge about analogs within the group. The EU (REACH) legislation clearly includes such in silico methods to be applied where appropriate to optimize hazard evaluation. However, their application in various areas of toxicity is still under development and discussion. Especially with regard to (Q)SAR, sufficient validation and documentation of the methods are mentioned as prerequisites. Pedersen et al. [84] estimate that the acceptance of (Q)SAR methods for reproductive and developmental toxicity endpoints is 10% and 25%, respectively, which is far below any other endpoint of toxicity. Another study showed that the vast majority of classified reproductive toxicants were not recognized by two existing (Q)SAR models [85]. These findings are not surprising in view of the limited database available to derive (Q)SAR models for reproductive toxicity and also understandable in view of the limited mechanistic knowledge in this area with its complex variety of mechanisms and windows of sensitivity in time throughout the reproductive cycle. Significant steps forward in methodology, development, and validation are necessary before these methods can be usefully employed in the area of reproductive toxicology hazard and risk assessment.

8.3   CONCLUSIONS

Regulatory reproductive toxicity testing has a history of nearly three decades of successful globally harmonized animal study designs and is currently undergoing a period of extensive revision and renewal. This is stimulated by developments such as the endocrine disrupter issue and the EU REACH legislation and by the realization that animal use in regulatory reproductive toxicity testing is relatively high. The existing testing paradigms having been in place for more than two decades warrants in itself a retrospective evaluation of their performance with the aims of improvement and refinement. Recent analyses of past performance have resulted in innovations in reproductive toxicity testing, by enhancing endpoint assessment and reducing animal use at the same time. In addition, efforts into the design and implementation of alternative tests have been intensified, although significant steps have to be taken before alternatives can replace current animal testing in regulatory toxicology. The OECD has published guidance documents on reproductive toxicity assessment (OECD GD 43) and on the validation of alternative methods (OECD GD 34) that serve as useful guides in the area. Integrated testing strategies take a step away from individual tests and toxicology domains in that they use testing schemes developed within specific areas of toxicology such as reproductive toxicology and attempt to integrate them on the level of toxicology as a whole. The aim is to achieve increased efficiency of testing through waiving of redundant tests based on results gathered in earlier stages of a tiered approach. Also, a combination of more endpoints in one test can improve efficiency, an example of which is the updated OECD TG 407 subchronic toxicity test with added parameters on endocrine organs. Alternatives can play a role at the earliest stages of testing, serving to prioritize and direct further testing. Testing strategy design should be critically based on retrospective analysis of existing data gathered over the last decades. In the coming years, these activities are expected to lead to new evidence-based proposals for integrated testing strategies. This is expected to result in significant efficiency gain and reduction and refinement of animal testing while retaining the current high standards of basic information requirements that form the indispensable basis for classification and labeling and risk assessment of chemicals.

REFERENCES

1. OECD Guidelines for the Testing of Chemicals—Section 4—Health effects. (2015). Available at http://www.oecd-ilibrary.org/environment/oecd-guidelines-for-the-testing-of-chemicals-section-4-health-effects_20745788.
2. European Chemicals Bureau, REACH. (2015). Available at http://echa.europa.eu/regulations/reach/legislation.
3. W. Lenz, *Kindliche Missbildungen nach Medikament wahrend der Graviditat*, Deutsch Med Wschr 86 (1961), pp. 2555–2556.
4. W.G. McBride, *Thalidomide and congenital abnormalities*, Lancet 2 (1961), p. 1358.
5. E. Carlsen, A. Giwercman, N. Keiding, and N.E. Skakkebaek, *Declining semen quality and increasing incidence of testicular cancer: Is there a common cause?* Environ. Health Perspect. 103 (1995), pp. 137–139.
6. T. Colborn, D. Dumanoski, and J. Peterson Myers, *Our Stolen Future*, Dutton Publishing, Boston, 1996.
7. OECD Chemicals Testing—Guidelines—Endocrine Disrupter Testing and Assessment. (2015). Available at http://www.oecd.org/env/ehs/testing/oecdworkrelatedtoendocrinedisrupters.htm.
8. K. Van der Jagt, S. Munn, J. Tørsløv, and J. de Bruijn, *Alternative approaches can reduce the use of test animals under REACH. Addendum to Assessment of additional testing needs under REACH Effects of (Q)SARS, risk based testing and voluntary industry initiatives*, JRC Report EUR 21405 EN, 2004.

9. G. Janer, W. Slob, B. Hakkert, T. Vermeire, and A.H. Piersma, *A retrospective analysis of developmental toxicity studies: What is the added value of the rabbit as an additional species?* Regul. Toxicol. Pharmacol. 50 (2008), pp. 206–217.

10. T.B. Knudsen, M.T. Martin, R.J. Kavlock, R.S. Judson, D.J. Dix, and A.V. Singh, *Profiling the activity of environmental chemicals in prenatal developmental toxicity studies using the U.S. EPA's ToxRefDB*, Reprod. Toxicol. 28 (2009), pp. 209–219.

11. G. Janer, B.C. Hakkert, A.H. Piersma, T. Vermeire, and W. Slob, *A retrospective analysis of the added value of the rat two-generation reproductive toxicity study versus the rat subchronic toxicity study*, Reprod. Toxicol. 24 (2007), pp. 103–113.

12. UNECE, Globally Harmonized System of Classification and Labelling of Chemicals (GHS), 2003. Available at http://www.unece.org/trans/danger/publi/ghs/ghs_welcome_e.html.

13. OECD, Report of the Initial Work Towards the Validation of the Rodent Uterotrophic Assay—Phase 1; and OECD Report of the Validation of the Rodent Uterotrophic Bioassay: Phase 2—Testing of Potent and Weak Oestrogen Agonists by Multiple Laboratories OECD Review documents 65 and 66, 2006. Available at http://www.oecd.org/document/30/0,2340,en_2649_34377_1916638_1_1_1_1,00.html.

14. L.E. Gray Jr., J. Furr, and J.S. Ostby, *Unit 16.9, Hershberger assay to investigate the effects of endocrine-disrupting compounds with androgenic or anti-androgenic activity in castrate-immature male rats*, in *Current Protocols in Toxicology, Chapter 16: Male Reproductive Toxicology*, L.G. Costa, J.C. Davila, D.A. Lawrence, D.J. Reed, and Y. Will, eds., John Wiley and Sons, New York, 2005, pp. 16.9.1–16.9.15.

15. H.P. Gelbke, A. Hofmann, J.W. Owens, and A. Freyberger, *The enhancement of the subacute repeat dose toxicity test OECD TG 407 for the detection of endocrine active chemicals: Comparison with toxicity tests of longer duration*, Arch. Toxicol. 81 (2007), pp. 227–250.

16. U. Hass, *The need for developmental neurotoxicity studies in risk assessment for developmental toxicity*, Reprod. Toxicol. 22 (2006), pp. 148–156.

17. R.L. Cooper, J.C. Lamb, S.M. Barlow, K. Bentley, A.M. Brady, N.G. Doerrer, D.L. Eisenbrandt, P.A. Fenner-Crisp, R.N. Hines, L.F. Irvine, C.A. Kimmel, H. Koeter, A.A. Li, S.L. Makris, L.P. Sheets, G. Speijers, and K.E. Whitby, *A tiered approach to life stages testing for agricultural chemical safety assessment*, Crit. Rev. Toxicol. 36 (2006), pp. 69–98.

18. G. Janer, A.H. Piersma, W. Slob, T. Vermeire, and B. Hakkert, *A retrospective analysis of the two-generation study: What is the added value of the second generation?* Reprod. Toxicol. 24 (2007), pp. 97–102.

19. A.H. Piersma, E. Rorije, M.E. Beekhuijzen, R. Cooper, D.J. Dix, B. Heinrich-Hirsch, M.T. Martin, E. Mendez, A. Muller, M. Paparella, D. Ramsingh, E. Reaves, P. Ridgway, E. Schenk, L. Stachiw, B. Ulbrich, and B.C. Hakkert, *Combined retrospective analysis of 498 rat multi-generation reproductive toxicity studies: On the impact of parameters related to F1 mating and F2 offspring*, Reprod. Toxicol. 31 (2011), pp. 392–401.

20. European Medicines Agency—CHMP, Draft Guideline on the need for non-clinical testing in juvenile animals on human pharmaceuticals for paediatric indications, September 2005.

21. N.A. Brown, *Teratogenicity testing in vitro: Status of validation studies*, Arch. Toxicol. Suppl. 11 (1987), pp. 105–114.

22. L.E. Gray Jr., J. Ostby, V. Wilson, C. Lambright, K. Bobseine, P. Hartig, A. Hotchkiss, C. Wolf, J. Furr, M. Price, L. Parks, R.L. Cooper, T.E. Stoker, S.C. Laws, S.J. Degitz, K.M. Jensen, M.D. Kahl, J.J. Korte, E.A. Makynen, J.E. Tietge, and G.T. Ankley, *Xenoendocrine disrupters-tiered screening and testing: Filling key data gaps*, Toxicology 181–182 (2002), pp. 371–382.

23. J.C. O'Connor, J.C. Cook, M.S. Marty, L.G. Davis, A.M. Kaplan, and E.W. Carney, *Evaluation of Tier I screening approaches for detecting endocrine-active compounds (EACs)*, Crit. Rev. Toxicol. 32 (2002), pp. 521–549.

24. A.H. Piersma, *Alternatives to animal testing in developmental toxicology*, Invited Rev. Toxicol. Pharmacol. 98 (2006), pp. 427–431.

25. B. Schenk, M. Weimer, S. Bremer, B. van der Burg, R. Cortvrindt, A. Freyberger, G. Lazzari, C. Pellizzer, A. Piersma, W.R. Schäfer, A. Seiler, H. Witters, and M. Schwarz, *The ReProTect Feasibility Study, a novel comprehensive in vitro approach to detect reproductive toxicants*, Reprod. Toxicol. 30 (2010), pp. 200–218.

26. A.H. Piersma, E. Genschow, A. Verhoef, M.Q.I. Spanjersberg, N.A. Brown, M. Brady, A. Burns, N. Clemann, A. Seiler, and H. Spielmann, *Validation of the postimplantation rat whole embryo culture test in the International ECVAM Validation study on three in vitro embryotoxicity tests*, ATLA 32 (2004), pp. 275–307.

27. E. Genschow, H. Spielmann, G. Scholz, I. Pohl, A. Seiler, N. Clemann, S. Bremer, and K. Becker, *Validation of the embryonic stem cell test in the international validation study on three in vitro embryotoxicity tests*, ATLA 32 (2004), pp. 209–244.

28. S. Scholz and I. Mayer, *Molecular biomarkers of endocrine disruption in small model fish*, Mol. Cell. Endocrinol. 293 (2008), pp. 57–70.

29. E.M. Johnson, *A subvertebrate system for rapid determination of potential teratogenic hazards*, J. Environ. Pathol. Toxicol. 4 (1980), pp. 153–156.

30. R.L. Schuler, B.D. Hardin, and R.W. Nicmeier, Drosophila *as a tool for the rapid assessment of chemicals for teratogenicity*, Teratog. Carcinog. Mutagen. 2 (1982), pp. 293–301.

31. J.N. Dumont, T.W. Schultz, M. Buchanan, and G. Kao, *Frog embryo teratogenesis assay: Xenopus (FETAX)—A short-term assay applicable to complex environmental mixtures*, in *Symposium on the Application of Short-Term Bioassays in the Analysis of Complex Environmental Mixtures*, M.D. Waters, S.S. Sandhu, J. Lewtas, L. Claxton, and S. Nesnow, eds., Plenum Press, New York, 1995, pp. 393–405.

32. R. Jelinek, *Use of chick embryo in screening for embryotoxicity*, Teratog. Carcinog. Mutagen. 2 (1982), pp. 255–261.

33. B.P. Schmid, *Teratogenicity testing of new drugs with the postimplantation embryo culture system*, in *In Vitro Embryotoxicity and Teratogenicity Tests*, F. Homburger and A.M. Goldberg, eds., S. Karger, Basel, 1985, pp. 46–57.

34. C.J. Van Leeuwen, E.M. Grootelaar, and G. Niebeek, *Fish embryos as teratogenicity screens: A comparison of embryotoxicity between fish and birds*, Ecotoxicol. Environ. Saf. 20 (1990), pp. 42–52.

35. N. Bournias-Vardiabasis and R.L. Teplitz, *Use of Drosophila embryo cell cultures as an in vitro teratogen assay*, Teratog. Carcinog. Mutagen. 2 (1982), pp. 333–341.

36. G.P. Daston, D. Baines, and J.E. Yonker, *Chick embryo neural retina cell culture as a screen for developmental toxicity*, Toxicol. Appl. Pharmacol. 109 (1991), pp. 352–366.

37. A.L. Wilk, J.H. Greenberg, E.A. Horigan, R.M. Pratt, and G.R. Martin, *Detection of teratogenic compounds using differentiating embryonic cells in culture*, In Vitro 16 (1980), pp. 269–276.

38. O.P. Flint and T.C. Orton, *An in vitro assay for teratogens with cultures of rat embryo midbrain and limb bud cells*, Toxicol. Appl. Pharmacol. 76 (1984), pp. 383–395.

39. R.M. Pratt and W.D. Willis, *In vitro screening assay for teratogens using growth inhibition of human embryonic cells*, Proc. Natl. Acad. Sci. U.S.A. 82 (1985), pp. 5791–4.

40. A.G. Braun, D.J. Emerson, and B.B. Nichinson, *Teratogenic drugs inhibit tumour cell attachment to lectin-coated surfaces*, Nature 282 (1979), pp. 507–509.

41. F. Welsch and D.B. Stedman, *Inhibition of metabolic cooperation between Chinese hamster V79 cells by structurally diverse teratogens*, Teratog. Carcinog. Mutagen. 4 (1984), pp. 285–301.

42. C.L. Mummery, C.E. van den Brink, P.T. van der Saag, and S.W. de Laat, *A short-term screening test for teratogens using differentiating neuroblastoma cells in vitro*, Teratology 29 (1984), pp. 271–279.

43. A.H. Piersma, L.A.G.J.M. van Aerts, A. Verhoef, J.M. Garbis-Berkvens, J.E. Robinson, J.H.J. Copius Peereboom-Stegeman, and P.W.J. Peters, *Biotransformation of cyclophosphamide in post-implantation rat embryo culture using maternal hepatocytes in coculture*, Pharmacol. Toxicol. 69 (1991), pp. 47–51.

44. G. Scholz, I. Pohl, E. Genschow, M. Klemm, and H. Spielmann, *Embryotoxicity screening using embryonic stem cells in vitro: Correlation to in vivo teratogenicity*, Cells Tissues Organs 165 (1999), pp. 203–211.

45. J. Legler, C.E. van den Brink, A. Brouwer, A.J. Murk, P.T. van der Saag, A.D. Vethaak, and B. van der Burg, *Development of a stably transfected estrogen receptor-mediated luciferase reporter gene assay in the human T47D breast cancer cell line*, Toxicol. Sci. 48 (1999), pp. 55–66.

46. E. Sonneveld, H.J. Jansen, J.A. Riteco, A. Brouwer, and B. van der Burg, *Development of androgen- and estrogen-responsive bioassays, members of a panel of human cell line-based highly selective steroid-responsive bioassays*, Toxicol. Sci. 83 (2005), pp. 136–148.

47. R. Dijkema, W.G. Schoonen, R. Teuwen, E. van der Struik, R.J. de Ries, B.A. van der Kar, and W. Olijve, *Human progesterone receptor A and B isoforms in CHO cells. I. Stable transfection of receptor and receptor-responsive reporter genes: Transcription modulation by (anti)progestagens*, J. Steroid Biochem. Mol. Biol. 64 (1998), pp. 147–156.

48. M. Hecker, J.L. Newsted, M.B. Murphy, E.B. Higley, P.D. Jones, R. Wu, and J.P. Giesy, *Human adenocarcinoma (H295R) cells for rapid in vitro determination of effects on steroidogenesis: Hormone production*, Toxicol. Appl. Pharm. 217 (2006), pp. 114–124.

49. D.A.T. New, *Development of explanted embryos in circulating medium*, J. Embryol. Exp. Morphol. 17 (1967), pp. 513–525.

50. D.A.T. New, *Whole embryo culture and the study of mammalian embryos during organogenesis*, Biol. Rev. 53 (1978), pp. 81–122.

51. N.A. Brown and S. Fabro, *Quantitation of rat embryonic development in vitro: A morphological scoring system*, Teratology 24 (1981), pp. 65–78.

52. G. Van Maele-Fabry, F. Delhaise, and J.J. Picard, *Morphogenesis and quantification of the development of post-implantation mouse embryos*, Toxicol. In Vitro 4 (1990), pp. 149–156.

53. G. Van Maele-Fabry and J.J. Picard, *Evaluation of the embryotoxic potential of ten chemicals in the whole mouse embryo culture*, Teratology 36 (1987), pp. 95–106.

54. R. Bechter and B. Schmid, *Teratogenicity in vitro—A comparative study of four antimycotic drugs using the whole embryo culture system*, Toxicol. In Vitro 1 (1987), pp. 11–15.

55. E. Genschow, H. Spielmann, G. Scholz, A. Seiler, N. Brown, A. Piersma, M. Brady, N. Clemann, H. Huuskonen, F. Paillard, S. Bremer, and K. Becker, *The ECVAM international validation study on in vitro embryotoxicity tests: Results of the definitive phase and evaluation of prediction models*, Altern. Lab. Anim. 30 (2002), pp. 151–176.

56. H. Spielmann, A. Seiler, S. Bremer, L. Hareng, T. Hartung, H. Ahr, E. Faustman, U. Hass, G. Moffat, H. Nau, P. Vanparys, A. Piersma, J. Riego Sintes, and J. Stewart, *The practical application of three validated in vitro embryotoxicity tests*, ATLA 34 (2006), pp. 527–538.

57. A.H. Piersma, G. Janer, A. Verhoef, G. Wolterink, and W. Slob, *Quantitative extrapolation of in vitro whole embryo culture embryotoxicity data to developmental toxicity in vivo using the Benchmark approach*, Toxicol. Sci. 101 (2008), pp. 91–100.

58. J.F. Robinson, V.A. van Beelen, A. Verhoef, M.F. Renkens, M. Luijten, M.H. van Herwijnen, A. Westerman, J.L. Pennings, and A.H. Piersma, *Embryotoxicant-specific transcriptomic responses in rat postimplantation whole-embryo culture*, Toxicol. Sci. 118 (2010), pp. 675–685.

59. H. Spielmann, I. Pohl, B. Döring, M. Liebsch, and F. Moldenauer, *The embryonic stem cell test, an in vitro embryotoxicity test using two permanent mouse cell lines: 3T3 fibroblasts and embryonic stem cells*, In Vitro Toxicol. 10 (1997), pp. 119–127.

60. M.A. Rudnicki and M.W. McBurney, *Cell culture methods and induction of differentiation of embryonal carcinoma cell lines*, in *Teratocarcinoma and Embryonic Stem Cells: A Practical Approach*, E. Robertson, ed., IRL Press, Washington, DC, 1987.

61. R. Chapin, D. Stedman, J. Paquette, R. Strecjk, S. Kumpf, and S. Deng, *Struggles for equivalence: In vitro developmental toxicity model evolution in pharmaceuticals in 2006*, Toxicol. In Vitro 21 (2007), pp. 1545–1551.

62. J.A. Paquette, S. Kumpf, R.D. Streck, J.J. Thomson, R.E. Chapion, and D.B. Stedman, *Assessment of the embryonic stem cell test and application and use in the pharmaceutical industry*, Birth Defects Res. B Dev. Reprod. Toxicol. 83 (2008), pp. 104–111.

63. P. Marx-Stoelting, E. Adriaens, H.J. Ahr, S. Bremer, B. Garthoff, H.P. Gelbke, A. Piersma, C. Pellizzer, U. Reuter, V. Rogiers, B. Schenk, S. Schwengberg, A. Seiler, H. Spielmann, M. Steemans, D.B. Stedman, P. Vanparys, J.A. Vericat, M. Verwei, F. van der Water, M. Weimer, and M. Schwarz, *A review of the implementation of the embryonic stem cell test (EST). The report and recommendations of an ECVAM/ReProTect workshop*, Altern. Lab. Anim. 37 (2009), pp. 313–328.

64. C. Pellizer, E. Bello, S. Adler, T. Hartung, and S. Bremer, *Detection of tissue-specific effects by methotrexate on differentiating mouse embryonic stem cells*, Birth Defects Res. B Dev. Reprod. Toxicol. 71 (2004), pp. 331–341.

65. T.C. Stummann, L. Hareng, and S. Bremer, *Embryotoxicity hazard assessment of methylmercury and chromium using embryonic stem cells*, Toxicology 242 (2007), pp. 130–143.

66. H. Terami, K. Hidaka, H. Shirai, H. Narumiya, T. Kuroyanagi, Y. Arai, H. Aburatani, and T. Morisaki, *Efficient capture of cardiogenesis-associated genes expressed in ES cells*, Biochem. Biophys. Res. Commun. 355 (2007), pp. 47–53.

67. M.X. Doss, J. Winkler, S. Chen, R. Hippler-Altenburg, I. Sotiriadou, M. Halbach, K. Pfannkuche, H. Liang, H. Schulz, O. Hummel, N. Hubner, R. Rottscheidt, R. Hescheler, J. Hescheler, and A. Sachinidis, *Global transcriptome analysis of murine embryonic stem cell-derived cardiomyocytes*, Genome Biol. 8 (2007), p. R56.

68. D.A.M. Van Dartel, J.L. Pennings, L.J. de la Fonteyne, M.H. van Herwijnen, J.H. van Delft, F.J. van Schooten, and A.H. Piersma, *Monitoring developmental toxicity in the embryonic stem cell test using differential gene expression of differentiation-related genes*, Toxicol. Sci. 116 (2010), pp. 130–139.

69. D.A.M. Van Dartel, J.L.A. Pennings, L.J.J. de la Fonteyne, K.J.J. Brauers, S. Claessen, J.H. van Delft, J.C.S. Kleinjans, and A.H. Piersma, *Evaluation of developmental toxicant identification using gene expression profiling in embryonic stem cell differentiation cultures*, Toxicol. Sci. 119 (2011), pp. 126–134.

70. C. Nusslein-Volhard, *Of flies and fishes*, Science 266 (1994), pp. 572–574.

71. Z. Lele and P.H. Krone, *The zebrafish as a model system in developmental, toxicological and transgenic research*, Biotechnol. Adv. 14 (1996), pp. 57–72.

72. K. Dooley and L.I. Zon, *Zebrafish: A model system for the study of human disease*, Curr. Opin. Genet. Dev. 10 (2000), pp. 252–256.

73. C.B. Kimmel, W.W. Ballard, S.R. Kimmel, B. Ullmann, and T.F. Schilling, *Stages of embryonic development of the zebrafish*, Dev. Dyn. 203 (1995), pp. 253–310.

74. W. Arthur, *The emerging conceptual framework of evolutionary developmental biology*, Nature 415 (2002), pp. 757–764.

75. T.F. Schilling and J. Webb, *Considering the zebrafish in a comparative context*, J. Exp. Zool. B Mol. Dev. Evol. 308 (2007), pp. 515–522.

76. T. Braunbeck and E. Lammer, *Fish embryo toxicity assays*, University of Heidelberg/ German Federal Environment Agency, Heidelberg/Dessau, 2006.

77. R. Nagel, *DarT: The embryo test with the Zebrafish* Danio rerio—*A general model in ecotoxicology and toxicology*, ALTEX 19 (2002), pp. 38–48.

78. F. Busquet, R. Nagel, F. von Landenberg, S.O. Mueller, N. Huebler, and T.H. Broschard, *Development of a new screening assay to identify proteratogenic substances using zebrafish* Danio rerio *embryo combined with an exogenous mammalian metabolic activation system (mDarT)*, Toxicol. Sci. 104 (2008), pp. 177–188.

79. A.P. Worth and M. Balls, *The importance of the prediction model in the validation of alternative tests*, Altern. Lab. Anim. 29 (2001), pp. 135–144.

80. W.L. Seng and K.A. Augustine, *Zebrafish: A predictive model for assessing developmental toxicity*, HESI Workshop on Alternative Assays for Developmental Toxicity, Cary, NC, 2007.

81. N. Ali, *Teratology in Zebrafish Embryos: A Tool for Risk Assessment*, Swedish University of Agricultural Sciences, Uppsala, 2007.

82. L. Yang, J.R. Kemadjou, C. Zinsmeister, M. Bauer, J. Legradi, F. Muller, M. Pankratz, J. Jakel, and U. Strahle, *Transcriptional profiling reveals barcode-like toxicogenomic responses in the zebrafish embryo*, Genome Biol. 8 (2007), p. R227.

83. L.I. Zon and R.T. Peterson, *In vivo drug discovery in the zebrafish*, Nat. Rev. Drug Discov. 4 (2005), pp. 35–44.

84. F. Pedersen, J. de Bruijn, S. Munn, and K. van Leeuwen, *Assessment of additional testing needs under REACH. Effects of (Q)SARs, risk based testing and voluntary industry initiatives*, EU-JRC-ECB, 2003.

85. L. Maslankiewicz, E.M. Hulzebos, T.G. Vermeire, J.J.A. Muller, and A.H. Piersma, *Can chemical structure predict reproductive toxicity?* RIVM Report no. 601200005/2005.

9 U.S. National Library of Medicine Resources for Computational Toxicology

Pertti J. Hakkinen

CONTENTS

ABSTRACT

The National Library of Medicine (NLM) offers online and downloadable resources accessible at no cost by global users. PubMed is NLM's web interface to the world's biomedical literature and to numerous sources of toxicological information. PubMed includes citations for journal articles, and these citations include links to full-text articles, when available, from PubMed Central or publisher websites.

NLM's Specialized Information Services Division (SIS) is responsible for extensive information resources and services in toxicology, environmental health, chemistry, and other topics. For example, NLM SIS's TOXNET (TOXicology Data NETwork) is a large set of widely used databases. TOXNET's TOXLINE database provides bibliographic information covering the toxicological, biochemical, pharmacological, and physiological effects of chemicals and incorporates citations from NLM's PubMed/MEDLINE and other sources.

Other TOXNET databases of key relevance to computational toxicology and related toxicology efforts are (1) DART (Developmental and Reproductive Toxicology), (2) HSDB (Hazardous Substances Data Bank), (3) IRIS (Integrated Risk Information System), and (4) ITER (International Toxicity Estimates of Risk). Also noteworthy for toxicologists and risk assessors is the CTD (Comparative Toxicogenomics Database).

As an example of how NLM's databases can be used in computational toxicology, a search for information relevant to Physiologically Based Pharmacokinetic (PBPK) modeling was conducted. The results show that NLM resources such as TOXLINE, HSDB, IRIS, and ITER provide access to information that is useful for PBPK modeling and related efforts.

Efforts are ongoing to consider adding additional resources, enhance existing resources, and keep up with the continuously changing ways to access information. For example, NLM developed an enhanced version of its ALTBIB—Resources for Alternatives to the Use of Live Vertebrates in Biomedical Research and Testing Web portal to provide easy and comprehensive access to publications in silico, in vitro, and refined animal testing methods, and easy access to methods that have been validated or undergoing validation in the United States and other parts of the world, for example, Canada, the European Union, Japan, and South Korea.

KEYWORDS

Databases, information, National Library of Medicine (NLM®), online, resources, TOXicology Data NETwork (TOXNET®)

9.1   INTRODUCTION

The (US) National Library of Medicine (NLM; http://www.nlm.nih.gov/) is the largest biomedical library in the world. Its mission is to collect, organize, preserve, and provide access to health-related information. NLM's wide range of online resources are accessible for free by global users. NLM's Specialized Information Services Division (SIS; http://sis.nlm.nih.gov/) is responsible for information resources and services in toxicology, environmental health, chemistry, and other topics. SIS databases and other resources related to toxicology and environmental health are accessed via the Environmental Health and Toxicology home page (http://sis.nlm.nih.gov/enviro.html) (Figure 9.1) [1,2].

9.2   SELECTED NLM RESOURCES FOR COMPUTATIONAL TOXICOLOGISTS AND OTHERS

PubMed (http://www.ncbi.nlm.nih.gov/pubmed) is NLM's free web interface to the world's biomedical literature and to additional sources of toxicological information. PubMed includes more than 24 million citations for biomedical articles from MEDLINE and life science journals. PubMed's citations include links to full-text articles, when available, from PubMed Central or publisher websites. Noteworthy for researchers and others is PubMed's "MyNCBI" (http://www.ncbi.nlm.nih.gov/sites/myncbi/), which allows users to store collections of search results for personal access or to share with others, and the ability to receive automatic updates for topic searches of interest. Also of interest to researchers and others is PubMed Commons (http://www.ncbi.nlm.nih.gov/pubmedcommons/faq/), released to the public in late

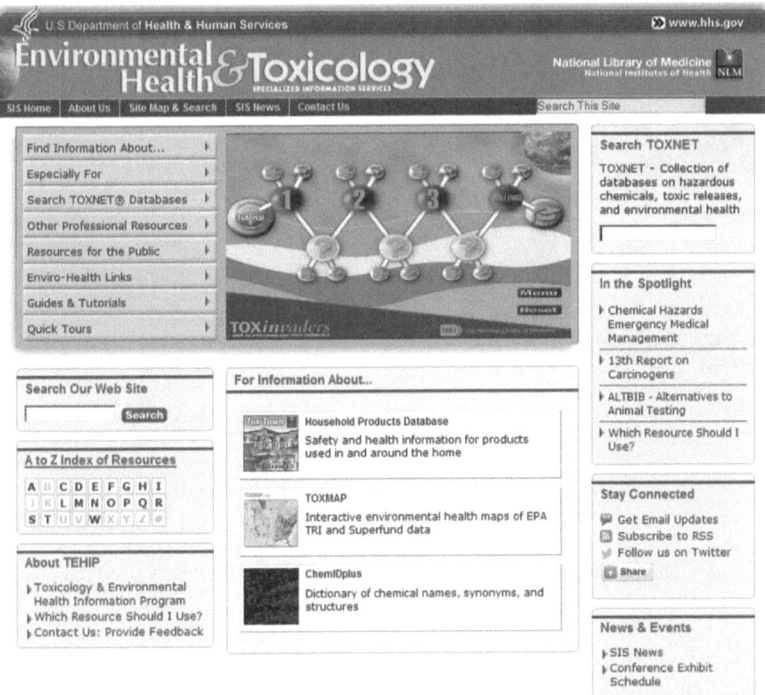

FIGURE 9.1   The home page for the NLM SIS Environmental Health and Toxicology collection of resources.

2013. PubMed Commons allows users with at least one article or any other item (including letters) in PubMed to write comments, for example, noting to others reading a PubMed citation where additional useful information is available.

SIS's TOXNET (TOXicology Data NETwork, http://sis.nlm.nih.gov/toxnet_faq .html and http://toxnet.nlm.nih.gov/) is a cluster of databases covering toxicology, hazardous chemicals, environmental health, and related areas (Figure 9.2). It is managed within the SIS Toxicology and Environmental Health Information Program. TOXNET's web interface is designed as an easy way to search databases of varying formats and content. The interface can be used to locate toxicology data, literature references, and other information for the chemical(s) of interest. It can also be used to search to identify chemicals that cause specific effects and offers a variety of ways to display and sort information.

TOXNET's TOXLINE database (http://www.nlm.nih.gov/pubs/factsheets/toxlinfs .html) provides bibliographic information covering the toxicological, biochemical, pharmacological, and physiological effects of chemicals and incorporates several million citations from NLM's PubMed/MEDLINE and other sources. The features offered by PubMed, such as saving PubMed citation searches, accessing full-text articles when available, and locating "similar articles" (renamed from "related citations" in April 2015) compared to TOXNET's TOXLINE, are such

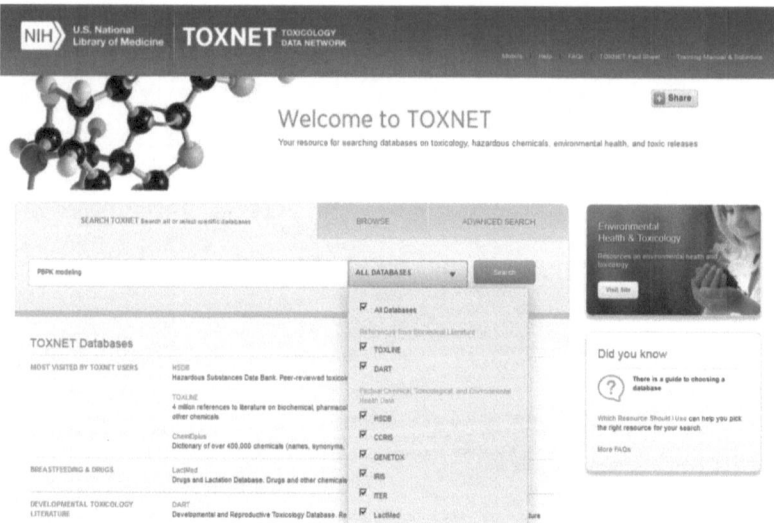

FIGURE 9.2 The home page for the NLM SIS TOXNET set of widely used databases.

that TOXLINE users may want to "turn off" the "include PubMed" option to avoid duplication of content and to plan to do both PubMed and TOXLINE searches.

DART (Developmental and Reproductive Toxicology; http://www.nlm.nih.gov /pubs/factsheets/dartfs.html) is a bibliographic database covering literature on teratology and other aspects of developmental and reproductive toxicology. It is managed by NLM and contains references to literature published since the early 1900s. Initially, DART was funded by the United States Environmental Protection Agency (EPA), the NIH National Institute of Environmental Health Sciences, the National Center for Toxicological Research of the United States Food and Drug Administration, and NLM. DART contains references to reproductive and developmental toxicology literature published since 1965. New citations in DART come only from PubMed, using a search profile (http://www.toxnet.nlm.nih.gov/help /newtoxnet/DARTCoreSearch.htm); however, DART includes historical content, for example, some meeting abstracts.

Hazardous Substances Data Bank (HSDB; http://www.nlm.nih.gov/pubs/fact sheets/hsdbfs.html) includes more than 5700 substances, with a focus on toxicology information [3]. In addition to toxicity data, HSDB provides information in the areas of emergency handling procedures, industrial hygiene, environmental fate, human exposure, detection methods, and regulatory requirements. The information is fully referenced and peer-reviewed by HSDB's Scientific Review Panel.

Integrated Risk Information System (IRIS; http://www.nlm.nih.gov/pubs/fact sheets/irisfs.html) includes data from the United States EPA in support of human health risk assessment, focusing on hazard identification and dose–response assessment. IRIS contains carcinogenic and noncarcinogenic health risk information on more than 550 chemical records, focusing on hazard identification and dose–response assessment information. Key data provided in IRIS are carcinogen classifications, unit risks, slope factors, oral reference doses, and inhalation reference concentrations. IRIS risk assessment data have been scientifically reviewed by EPA scientists and represents EPA consensus.

International Toxicity Estimates of Risk (ITER; http://www.nlm.nih.gov/pubs /factsheets/iterfs.html) contains data in support of human health risk assessments. It contains more than 680 chemical records and provides a comparison of international risk assessment information in a side-by-side format and explains differences in risk values derived by different organizations. ITER data, focusing on hazard identification and dose–response assessment, are extracted from each organization's assessment and contains links to the source documentation. These organizations include the US EPA, the US Agency for Toxic Substances and Disease Registry, the (US) State of Texas Commission on Environmental Quality, NSF International, Health Canada, the Dutch National Institute for Public Health and the Environment, and the International Agency for Research on Cancer. ITER also includes risk values from independent parties that have undergone peer review.

9.3 OTHER NLM RESOURCES TO CONSIDER

NLM has developed an enhanced version of its ALTBIB Web portal (http://toxnet .nlm.nih.gov/altbib.html) to provide access to publications in silico, in vitro, and refined animal testing methods. ALTBIB offers a free searchable bibliographic collection on alternatives to animal testing including being searchable by term(s) or by 1 of 15 categories, (e.g., pharmacokinetic/mechanistic studies) and reproductive/ developmental toxicity. Included are citations/links to articles, books, and technical reports, as well as links to online "3Rs" (reduction, refinement, and replacement), animal testing alternatives, and other databases.

ChemIDplus (http://chem.sis.nlm.nih.gov/chemidplus/chemidlite.jsp) contains several hundred thousand chemical records, most of which include chemical structures. ChemIDplus is searchable by name, synonym, CAS registry number, molecular formula, classification code, locator code, structure, and physical properties. An enhanced structure display is available in ChemIDplus Advanced and it also includes a very useful structure comparison tool to help identify toxicology and other information for chemicals with a similar structure.

Haz-Map (http://www.nlm.nih.gov/pubs/factsheets/hazmap.html and http:// hazmap.nlm.nih.gov/) is an occupational toxicology database developed from expert review, extraction, and compilation of published information. It is designed primarily for health and safety professionals and also for consumers and others seeking information about the health effects of exposure to chemicals and biologicals at home and at work. Haz-Map links jobs and hazardous tasks with occupational diseases and their symptoms. The chemicals and biological agents in the database are related to industrial processes and other activities such as hobbies. The occupational diseases and their symptoms are associated with hazardous job tasks. The categories of chemicals include those identified as "reproductive toxins."

The most recent addition to TOXNET (in early 2011) is the Comparative Toxicogenomics Database (CTD; http://www.nlm.nih.gov/pubs/factsheets/ctdfs.html) [4]. The CTD contains manually curated data describing cross-species chemical–gene/protein interactions and chemical– and gene–disease relationships. CTD users can perform several types of searches, for example, (1) browse relationships among chemicals, and obtain detailed information about them, including structure, toxicology data and related genes, diseases, pathways, and references, and (2) browse relationships among diseases, and obtain detailed information about them, including related chemicals, genes, pathways, and references. Further examples of potential CTD searches include finding which (1) human diseases are associated with a gene/protein, (2) human diseases are associated with a chemical, (3) genes/proteins interact with a chemical, (4) chemicals interact with a gene/protein, (5) references report a chemical–gene/protein interaction, and (6) cellular functions are affected by a chemical. Users can also easily conduct their CTD search strategy against other databases, for example, HSDB, TOXLINE, and ChemIDplus. The CTD has been enhanced in recent years to include increased data content, a new "Pathway View" visualization tool, enhanced curation practices, pilot chemical-phenotype results, and an extensive curated exposure data set.

The National Center for Biotechnology Information (NCBI; http://www.ncbi.nlm.nih.gov) is another part of the NLM and offers access to numerous other resources. NCBI's "Entrez, The Life Sciences Search Engine" (http://www.ncbi.nlm.nih.gov/sites/gquery) provides "search all" access to these resources, including PubChem (http://pubchem.ncbi.nlm.nih.gov/). PubChem provides access to extensive information (millions of records) on the biological activities of small molecules, including substance information, compound structures, and BioActivity data in three primary databases, PubChem Substance, PubChem Compound and PubChem Bioassay. PubChem's information comes from numerous depositors/resources and incorporates some of the content of TOXNET's databases.

9.4   CASE STUDY OF NLM DATABASE RESULTS RELEVANT TO PBPK MODELING

As an example of how NLM's databases can be used in computational toxicology, a search for information relevant to Physiologically Based Pharmacokinetic (PBPK) modeling was conducted using physiological parameters as the topic of interest.

The focus was on the types of information relevant to computational methods for reproductive and developmental toxicology available in NLM's databases. As shown below, searching TOXNET for "PBPK modeling" (the search included these words without the quotation marks) retrieved more than 100 TOXLINE (including PubMed) citations, along with information in the DART and HSDB databases (Figures 9.3 through 9.6). The results for the IRIS and ITER databases are not shown.

9.5   RECOMMENDATIONS AND FUTURE RESOURCES

The features offered by PubMed (e.g., the abilities to save the PubMed citation searches, to easily get full-text access when available, and to be able to easily identify and look at

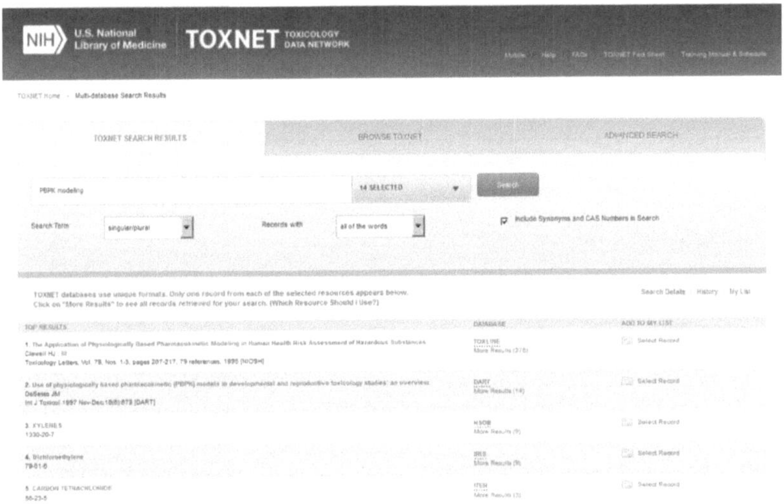

FIGURE 9.3   An example of the results from searching TOXNET for "PBPK modeling."

FIGURE 9.4   An example of the results from searching TOXNET's TOXLINE for "PBPK modeling."

"similar articles") compared to TOXNET's TOXLINE are such that users of TOXLINE may want to "turn off" the "include PubMed" option to avoid duplication of content in the TOXLINE searches and to plan to do both PubMed and TOXLINE searches.

NLM's online resources in toxicology and related topics are accessible at no cost by global users. The large number of databases in NLM SIS's TOXNET (http://toxnet.nlm.nih.gov/) can be used to identify and access bibliographic information covering the toxicological and other health and environmental effects of chemicals, including PBPK-related information.

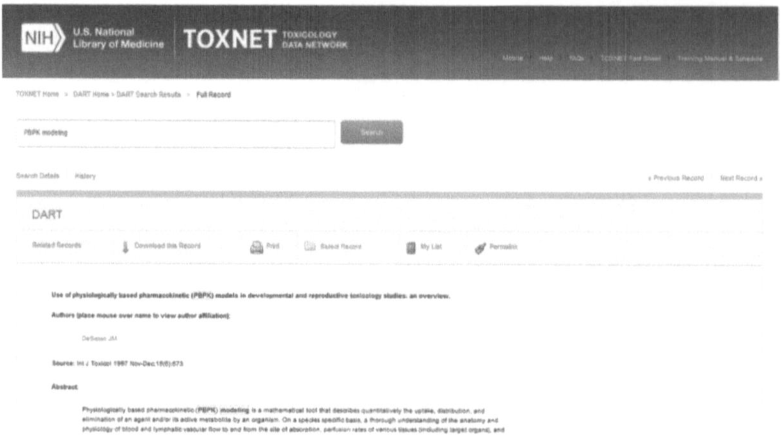

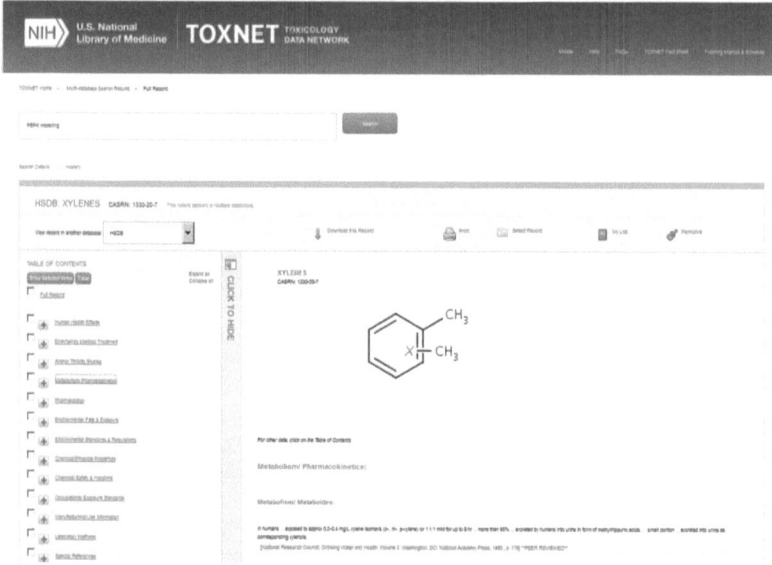

FIGURE 9.5   An example of the results from searching TOXNET's DART for "PBPK modeling."

FIGURE 9.6   An example of the results from searching TOXNET's HSDB for "PBPK modeling."

The databases in TOXNET will continue to be enhanced, for example, by adding images to HSDB of a chemical's metabolism- and toxicity-related pathways. In addition, there are new databases and other types of information resources that provide access to the latest information from the US EPA, other organizations in the United States, and beyond. These databases might be considered for addition to TOXNET's suite of databases. One resource is the Virtual Embryo (v-Embryo) project from the US EPA (http://www.epa.gov/ncct/v-Embryo/). Other examples from the United

States include "Tox21" (http://www.epa.gov/ncct/Tox21/), "ToxCast" (http://www.epa.gov/ncct/toxcast/), "ACToR" (Aggregated Computational Toxicology Resource; http://actor.epa.gov/actor/faces/ACToRHome.jsp), "NexGen" (Next Generation of Risk Assessment; http://www.epa.gov/risk/nexgen/), and "HERO" (Health and Environmental Research Online; http://www.epa.gov/hero/). Examples from the European Union include databases and other information resources from the European Chemicals Agency (http://echa.europa.eu/).

An additional challenge is the need to keep up with the ways to access information via smartphones, tablets, and other devices. For example, users can now access some TOXNET databases via TOXNET Mobile (http://toxnet.nlm.nih.gov/pda/). Users of smartphones, tablets, and other devices can stay updated about NLM's apps and websites optimized for mobile devices via http://www.nlm.nih.gov/mobile/index.html. Finally, users can learn about NLM databases via e-mail updates (http://sis.nlm.nih.gov/enviro/envirolistserv.html) and Twitter (https://twitter.com/nlm_sis).

ACKNOWLEDGMENTS

The author acknowledges the efforts of the staff of NLM SIS and SIS contractors to develop and maintain the many resources noted in this chapter.

REFERENCES

1. C. Hochstein, S. Arnesen, and J. Goshorn, *Environmental health and toxicology resources of the United States National Library of Medicine*, Med. Ref. Serv. Q. 26 (2007), pp. 21–45.
2. I. Laamanen, J. Verbeek, G. Franco, M. Lehtola, and M. Luotamo, *Finding toxicological information: An approach for occupational health professionals*, J. Occup. Med. Toxicol. 3 (2008), p. 18.
3. G.C. Fonger, P. Hakkinen, S. Jordan, and S. Publicker, *The National Library of Medicine's (NLM) Hazardous Substances Data Bank (HSDB): Background, recent enhancements and future plans*, Toxicology 325 (2014), pp. 209–216.
4. A.P. Davis, C.J. Grondin, K. Lennon-Hopkins, C. Saraceni-Richards, D. Sciaky, B.L. King, T.C. Wiegers, and C.J. Mattingly, *The Comparative Toxicogenomics Database's 10th year anniversary: Update 2015*, Nuclei Acids Res. 43 (2015), pp. D914–D920.

10 Modeling Endocrine Regulation of the Menstrual Cycle Using Delay Differential Equations*

Leona A. Harris and James F. Selgrade

CONTENTS

ABSTRACT

This chapter develops a mathematical model describing blood levels of five hormones important for regulating the menstrual cycle of adult women. The resulting system of 13 nonlinear, delay, differential equations with 44 parameters correctly predicts the serum concentrations of ovarian and pituitary hormones found in the

---

* Reprinted from *Mathematical Biosciences*, 257, Leona A. Harris and James F. Selgrade, Modeling endocrine regulation of the menstrual cycle using delay differential equations, 11–22, Copyright 2014, with permission from Elsevier.

biological literature for normally cycling women. In addition to this normal cycle, the model exhibits another stable cycle that may describe a biologically feasible "abnormal" condition such as polycystic ovarian syndrome. Model simulations illustrate how one cycle can be perturbed to the other cycle. This model may be used to test the effects of external hormone therapies on abnormally cycling women as well as the effects of exogenous compounds on normally cycling women. Sensitive parameters are identified and bifurcations in model behavior with respect to parameter changes are examined. Modeling various aspects of menstrual cycle regulation should be helpful in predicting successful hormone therapies, in studying the phenomenon of cycle synchronization, and in understanding many factors affecting the aging of the female reproductive endocrine system.

KEYWORDS

Pituitary, ovaries, follicle, estradiol, progesterone

10.1   INTRODUCTION

Complex endocrine signaling between the ovaries and the hypothalamus and pituitary glands is crucial for regulating and maintaining the female reproductive system of many mammals and birds. Abnormal levels of reproductive hormones often result in cycle irregularities. For instance, polycystic ovarian syndrome (PCOS), a leading cause of infertility in women [1–3], is usually associated with hormonal imbalances. Many PCOS women exhibit high androgen and low progesterone levels and their estrogen fluctuates very little during the month at levels that may be contraceptive [3]. Another example pertains to the observation that the breeding of dairy cows to maximize milk production is concurrent with a decrease in bovine fertility [4–7]. There is evidence that high-milk-yield cows have lower amounts of progesterone and luteinizing hormone (LH) than cows that were not genetically engineered. Also, there is concern [8–11] that environmental substances with estrogenic activity may disrupt the sexual endocrine system and, hence, may contribute to the increased incidence of breast cancer [12], to declines in sperm counts [13], and to developmental abnormalities [14]. Mathematical models may be used to simulate the effects of exogenous compounds and hormonal treatments on the reproductive endocrine system.

The fact that the hypothalamus and pituitary glands are essential to the control of the female reproductive cycle was not known until the 20th century (see Greep [15]). Much research (e.g., see Refs. [16–19]) has been done to understand the physiological mechanisms involved in the regulation of the menstrual and estrous cycles. However, many aspects are not completely understood because of experimental difficulties in determining these mechanisms especially at the level of the hypothalamus and pituitary. Modeling various aspects of menstrual and estrous cycle regulation may be helpful in understanding the roles of the many components of the reproductive endocrine system and may assist the experimentalist by indicating directions of investigation.

Most mathematical models of cycle regulation track blood levels of hormones produced by the brain and the ovaries. Follicle-stimulating hormone (FSH) and LH, which are produced by the pituitary gland responding to signaling from the hypothalamus, initiate the development of ovarian follicles and promote ovulation and the formation of the corpus luteum (see Refs. [16,18,19]). Simultaneously, at least three ovarian hormones, estradiol (E_2), progesterone (P_4), and inhibin (Inh), affect the synthesis and release of LH and FSH (see Refs. [20–22]). One of the early models of the female reproductive cycle was developed by Schwartz [23] to describe the rat estrous cycle. Similar to humans, a surge in LH leads to ovulation but rats ovulate at night; thus, Schwartz's model contains a 24-h clock to force the right timing of ovulation. Another early model was published by Bogumil et al. [24,25], which consists of 34 algebraic and ordinary differential equations. In order to produce the LH surge, their model assumed that the pituitary produced "tonic" and "surge" amounts of LH. They also expressed an LH surge threshold in terms of convolution integrals to weight more heavily recent concentrations of E_2 and P_4. Subsequent models of cycle regulation include those of McIntosh and McIntosh [26] and Plouffe and Luxenberg [27]. For articles that review the literature on mathematical models of the menstrual cycle and the estrus cycle, see Chávez-Ross [28] and Vetharaniam et al. [29]. All of these models describe some biological mechanisms but also many contain artificial features such as clocks or convolution integrals.

Over the last decade, we have developed and analyzed a mechanistic, deterministic, mathematical model [30–36] that predicts average serum concentrations of FSH, LH, E_2, P_4, and Inh that agree with data in the biological literature for normally cycling adult women (McLachlan et al. [37]). Because of the interplay between the brain and ovaries, this system may be described as dual control. Hence, the modeling procedure is divided into three distinct steps. First, we derive a linear system of ordinary differential equations for the synthesis and release of FSH and LH in the pituitary, which respond to the signaling of the ovarian hormones E_2, P_4, and Inh. The McLachlan data [37] are used to obtain explicit time-periodic input functions for serum levels of E_2, P_4, and Inh, and the unknown state variables in the system of differential equations are FSH and LH. Then, the parameters of this system are estimated from the McLachlan data for FSH and LH using a numerical optimization routine such as Nelder–Mead [38,39] with least squares. The second step reverses this process by developing a model for the monthly cyclic changes in the ovarian hormones E_2, P_4, and Inh under the influence of the pituitary hormones FSH and LH. This linear system of differential equations for the ovarian hormones contains parameters and time-periodic input functions for FSH and LH. Parameter identification is performed on this system using the data from McLachlan et al. [37] for E_2, P_4, and Inh. With a complete set of parameters determined, the final step is to merge these two linear systems into one system, which is highly nonlinear because all the variables are considered as state variables.

As an illustration of this process, suppose that experimental data over a span of time, t, are available for two state variables x and y, where $x = x(t)$ and $y = y(t)$. First, we derive a single differential equation describing the rate at which the state variable x is changing with respect to t based on known biological interactions between x, y, and $\frac{dx}{dt}$. The resulting equation contains an explicit input function $y(t)$, derived empirically from the data set, and contains unknown parameters values, $p_1, p_2, \ldots, p_n$. Equation 10.1 gives an example of such a differential equation with two parameters p_1 and p_2:

$$\frac{dx}{dt} = \frac{p_1 y(t)}{p_2 + y(t)} x. \tag{10.1}$$

Notice that the differential equation in Equation 10.1 is linear in the state variable x and has a time-dependent coefficient function $\dfrac{p_1 y(t)}{p_2 + y(t)}$. The existing experimental data for x are used to estimate the parameters p_1 and p_2 by applying a parameter identification numerical routine in conjunction with numerically solving the differential equation for $x = x(t)$.

The procedure is then reversed by introducing a differential equation for the state variable y that contains an explicit empirical approximation for $x(t)$ derived from the data and that contains unknown parameters, say p_3 and p_4. An example of such a differential equation is given by Equation 10.2:

$$\frac{dy}{dt} = p_3 x(t) y + p_4 x^2(t). \tag{10.2}$$

Here, experimental data for y and a parameter identification routine are used to estimate p_3 and p_4 and numerically solve for $y = y(t)$. With these estimates for the four model parameters, p_1, p_2, p_3, and p_4, Equations 10.1 and 10.2 are then merged together to create the following system of differential equations in Equation 10.3, which describes the rates at which the state variables x and y are changing with respect to time:

$$\frac{dx}{dt} = \frac{p_1 xy}{p_2 + y}$$
$$\frac{dy}{dt} = p_3 xy + p_4 x^2. \tag{10.3}$$

Notice that the system of differential equations in Equation 10.3 is nonlinear in two state variables x and y as compared to the single, linear differential equations in Equations 10.1 and 10.2. It is important to note here that in order to fit this system of differential equations to the existing data for x and y simultaneously, it may be necessary to reestimate all four parameters in Equation 10.3, but the estimates already obtained serve as a good starting place.

After discussing biological background, we present the three components of our menstrual cycle model in detail.

10.2   BIOLOGICAL PRELIMINARIES

Typically, a woman is born with 500,000 to 700,000 primordial follicles, and this number decreases because of atresia, with an increasing decay rate as the woman ages (e.g., see Hansen et al. [40]). During her reproductive life, only a small number of these follicles develop to ovulatory status before the onset of menopause, which occurs at an average age of 51. The length of a normal menstrual cycle (Figure 10.1) for an adult woman is 28 days on average but may range from 25 to 35 days (Ojeda [41]). The cycle is divided into the follicular phase (roughly 14 days), ovulation,

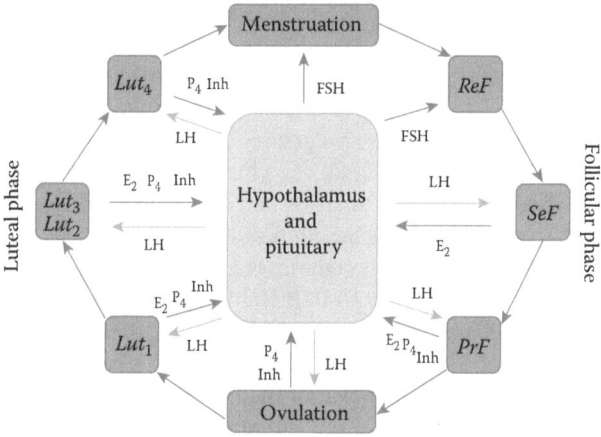

FIGURE 10.1 The follicular and luteal phases of the menstrual cycle. The outer ring depicts various stages of the ovary during a monthly cycle. *ReF*, *SeF*, and *PrF* represent the recruited, secondary, and primary follicles and Lut_i, $i = 1...4$, represent the corpus luteum. Directed arrows indicate hormonal actions.

and the luteal phase (roughly 14 days). The brain regulates ovarian cycling via the hypothalamus and the pituitary glands. The hypothalamus produces gonadotropin-releasing hormone (GnRH), which modulates the pituitary's secretion of the gonado-tropin hormones FSH and LH (see Clayton et al. [42]). To simplify our model, we lump the effects of the hypothalamus and the pituitary together and just consider the synthesis and release of FSH and LH. These hormones are secreted in a pulsatile pattern on the time scale of minutes but, because the ovaries respond to average daily blood levels (Odell [43]), our model tracks average daily gonadotropin concentrations in the blood. As part of its normal function, the ovary produces E_2, P_4, and Inh, which control the pituitary's synthesis and release of the gonadotropin hormones during the various stages of the cycle (see Figure 10.1).

The follicular phase of the cycle begins with the first day of menstrual flow, when blood levels of FSH rise and promote the recruitment and growth of 6 to 12 immature follicles. As these follicles develop by adding layers of granulosa cells (Odell [43]), the production of E_2 increases. During the second third of the follicular phase, typically a single dominant follicle is selected to continue its development and ultimately to release its ovum and the remaining follicles begin to atrophy. We do not model the process of follicle selection because the biological mechanism is not understood. As the ovaries pass into the primary follicular stage, the dominant follicle grows more rapidly and produces E_2 in large amounts. During the first two-thirds of the follicular phase, LH levels are roughly constant, but E_2 primes the pituitary for gonadotropin synthesis and, 1 day after E_2 reaches its maximum, LH peaks at approximately 10 times its early follicular concentration. This rapid rise and fall of LH over a period of 5 days is referred to as the LH surge and is necessary for ovulation. The day of the LH peak is considered the midpoint of the menstrual cycle and hormone data are usually centered at the day of the LH surge before averaging is done or comparisons are made. After a significant decrease during the primary follicular stage, FSH also surges concurrently with LH.

Ovulation occurs within a day after the LH surge [41] and, hence, the dominant follicle is transformed into the corpus luteum. The corpus luteum ("yellow body") secretes hormones in preparation for pregnancy and is characterized by increased fat storage in the theca and granulosa cells. P_4, which is low during the follicular phase, begins to rise several days before ovulation and continues to increase to a maximum midway through the luteal phase. The Inh profile is similar to that of P_4. During the luteal phase, P_4 and Inh inhibit the synthesis of LH and FSH, respectively, so that no immature follicles begin to grow [37,44]. If fertilization does not occur, then the corpus luteum decreases in size, decreases in hormone secretion, and becomes inactive by the end of the month. The decline of the corpus luteum results in a decrease in P_4 and Inh and, consequently, the removal of the inhibition on LH and FSH synthesis. The resulting gradual rise in FSH at the end of the month promotes the growth of a new cohort of immature follicles and initiates the next cycle.

10.3   MODEL DEVELOPMENT

Our modeling approach is divided into three components: the pituitary model, the ovarian model and the merged model. The pituitary model describes the production of the pituitary hormones LH and FSH during the menstrual cycle in response to the circulating ovarian hormones E_2, P_4, and Inh (inputs to the model). The ovarian model describes follicular and luteal development during the menstrual cycle and the production of the ovarian hormones in response to the pituitary hormones LH and FSH (inputs to the model). Each of these models are linear systems of differential equations with time-dependent coefficients (inputs) that are derived empirically from existing clinical data. The third component of the modeling process involves merging the pituitary and ovarian models together, creating a 13-dimensional, highly nonlinear, autonomous (time-independent) system of differential equations that describes the stages of the menstrual cycle and the interactions of all five hormones during the menstrual cycle while eliminating the use of input functions derived from clinical data.

10.3.1   Pituitary Model: Systems of Differential Equations

The pituitary model, first developed by Schlosser and Sclgrade [33], describes the synthesis, release, and clearance of LH and FSH on the basis of the pituitary's response to circulating levels of the ovarian hormones E_2, P_4, and Inh. The model consists of two systems, the LH system and the FSH system, of ordinary differential equations with time-dependent coefficients. Each system is linear in its state variables; however, the time-dependent coefficients are nonlinear functions of the ovarian hormones. Functions that approximate clinical study data (McLachlan et al. [37]) for the daily mean serum levels of E_2, P_4, and Inh during the menstrual cycle of 33 normally cycling women are used as inputs to the pituitary systems in order to predict the serum levels of LH and FSH during that cycle. Because the McLachlan data contain hormone values for 31 consecutive days, we assume a menstrual cycle period of 31 days and use the following input functions to approximate the ovarian hormone profiles over two menstrual periods:

$$E_2(t) = 62.5 + 230e^{-\frac{(t-14)^2}{5}} + 115e^{-\frac{(t-23)^2}{20}} + 230e^{-\frac{(t-45)^2}{5}} + 115e^{-\frac{(t-54)^2}{20}} \quad (10.4)$$

$$P_4(t) = 0.8 + 52.24e^{-\frac{(t-22)^2}{19.15}} + 52.24e^{-\frac{(t-53)^2}{19.15}} \quad (10.5)$$

$$Inh(t) = 290 + 1401.5e^{-\frac{(t-22)^2}{15}} + 1401.5e^{-\frac{(t-53)^2}{15}}. \quad (10.6)$$

The ovarian input functions are graphed against the ovarian hormone data from McLachlan et al. [37] over two menstrual cycles in Figure 10.2. In the McLachlan data, the follicular phase E_2 peak occurred at day 14 and the luteal peak occurred at day 23. To produce these elevations in $E_2(t)$, we use negative exponential functions where the exponents are translated to days 14 and 45 for the follicular phases and translated to days 23 and 54 for the luteal phases of two cycles. The input functions $P_4(t)$ and $Inh(t)$ are constructed in a similar manner.

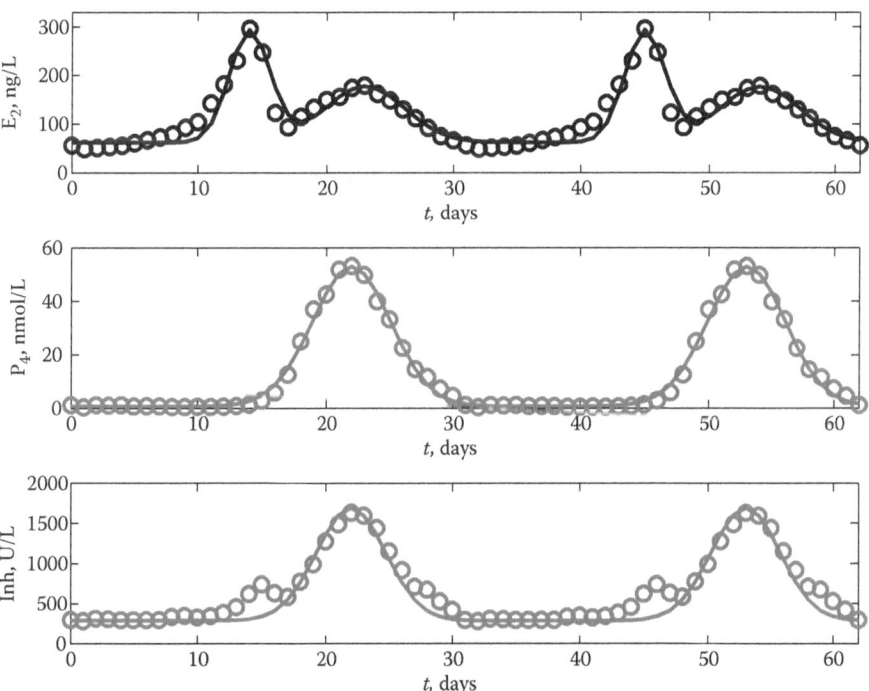

FIGURE 10.2 Ovarian input functions. Open circles represent daily mean serum levels of estradiol, progesterone, and inhibin of 33 normally cycling women as measured by McLachlan et al. [37]. Time-dependent functions (solid curves) approximating these values over two menstrual cycles are used as inputs to the LH and FSH systems.

The pituitary systems of differential equations model the synthesis, release, and clearance of LH and FSH, in response to stimulatory and inhibitory effects of the ovarian hormones. The schematic diagram in Figure 10.3 illustrates the effects of circulating levels of E_2, P_4, and Inh and outlines two major modeling assumptions: (1) LH and FSH synthesis occurs in the pituitary and (2) LH and FSH are held on reserve in the pituitary in what we call the "reserve pool" awaiting release into the bloodstream.

The LH system of differential equations has two state variables, RP_{LH}, representing the amount of LH in the reserve pool awaiting release into the bloodstream, and LH, representing the concentration of LH in the blood. In the model, the synthesis and release rates of LH are described as rational functions of ovarian hormones in which stimulatory effects appear in the numerators and inhibitory effects appear in the denominators.

It has been shown that high blood levels of estradiol promote rapid LH synthesis; therefore, the numerator of the LH synthesis term contains a Hill function (see Equation 10.9) to reflect estradiol's stimulatory effect on LH. This effect is most evident in the late follicular phase of the menstrual cycle when large amounts of estradiol are secreted by the dominant follicle, inducing the LH surge. This Hill function was selected because it increases rapidly as estradiol concentrations vary within a range of 200 and 600 pg/mL during the late follicular phase. This range includes normal and elevated levels of estradiol [33] and therefore the model can be used to monitor the effects of administering exogenous estrogens to existing estradiol levels. The exponent in the Hill function, called the Hill coefficient, was chosen to be $h = 8$ so that the Hill function begins increasing around 200 pg/mL and reaches its maximum around 600 pg/mL. It can easily be shown that if the Hill coefficient is $h = 9$, the synthesis rate increases too rapidly, and if $h = 7$, the increase is not rapid enough. During the luteal phase of the cycle, estradiol blood levels peak for a second time; however, this peak is not as substantial as the late follicular phase peak. It is believed that during this time, progesterone blood levels inhibit LH synthesis [45]. The period between changes in estradiol and progesterone blood levels and changes

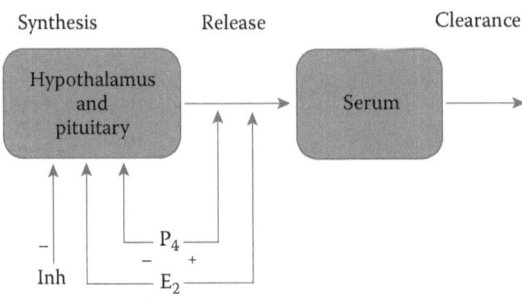

FIGURE 10.3 The ovarian hormones control the synthesis and release of LH and FSH in the brain. Plus (+) arrows indicate stimulation and minus (−) arrows indicate inhibition. (With kind permission from Springer Science + Business Media: *Bulletin of Mathematical Biology*, "Multiple stable periodic solutions in a model for hormonal control of the menstrual cycle," volume 65, 2003, pages 157–173, Leona Harris Clark, Paul M. Schlosser, and James F. Selgrade, Figure 1.)

in the synthesis rate of LH is captured by incorporating time delays, δ_E and δ_P, into the input functions $E_2(t)$ and $P_4(t)$, which appear in the LH synthesis term.

It has also been shown that estradiol and progesterone have similar effects on the release of LH and FSH into the bloodstream. A study by Chang and Jaffe [46] showed that progesterone stimulates the release of LH and FSH when estradiol blood levels are in a normal range during the late follicular phase. Tsai and Yen [47] demonstrated that blood levels of LH and FSH decline after the administration of ethinyl estradiol. This suggests that estradiol inhibits the release of LH and FSH into circulation. Finally, the clearance rate of LH is assumed to be proportional to LH blood levels. Therefore, the equations that govern the synthesis, release, and clearance of LH have the form

$$\frac{d}{dt}RP_{LH} = syn_{LH}(E_2, P_4) - rel_{LH}(E_2, P_4, RP_{LH}) \tag{10.7}$$

$$\frac{d}{dt}LH = \frac{1}{v}rel_{LH}(E_2, P_4, RP_{LH}) - clear_{LH}(LH) \tag{10.8}$$

where

$$syn_{LH}(E_2, P_4) = \frac{V_{0,LH} + \dfrac{V_{1,LH}E_2(t - d_E)^8}{Km_{LH}^8 + E_2(t - d_E)^8}}{1 + P_4(t - d_P)/Ki_{LH,P}}, \tag{10.9}$$

$$rel_{LH}(E_2, P_4, RP_{LH}) = \frac{k_{LH}[1 + c_{LH,P}P_4(t)]RP_{LH}}{1 + c_{LH,E}E_2(t)}, \tag{10.10}$$

$$clear_{LH}(LH) = a_{LH}LH. \tag{10.11}$$

The compartmental structure of the FSH system of differential equations is identical to that of the LH system of differential equations with state variables, RP_{FSH}, representing the amount of FSH in the reserve pool, and FSH, representing the concentration of FSH in the blood. However, there are variations in the synthesis and release terms because FSH responds differently to the ovarian hormones. There is evidence that inhibin has an inhibitory effect on FSH synthesis [37,48–50] and, as with $E_2(t)$ and $P_4(t)$ in the LH synthesis term, a time delay δ_{Inh} is used in the input function $Inh(t)$, which appears in the denominator of the FSH synthesis term (see Equation 10.14).

Recall that estradiol and progesterone have similar effects on the release of LH and FSH into the bloodstream: estradiol inhibits the release of LH and FSH and progesterone stimulates the release of LH and FSH. Tsai and Yen [47] also showed that estradiol has a greater inhibitory effect on FSH release. In addition, the preovulatory decline in FSH blood levels, not present in the LH profile, provides further evidence of the greater inhibitory effect of rising estradiol levels in the late follicular phase of the cycle. Therefore, a second-order inhibitory effect of estradiol on FSH release is used in the FSH system of differential equations instead of the first-order effect used in the LH equations [33,36]. Finally, the clearance rate of FSH is assumed to be proportional to FSH blood levels. Therefore, the equations that govern the synthesis, release, and clearance of FSH are given by

$$\frac{d}{dt} RP_{FSH} = syn_{FSH}(Inh) - rel_{FSH}(E_2, P_4, RP_{FSH}) \tag{10.12}$$

$$\frac{d}{dt} FSH = \frac{1}{v} rel_{FSH}(E_2, P_4, RP_{FSH}) - clear_{FSH}(FSH) \tag{10.13}$$

where

$$syn_{FSH}(Inh) = \frac{V_{FSH}}{1 + Inh(t - d_{Inh})/Ki_{FSH,Inh}} \tag{10.14}$$

$$rel_{FSH}(E_2, P_4, RP_{FSH}) = \frac{k_{FSH}[1 + c_{FSH,P} P_4(t)] RP_{FSH}}{1 + c_{FSH,E}(E_2(t))^2} \tag{10.15}$$

$$clear_{FSH}(FSH) = a_{FSH} FSH. \tag{10.16}$$

10.3.2 Ovarian Model: System of Differential Equations and Auxiliary Equations

The ovarian model, first developed by Selgrade and Schlosser [36], describes nine stages in the monthly development of the ovary and the production of the ovarian hormones E_2, P_4, and Inh. The model consists of a linear, time-dependent system of nine ordinary differential equations that represent the active capacities of follicular and luteal tissue to produce hormones under the influence of the pituitary hormones. Here, "active" means actively growing and secreting hormones. The follicular phase of the menstrual cycle is divided into three stages: the recruited follicular stage ReF, the secondary follicular stage SeF, and the primary follicular stage PrF. Ovulation and luteinization are represented by two ovulatory follicular stages: Ov_1 and Ov_2. The luteal phase of the cycle is represented by four stages of luteal development: Lut_i where $i = 1\ldots4$.

The pituitary hormones stimulate the growth of follicular tissue within a stage and the transfer of follicular tissue from one stage to the next as indicated in Figure 10.4. The capacity to produce hormones at each stage of the cycle is assumed to be proportional to the mass of the ovarian follicles or corpus lutea at that stage, and therefore, the schematic diagram of the ovarian model in Figure 10.4 also illustrates the stages of luteal tissue development and the production of E_2, P_4, and Inh by the secondary follicles, primary follicle, and the corpus luteum.

Functions $LH(t)$ and $FSH(t)$, which approximate the data by McLachlan et al. [37] for the daily mean serum levels of LH and FSH, are used as inputs to the ovarian system in Equations 10.17 through 10.25 in order to predict the serum levels of E_2, P_4, and Inh during that cycle. These functions are linear combinations of exponential functions (see Ref. [30]) and are similar to those in Section 10.3.1 for the ovarian

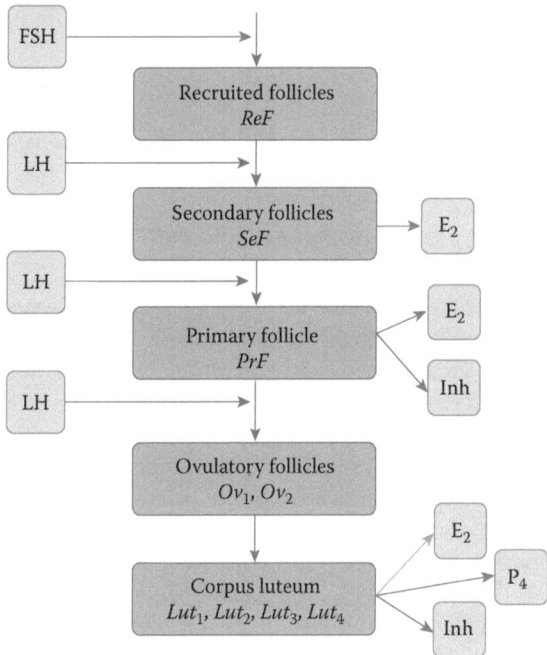

FIGURE 10.4 The compartments represent stages of follicular and luteal development during one menstrual cycle. FSH and LH promote the growth and transition between stages as indicated on the left and the ovarian hormones are secreted during the ovarian stages as indicated on the right.

hormones, that is, Equations 10.4 through 10.6. The following system of ordinary differential equations describes the ovarian model:

$$\frac{d}{dt}ReF = bFSH(t) + [c_1FSH(t) - c_2(LH(t))^\alpha]ReF \tag{10.17}$$

$$\frac{d}{dt}SeF = c_2(LH(t))^\alpha ReF + [c_3(LH(t))^\beta - c_4LH(t)]SeF \tag{10.18}$$

$$\frac{d}{dt}PrF = c_4LH(t)SeF - c_5(LH(t))^\gamma PrF \tag{10.19}$$

$$\frac{d}{dt}Ov_1 = c_5(LH(t))^\gamma PrF - d_1Ov_1 \tag{10.20}$$

$$\frac{d}{dt}Ov_2 = d_1Ov_1 - d_2Ov_2 \tag{10.21}$$

$$\frac{d}{dt}Lut_1 = d_2Ov_2 - k_1Lut_1 \tag{10.22}$$

$$\frac{d}{dt}Lut_2 = k_1 Lut_1 - k_2 Lut_2 \tag{10.23}$$

$$\frac{d}{dt}Lut_3 = k_2 Lut_2 - k_3 Lut_3 \tag{10.24}$$

$$\frac{d}{dt}Lut_4 = k_3 Lut_3 - k_4 Lut_4. \tag{10.25}$$

The first term $bFSH(t)$ in Equation 10.17 initiates the recruitment and growth of inactive antral follicles. During the follicular phase of the cycle, follicular growth rates and transfer rates are assumed to be proportional to $FSH(t)$ and powers of $LH(t)$ as indicated by Equations 10.17 through 10.20. The transition from the secondary follicular stage to the primary follicular stage depends on LH serum levels as indicated in Equations 10.18 and 10.19 and corresponds to the selection of the dominant follicle. Since ovulation and luteinization of the primary follicle are processes that are not instantaneous events [43], they are represented by two stages of ovulatory follicular development, Ov_1 and Ov_2. Little hormone production is assumed during this time. Finally, the model divides the luteal phase of the cycle into four stages represented by Equations 10.22 through 10.25 and reflects the corpus luteum as the primary source of P_4 and Inh production.

Because the clearance of the ovarian hormones from the blood is rapid compared to the clearance of the pituitary hormones, we assume that the blood levels of the ovarian hormones are at quasi-steady state [51] and their concentrations are modeled as linear combinations of the appropriate ovarian stages of the cycle. The following three auxiliary equations represent the serum levels of the ovarian hormones:

$$E_2 = e_0 + e_1 SeF + e_2 PrF + e_3 Lut_4 \tag{10.26}$$

$$P_4 = p_0 + p_1 Lut_3 + p_2 Lut_4 \tag{10.27}$$

$$Inh = h_0 + h_1 PrF + h_2 Lut_3 + h_3 Lut_4. \tag{10.28}$$

Because of the form of Equations 10.26 through 10.28, the effect of an exogenous ovarian hormone on the menstrual cycle may be simulated by adding a function representing an amount of that hormone to the appropriate equation.

10.3.3 Merged Model

The third and final step of the modeling process, as developed by Harris-Clark et al. [31], is to merge the pituitary model and ovarian model together to create a single 13-dimensional system of nonlinear, delay differential equations (Equations 10.29 through 10.41) with three auxiliary equations (Equations 10.26 through 10.28). The merged system has the form

$$\frac{d}{dt}RP_{LH} = \frac{V_{0,LH} + \dfrac{V_{1,LH}E_2(t-d_E)^8}{Km_{LH}^8 + E_2(t-d_E)^8}}{1 + P_4(t-d_P)/Ki_{LH,P}} - \frac{k_{LH}[1+c_{LH,P}P_4]RP_{LH}}{1+c_{LH,E}E_2} \tag{10.29}$$

$$\frac{d}{dt}LH = \frac{1}{v}\frac{k_{LH}[1+c_{LH,P}P_4]RP_{LH}}{1+c_{LH,E}E_2} - a_{LH}LH \tag{10.30}$$

$$\frac{d}{dt}RP_{FSH} = \frac{V_{FSH}}{1+Inh(t-d_{Inh})/Ki_{FSH,Inh}} - \frac{k_{FSH}[1+c_{FSH,P}P_4]RP_{FSH}}{1+c_{FSH,E}E_2^2} \tag{10.31}$$

$$\frac{d}{dt}FSH = \frac{1}{v}\frac{k_{FSH}[1+c_{FSH,P}P_4]RP_{FSH}}{1+c_{FSH,E}E_2^2} - a_{FSH}FSH \tag{10.32}$$

$$\frac{d}{dt}ReF = bFSH + [c_1FSH - c_2LH^\alpha]ReF \tag{10.33}$$

$$\frac{d}{dt}SeF = c_2LH^\alpha ReF + [c_3LH^\beta - c_4LH]SeF \tag{10.34}$$

$$\frac{d}{dt}PrF = c_4LHSeF - c_5LH^\gamma PrF \tag{10.35}$$

$$\frac{d}{dt}Ov_1 = c_5LH^\gamma PrF - d_1Ov_1 \tag{10.36}$$

$$\frac{d}{dt}Ov_2 = d_1Ov_1 - d_2Ov_2 \tag{10.37}$$

$$\frac{d}{dt}Lut_1 = d_2Ov_2 - k_1Lut_1 \tag{10.38}$$

$$\frac{d}{dt}Lut_2 = k_1Lut_1 - k_2Lut_2 \tag{10.39}$$

$$\frac{d}{dt}Lut_3 = k_2Lut_2 - k_3Lut_3 \tag{10.40}$$

$$\frac{d}{dt}Lut_4 = k_3Lut_3 - k_4Lut_4 \tag{10.41}$$

where the ovarian hormone functions E_2, P_4, and Inh in Equations 10.29 through 10.32 are linear combinations of the ovarian state variables, as defined by the auxiliary equations (Equations 10.26 through 10.28). The pituitary hormone functions LH and FSH in Equations 10.33 through 10.36 are the pituitary state variables represented by Equations 10.30 and 10.32. Therefore, the merged system is an autonomous system of differential equations since there are no time-dependent inputs to the system of differential equations as there were in the unmerged, pituitary, and ovarian models. In addition, the merged system is nonlinear because many of the equations involve nonlinear functions of the state variables as opposed to the unmerged models where the nonlinearities appeared in the time-dependent coefficients of the linear differential equations. Finally, the merged system involves delay differential equations as the ovarian state variables are delayed in the LH and FSH synthesis terms (see Equations 10.29 and 10.31).

10.4   PARAMETER ESTIMATION AND MODEL SIMULATIONS

In order to study the dynamical behavior of the merged model, estimates of the 44 model parameters are obtained either from the literature or through a parameter estimation scheme. The only known model parameters are the clearance rates for LH and FSH and the blood volume v, and the remaining 41 model parameters were estimated using daily mean serum levels of LH, FSH, estradiol, progesterone, and inhibin of 33 normally cycling women as measured by McLachlan et al. [37]. To estimate the 15 unknown pituitary parameters in Equations 10.7 through 10.15, Harris-Clark et al. [31] applied the Nelder–Mead Method in MATLAB® to a least squares cost function in order to fit the pituitary model to the LH and FSH data in McLachlan et al. [37]. To estimate the ovarian model parameters, Harris-Clark et al. [31] estimated the 15 ovarian system parameters in Equations 10.17 through 10.25 and the 4 estradiol parameters in Equation 10.26 using the Nelder–Mead Method and a least squares cost function that fit the ovarian system and the estradiol auxiliary equation to the E_2 data in McLachlan et al. [37]. Then, the remaining 7 parameters in the auxiliary equations for progesterone and inhibin (Equations 10.27 and 10.28) were estimated using separate least squares cost functions for P_4 and Inh. For a complete description of the parameter estimation process, refer to Harris-Clark et al. [31] and Harris [30].

Once the pituitary and ovarian models are merged together, the 44 parameters obtained in the preceding two steps are then used as estimates of the merged model parameters. Numerical simulations of the merged model were run in MATLAB using the delay differential equation solver *dde23* to analyze the model output. These simulations are discussed in detail in Refs. [31,35] and will be described briefly here. Using appropriate initial conditions, we observe the existence of two locally asymptotically stable periodic solutions for the same set of parameter values. One is a large-amplitude solution with a period of 29.5 days that approximates the McLachlan data for normally cycling women. See Figure 10.5 for graphs of the model simulations of E_2 and LH as compared to the data. We refer to this solution as the normal cycle. The second is a smaller-amplitude solution that has a period of 24 days and represents an abnormal menstrual cycle. Because there is no LH surge, the abnormal cycle is anovulatory and its acyclic E_2 profile suggests the possibility of PCOS [3]. See the dashed LH and E_2 curves in Figures 10.6 and 10.7 that compare the hormone profiles of the normal and abnormal cycles over 120 days.

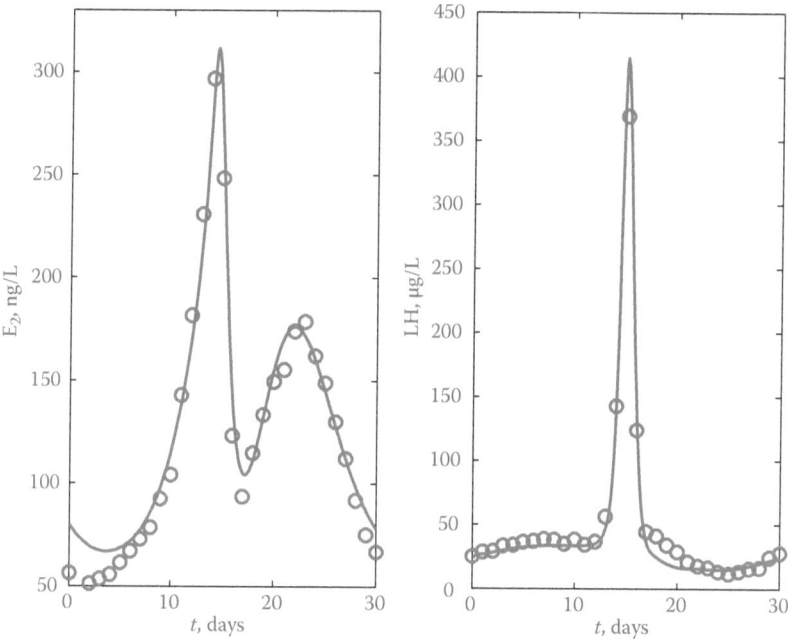

FIGURE 10.5   Model simulations of the normal cycle of the merged model (solid curves) in Equations 10.29 through 10.41 as compared to the clinical data (open circles) in McLachlan et al. [37]. (With kind permission from Springer Science + Business Media: *Bulletin of Mathematical Biology*, "Multiple stable periodic solutions in a model for hormonal control of the menstrual cycle," volume 65, 2003, pages 157–173, Leona Harris Clark, Paul M. Schlosser, and James F. Selgrade, Figure 3.)

10.5   SENSITIVITY ANALYSIS AND BIFURCATION ANALYSIS: PERTURBING THE MODEL PARAMETERS

Since the Nelder–Mead Method was used to search for a parameter set that minimized the least squares cost functions locally, it is quite possible that other parameter sets exist that fit the data well. As such, it is important to determine how sensitive the model is to changes in the model parameters. A local sensitivity analysis of the model parameters was performed by Selgrade et al. [35] to determine the effects of small variations in the model parameters on model outputs. In this analysis, normalized sensitivity coefficients were measured by discrete changes in a model output relative to the output value divided by changes in a model parameter relative to the parameter value. For example, if the original value of the parameter p is increased by 1% and a model output is denoted by a function of p, $MO(p)$, then the normalized sensitivity coefficient is computed according to the following formula:

$$S(p) = \frac{\Delta MO}{MO}\frac{p}{\Delta p} = \frac{MO(1.01p) - MO(p)}{MO(p)}\frac{p}{0.01p} = 100\frac{MO(1.01p) - MO(p)}{MO(p)}.$$

$$(10.42)$$

This coefficient approximates the partial derivative of some model output, a function of the model state variables, with respect to a model parameter that is normalized so that comparisons may be made across model outputs and across model parameters.

Selgrade et al. [35] decided to use the height of the E_2 midcycle peak along the normal cycle as the model output in this analysis because a significant follicular phase rise in E_2 stimulates the secretion of LH and causes the LH surge needed for ovulation and normal ovarian function. After computing normalized sensitivity coefficients for the 44 model parameters, Selgrade et al. [35] found that with respect to the E_2 midcycle peak, there were six parameters most sensitive to small variations: α, Km_{LH}, c_2, V_{FSH}, c_1, and $V_{0,LH}$. To further study the impact of perturbing sensitive model parameters, a bifurcation analysis was performed to determine the effects that variations of the parameter values have on the existence of the two locally asymptotically stable periodic solutions observed in Figures 10.6 and 10.7. Selgrade et al. [35] chose the parameter Km_{LH} for this analysis because of its physiological significance.

The bifurcation diagram for the merged model is shown in Figure 10.8, where the vertical axis denotes the difference between the maximum and the minimum

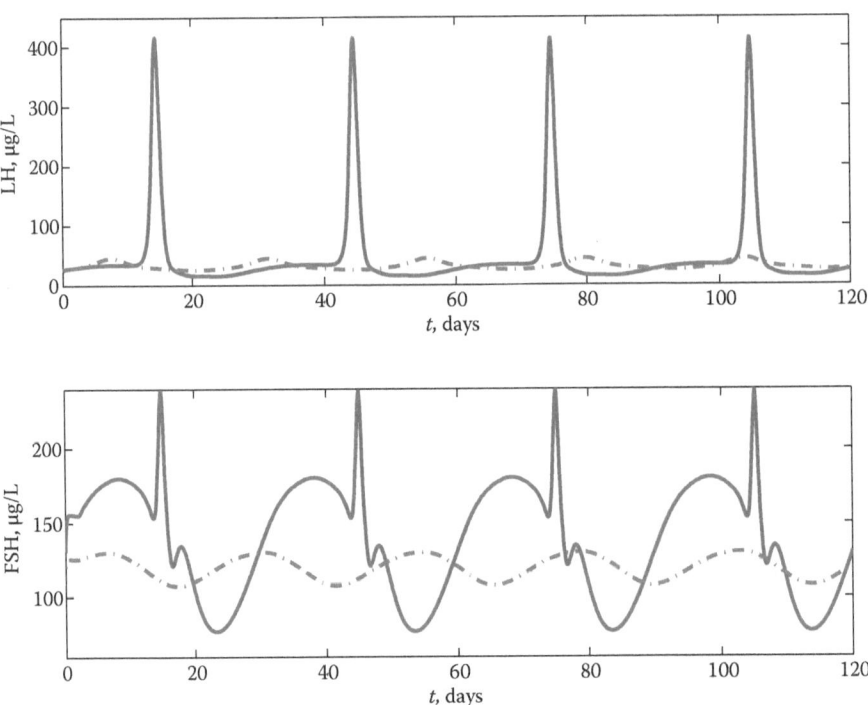

FIGURE 10.6 Profiles of the pituitary hormones for the normal (solid curves) and abnormal (dashed curves) cycles. Notice that the abnormal cycle has no LH surge and therefore is anovulatory. (With permission from *Bulletin of Mathematical Biology*, "Multiple stable periodic solutions in a model for hormonal control of the menstrual cycle," volume 65, 2003, pages 157–173, Leona Harris Clark, Paul M. Schlosser, and James F. Selgrade, Figure 4.)

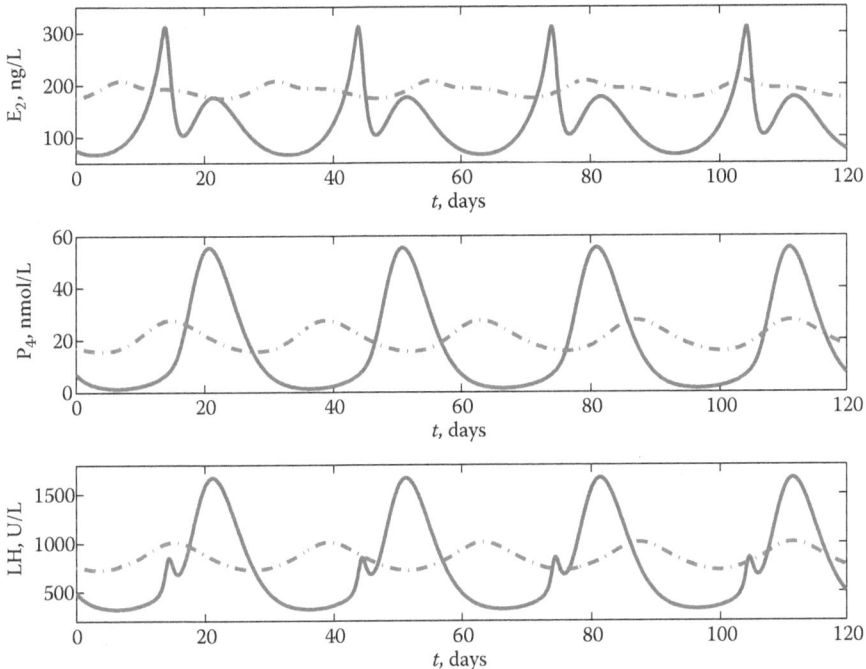

FIGURE 10.7 Ovarian hormone profiles for the normal (solid curves) and abnormal (dashed curves) cycles. (With kind permission from Springer Science + Business Media: *Bulletin of Mathematical Biology*, "Multiple stable periodic solutions in a model for hormonal control of the menstrual cycle," volume 65, 2003, pages 157–173, Leona Harris Clark, Paul M. Schlosser, and James F. Selgrade, Figure 5.)

of the first state variable, RP_{LH}, along a periodic solution or at an equilibrium (a solution where all state variables are constant in time). Hence, this difference is a measure of the amplitude of the periodic solution or is zero at an equilibrium. For a detailed description of the tedious method that was used to track the positions of stable and unstable periodic solutions as the parameter Km_{LH} is varied, refer to Selgrade et al. [35]. The bifurcation diagram has a closed loop of stable and unstable cycles (periodic solutions) where the upper half of the loop (solid curve) represents stable large-amplitude cycles and the lower half (dashed curve) represents unstable cycles. Saddle-node bifurcations occur at $Km_{LH} \approx 270$ and $Km_{LH} \approx 770$ where the stable and unstable cycles coalesce. The horizontal axis in Figure 10.8 represents a curve of equilibria. Along this axis, a supercritical Hopf bifurcation occurs at $Km_{LH} \approx 265$, resulting in a stable small-amplitude periodic solution and an unstable equilibrium solution. The branch of stable small-amplitude cycles continues through $Km_{LH} \approx 1500$ and then disappears (not shown on graph). The value $Km_{LH} = 360$ is the parameter value that fits the McLachlan data best. When $Km_{LH} = 360$, the normal cycle is indicated by an asterisk (*) in Figure 10.8 but there also exists a stable, small-amplitude abnormal cycle. Hence, when $Km_{LH} = 360$, a woman has the possibility of

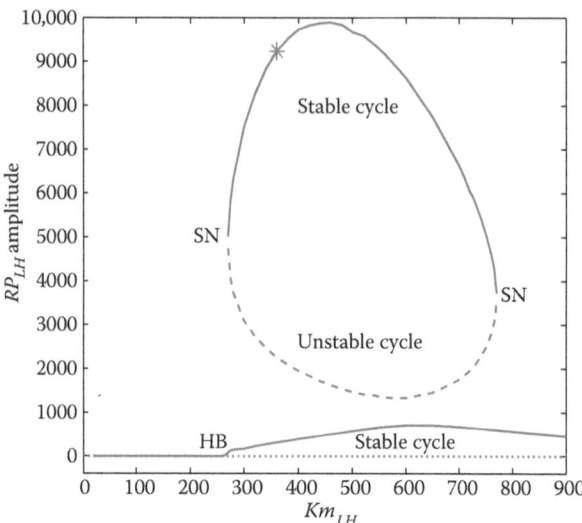

FIGURE 10.8 This bifurcation diagram plots cycle amplitude against the parameter Km_{LH}. The horizontal axis represents equilibria. A solid curve indicates stable cycles or equilibria and a dashed curve represents unstable cycles or equilibria. HB and SN denote Hopf and saddle-node bifurcations. The asterisk (*) indicates the normal cycle when $Km_{LH} = 360$. (Reprinted from Figure 8 in the *Journal of Theoretical Biology*, volume 260, J.F. Selgrade, L.A. Harris, and R.D. Pasteur, "A model for hormonal control of the menstrual cycle: Structural consistency but sensitivity with regard to data," pages 572–580, 2009, with permission from Elsevier.)

having a normal menstrual cycle or an abnormal menstrual cycle depending on her initial hormone levels. In fact, there are two stable periodic solutions that exist for every Km_{LH} value between 270 and 770 (as seen in Figure 10.8). Therefore, the initial hormone levels of a woman with a Km_{LH} in this range will determine whether she will cycle normally or abnormally. When Km_{LH} has a value outside this range, the amount of RP_{LH} is too low to produce an LH surge and, therefore, the woman will have only an anovulatory cycle.

10.6 EXOGENOUS EXPOSURE OF OVARIAN HORMONES

In Section 10.4, we observe that the merged model produces two asymptotically stable periodic solutions for the same set of model parameters, a large-amplitude cycle (normal cycle) fitting the McLachlan data for normally cycling women and a small-amplitude cycle (abnormal cycle) that resembles the hormone profiles of women with menstrual cycle irregularities, possibly PCOS. Since a specific set of parameters represents the behavior of an individual woman, Figures 10.6 through 10.8 indicate that a woman's initial hormone levels will determine whether she will cycle normally or abnormally. By perturbing one parameter and keeping all of the other parameters

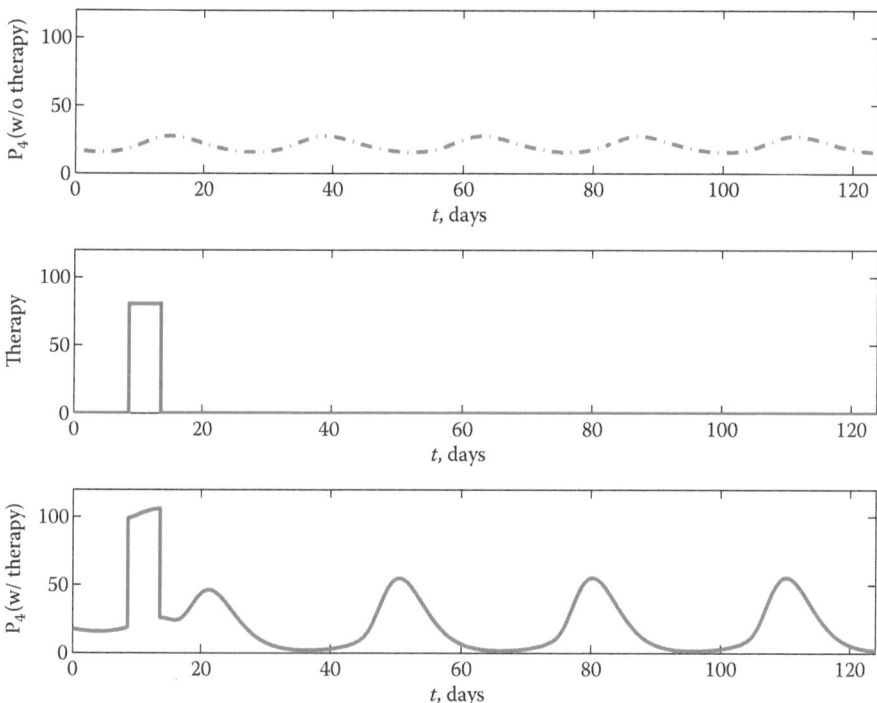

FIGURE 10.9 The upper graph is the P_4 profile of the abnormal cycle. The middle graph is a P_4 treatment of 80 nmol/L for the first 5 days of the luteal phase of the first cycle. The lower curve graphs P_4 with this treatment to show that normal P_4 levels are restored after one cycle. (With permission from *Bulletin of Mathematical Biology*, "Multiple stable periodic solutions in a model for hormonal control of the menstrual cycle," volume 65, 2003, pages 157–173, Leona Harris Clark, Paul M. Schlosser, and James F. Selgrade, Figure 6.)

fixed, we also observe that women with similar hormone profiles can also have a normal or an abnormal cycle (see Figure 10.9). These results lead us to the following questions: Can the abnormal cycle be perturbed into the normal cycle by applying some exogenous exposure of ovarian hormones (while keeping the parameter values fixed)? Similarly, can the normal cycle be perturbed into the abnormal cycle?

10.6.1 PCOS AND PROGESTERONE TREATMENT

PCOS, a menstrual cycle abnormality that is a leading cause of infertility in women [1–3], is usually associated with abnormal hormone profiles. Many PCOS women exhibit high androgen levels and low progesterone levels [3]. For example, low progesterone during the luteal phase permits more LH secretion at the expense of FSH secretion because of very rapid pulsing of GnRH that affects the pituitary's synthesis and release of the gonadotropins (see Marshall et al. [52]). Assuming that the abnormal cycle (dashed curves in Figure 10.7) of our model represents PCOS, a progesterone treatment may be tested in the setting of this model by trying to perturb the abnormal cycle to the normal cycle (solid curves in Figure 10.7) with exogenous P_4. In fact, the administration of exogenous P_4 was implemented by Harris-Clark et al. [31] by adding a constant term to the progesterone auxiliary equation (Equation 10.27). The progesterone therapy shown in Figure 10.9 adds 80 nmol/L of P_4 to (10.27) for 5 days at the beginning of the luteal phase of the abnormal cycle (from day 8 to day 13 of the cycle) and results in normal serum levels of all five hormones by the next cycle.

10.6.2   ENDOCRINE DISRUPTION

The model described in this chapter can also be used to simulate the effects of exogenous substances on normal menstrual cycle behavior. There are concerns that environmental chemicals with estrogenic activity can disrupt the reproductive endocrine system and may contribute to the increased incidence of breast cancer [12], declines in sperm counts [13], and developmental abnormalities [14]. To test whether the normal cycle of our model can be perturbed (disrupted) into the abnormal cycle because of exposure to exogenous estrogen, a constant term can be added to the estradiol auxiliary equation (Equation 10.26). Figure 10.10 shows that the administration of 50 ng/L of E_2 for one complete cycle (~30 days) starting at day 6 of the follicular

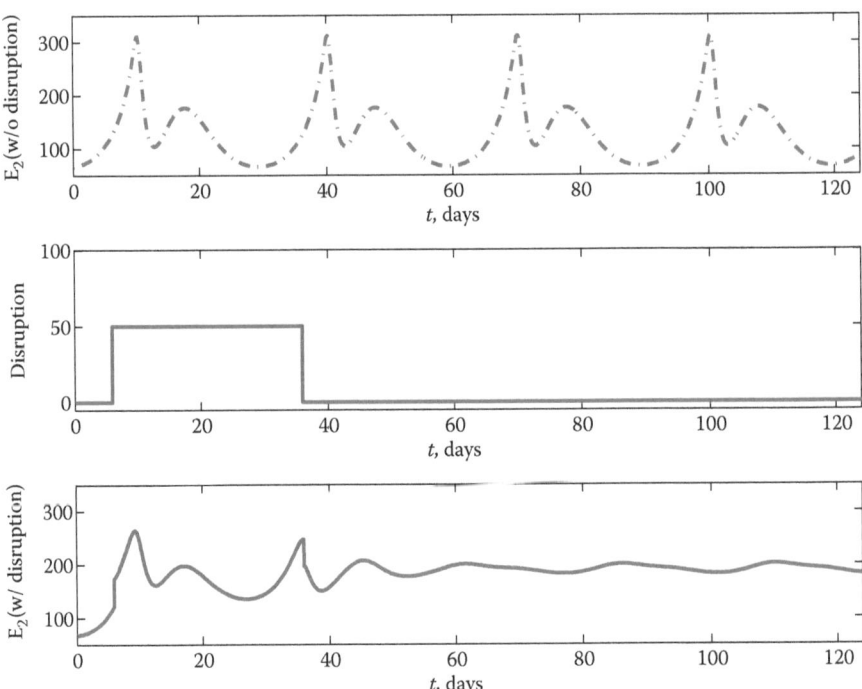

FIGURE 10.10   The upper graph shows the normal E_2 profile. The middle graph depicts a 30-day exposure to 50 ng/L of exogeneous E_2 starting on day 6 of the normal cycle. The lower graph shows that this exposure perturbs the normal cycle into the abnormal cycle.

phase of the normal cycle results in E_2 hormone levels that are too low to produce an LH surge and, hence, disrupts the normal menstrual cycle. A more complicated estrogen disruption was carried out by Harris-Clark et al. [31], which also perturbed the normal cycle to the abnormal cycle.

10.7   SUMMARY AND DISCUSSION

The mathematical model presented here describes the biological mechanisms pertinent to hormonal control of the menstrual cycle of adult women. Average daily blood levels of five essential hormones are tracked. Because the biological system is dual control, the model may be decomposed into two submodels—one submodel for the pituitary hormones LH and FSH under the control of only ovarian hormones and the other submodel for the ovarian hormones E_2, P_4, and Inh under the control of only pituitary hormones. Each submodel is linear in its state variables with time-dependent input functions for the control variables estimated from data in the literature (e.g., McLachlan et al. [37]). Parameter identification is performed on each submodel separately. These parameter estimates are good starting values for parameter identification for the merged model. The final merged model is a system of 13 nonlinear differential equations with three discrete time delays representing time lags in the pituitary's synthesis response to changes in ovarian hormone levels. Simulations of the merged model provide an excellent approximation to the hormone data in McLachlan et al. [37] for normally cycling women (see Figure 10.5).

Surprisingly, this model with the parameters that fit the McLachlan data best also has another stable periodic solution, which we refer to as the abnormal cycle. Because of a lack of an LH surge, the abnormal cycle is anovulatory and hormone profiles are reminiscent of PCOS. In fact, the acyclic E_2 level of approximately 200 ng/L may be contraceptive (see Figure 10.7). We illustrate how exogenous ovarian hormones can be used to perturb one stable cycle to the other. Although the model with this parameter set is bistable, multiple simulations indicate that the state space region of initial conditions giving solutions that approach the normal cycle is much larger than the region of initial conditions approaching the abnormal cycle.

Biological data are inherently variable. Data collected by Welt et al. [53] for the same five hormones in McLachlan et al. [37] has been used by Pasteur [32] to estimate parameters for the model described above. The McLachlan and Welt data sets are somewhat different; hence, the resulting parameter sets are different. Pasteur's simulations of the model using the Welt data exhibit only one stable periodic solution and it fits the Welt data for normally cycling women. Selgrade et al. [35] explained this apparent inconsistency by showing that changing the value of Km_{LH} in the Welt system resulted in the Welt model exhibiting two stable cycles like the McLachlan model. As discussed in Section 10.5, model output is sensitive to changes in the parameter Km_{LH}. Small changes in a sensitive parameter may result in changes to the asymptotic behavior of model solutions because of the occurrence of bifurcations. Hence, sensitivity analysis and bifurcation analysis are essential to understanding and using a mathematical model.

What does this model say about the menstrual cycle of individual women? Depending on an individual's parameters, she may cycle normally after a length of time regardless of her initial hormone levels. Her menstrual cycle corresponds to the Welt parameter set. On the other hand, another woman may have two possible menstrual cycles depending on initial hormone levels and one of these cycles is anovulatory. Her menstrual cycle corresponds to the McLachlan parameter set. If her cycle is anovulatory, then we demonstrate how the administration of exogenous hormones may perturb it to the ovulatory cycle.

Finally, this model may be refined by including additional important reproductive hormones. Welt et al. [53] collected data for two types of inhibin, Inh A and Inh B. Both inhibit FSH synthesis, but Inh B is a good indicator of ovarian aging and would be useful for extending the model to older reproductive women in the age range 35–45 years. Another refinement would be to separate the functions of the pituitary and hypothalamus in order to describe the role of GnRH, because many cycle abnormalities involve irregular GnRH pulsing. However, the time scale for GnRH pulsing is that of minutes and hours. The present model is on a time scale of days and months and, therefore, incorporating multiple time scales would complicate the model significantly.

ACKNOWLEDGMENT

Research was supported by National Science Foundation grant DMS-0920927.

REFERENCES

1. F. Alvarez-Blasco, J.I. Botella-Carretero, J.L. San Millan, and H.F. Escobar-Morreale, *Prevalence and characteristics of the polycystic ovary syndrome in overweight and obese women*, Arch. Intern. Med. 166 (2006), pp. 2081–2086.
2. R. Azziz, K.S. Woods, R. Reyna, T.J. Key, E.S. Knochenhauser, and B.O. Yildiz, *The prevalence and features of the polycystic ovary syndrome in an unselected population*, J. Clin. Endocrinol. Metab. 89 (2004), pp. 2745–2749.
3. S.S.C. Yen, *Polycystic ovarian syndrome (hyperandrogenic chronic anovulation)*, in *Reproductive Endocrinology. Physiology, Pathophysiology and Clinical Management*, 4th ed., S.S.C. Yen, R.B. Jaffe, and R.L. Barbieri, eds., W.B. Saunders Co., Philadelphia, PA, 1999, pp. 436–478.
4. J.E. Pryce, M. Royal, P.C. Garnsworthy, and I.L. Mao, *Fertility in the high-producing dairy cow*, Livest. Prod. Sci. 86 (2004), pp. 125–135.
5. M.D. Royal, *Genetic variation in endocrine parameters of fertility in dairy cattle*, PhD thesis, University of Nottingham, 1999.
6. M.D. Royal, A.O. Darwash, A.P.F. Flint, R. Webb, J.A. Woolliams, and G.E. Lamming, *Declining fertility in dairy cattle: Changes in traditional and endocrine parameters of fertility*, Anim. Sci. 70 (2000), pp. 487–501.
7. R.F. Veerkamp, B. Beerda, and T. van der Lende, *Effects of genetic selection for milk yield on energy balance, levels of hormones, and metabolites in lactating cattle, and possible links to reduced fertility*, Livest. Prod. Sci. 83 (2003), pp. 257–275.
8. G.P. Daston, J.W. Gooch, W.J. Breslin, D.L. Shuey, A.I. Nikiforov, T.A. Fico, and J.W. Gorsuch, *Environmental estrogens and reproductive health: A discussion of the human and environmental data*, Reprod. Toxicol. 11 (1997), pp. 465–481.
9. J.A. McLachlan and K.S. Korach, *Symposium on estrogens in the environment, III*, Environ. Health Perspect. 103 (1995), pp. 3–4.
10. S.H. Safe, *Environmental and dietary estrogens and human health: Is there a problem?* Environ. Health Perspect. 103 (1995), pp. 346–351.

11. L. Wright, *Silent sperm*, The New Yorker (1996), pp. 42–55.
12. D.L. Davis, H.L. Bradlow, M. Wolff, T. Woodruff, D.G. Hoel, and H. Anton-Culver, *Medical hypothesis: Xenoestrogens as preventable causes of breast cancer*, Environ. Health Perspect. 101 (1993), pp. 372–377.
13. R.M. Sharpe and N.E. Skakkebaek, *Are oestrogens involved in falling sperm counts and disorders of the male reproductive tract?* Lancet 341 (1993), pp. 1392–1395.
14. J.A. McLachlan, *Estrogens in the Environment II: Influences on Development*, Elsevier North Holland, New York, 1985.
15. R.O. Greep, *History of research on anterior hypophysial hormones*, in *Handbook of Physiology. Section 7: Endocrinology, Vol. IV. The Pituitary Gland, Part 2*, E. Knobil and W.H. Sawyer, eds., American Physiological Society, Washington, DC, 1974, pp. 1–27.
16. J. Hotchkiss and E. Knobil, *The menstrual cycle and its neuroendocrine control*, in *The Physiology of Reproduction*, 2nd ed., E. Knobil and J.D. Neill, eds., Raven Press, Ltd., New York, 1994, pp. 711–750.
17. S.M. Moenter, R.C. Brand, and F.J. Karsch, *Dynamics of gonadotropin hormone secretion during the GnRH surge: Insights into the mechanisms of GnRH surge induction*, Endocrinology 130 (1992), pp. 2978–2984.
18. S.S.C. Yen, *The human menstrual cycle: Neuroendocrine regulation*, in *Reproductive Endocrinology. Physiology, Pathophysiology and Clinical Management*, 4th ed., S.S.C. Yen, R.B. Jaffe and R.L. Barbieri, eds., W.B. Saunders Co., Philadelphia, PA, 1999, pp. 191–217.
19. A.J. Zeleznik and D.F. Benyo, *Control of follicular development, corpus luteum function, and the recognition of pregnancy in higher primates*, in *The Physiology of Reproduction*, 2nd ed., E. Knobil and J.D. Neill, eds., Raven Press, Ltd., New York, 1994, pp. 751–782.
20. F.J. Karsch, D.J. Dierschke, R.F. Weick, T. Yamaji, J. Hotchkiss, and E. Knobil, *Positive and negative feedback control by estrogen of luteinizing hormone secretion in the rhesus monkey*, Endocrinology 92 (1973), pp. 799–804.
21. J.H. Liu and S.S.C. Yen, *Induction of midcycle gonadotropin surge by ovarian steroids in women: A critical evaluation*, J. Clin. Endocrinol. Metab. 57 (1983), pp. 797–802.
22. C.F. Wang, B.L. Lasley, A. Lein, and S.S.C. Yen, *The functional changes of the pituitary gonadotrophs during the menstrual cycle*, J. Clin. Endocrinol. Metab. 42 (1976), pp. 718–728.
23. N.B. Schwartz, *Cybernetics of mammalian reproduction*, in *Mammalian Reproduction*, H. Gibian and E.J. Plotz, eds., Springer-Verlag, Berlin, 1970, pp. 97–111.
24. R.J. Bogumil, M. Ferin, J. Rootenberg, L. Speroff, and R.L. Vande Wiele, *Mathematical studies of the human menstrual cycle. I: Formulation of a mathematical model*, J. Clin. Endocrinol. Metab. 35 (1972), pp. 126–143.
25. R.J. Bogumil, M. Ferin and R.L. Vande Wiele, *Mathematical studies of the human menstrual cycle. II: Simulation performance of a model of the human menstrual cycle*, J. Clin. Endocrinol. Metab. 35 (1972), pp. 144–156.
26. J.E.A. McIntosh and R.P. McIntosh, *Mathematical Modeling and Computers in Endocrinology*, Springer-Verlag, Berlin, 1980.
27. L. Plouffe Jr. and S.N. Luxenberg, *Biological modeling on a microcomputer using standard spreadsheet and equation solver programs: The hypothalamic-pituitary-ovarian axis as an example*, Comput. Biomed. Res. 25 (1992), pp. 117–130.
28. A. Chávez-Ross, *Follicle selection dynamics in the mammalian ovary*, PhD thesis, University College London, 1999.
29. I. Vetharaniam, A.J. Peterson, K.P. McNatty, and T.B. Soboleva, *Modelling female reproductive function in farmed animals*, Anim. Reprod. Sci. 122 (2010), pp. 164–173.

30. L.A. Harris, *Differential equation models for the hormonal regulation of the menstrual cycle*, PhD thesis, North Carolina State University, Raleigh, NC, 2001. Available at http://www.lib.ncsu.edu/resolver/1840.16/4693.

31. L. Harris-Clark, P.M. Schlosser, and J.F. Selgrade, *Multiple stable periodic solutions in a model for hormonal control of the menstrual cycle*, Bull. Math. Biol. 65 (2003), pp. 157–173.

32. R.D. Pasteur, *A multiple-inhibin model for the human menstrual cycle*, PhD thesis, North Carolina State University, Raleigh, NC, 2008. Available at http://www.lib.ncsu.edu/resolver/1840.16/5587.

33. P.M. Schlosser and J.F. Selgrade, *A model of gonadotropin regulation during the menstrual cycle in women: Qualitative features*, Environ. Health Perspect. 108 (2000), pp. 873–881.

34. J.F. Selgrade, *Bifurcation analysis of a model for hormonal regulation of the menstrual cycle*, Math. Biosci. 225 (2010), pp. 108–114.

35. J.F. Selgrade, L.A. Harris, and R.D. Pasteur, *A model for hormonal control of the menstrual cycle: Structural consistency but sensitivity with regard to data*, J. Theor. Biol. 260 (2009), pp. 572–580.

36. J.F. Selgrade and P.M. Schlosser, *A model for the production of ovarian hormones during the menstrual cycle*, Fields Inst. Commun. 21 (1999), pp. 429–446.

37. R.I. McLachlan, N.L. Cohen, K.D. Dahl, W.J. Bremner, and M.R. Soules, *Serum inhibin levels during the periovulatory interval in normal women: Relationships with sex steroid and gonadotrophin levels*, Clin. Endocrinol. 32 (1990), pp. 39–48.

38. J.A. Nelder and R. Mead, *A simplex method for function minimization*, Comput. J. 7 (1965), pp. 308–313.

39. M.H. Wright, *Direct search methods: Once scorned, now respectable*, in *Numerical Analysis 1995. Proceedings of the 1995 Dundee Biennial Conference in Numerical Analysis*, D.F. Griffiths and G.A. Watson, eds., Pitman Research Notes in Mathematics, Addison Wesley Longman, London, 1996, pp. 191–208.

40. K.R. Hansen, N.S. Knowlton, A.C. Thyer, J.S. Charleston, M.R. Soules, and N.A. Klein, *A new model of reproductive aging: The decline in ovarian non-growing follicle number from birth to menopause*, Hum. Reprod. 23 (2008), pp. 699–708.

41. S.R. Ojeda, *Female reproductive function*, in *Textbook of Endocrine Physiology*, 2nd ed., J.E. Griffin and S.R. Ojeda, eds., Oxford University Press, Oxford, 1992, pp. 134–188.

42. R.N. Clayton, A.R. Solano, A. Garcia-Vela, M.L. Dufau, and K.J. Catt, *Regulation of pituitary receptors for gonadotropin-releasing hormone during the rat estrous cycle*, Endocrinology 107 (1980), pp. 699–706.

43. W.D. Odell, *The reproductive system in women*, in *Endocrinology*, L.J. DeGroot, ed., Grune & Stratton, New York, 1979, pp. 1383–1400.

44. I.J. Clarke and J.T. Cummins, *Direct pituitary effects of estrogen and progesterone on gonadotropin secretion in the ovariectomized ewe*, Neuroendocrinology 39 (1984), pp. 267–274.

45. J.F. Selgrade, *Modeling hormonal control of the menstrual cycle*, Comments Theor. Biol. 6 (2001), pp. 79–101.

46. R.J. Chang and R.B. Jaffe, *Progesterone effects on gonadotropin release in women pretreated with estradiol*, J. Clin. Endocrinol. Metab. 47 (1978), pp. 119–125.

47. C.C. Tsai and S.S.C. Yen, *The effects of ethinyl estradiol administration during early follicular phase of the cycle on the gonadotropin levels and ovarian function*, J. Clin. Endocrinol. 33 (1971), pp. 917–923.

48. N.P. Groome, P. Illingsworth, M. O'Brien, R. Pai, F. Rodger, J. Mather, and A. McNeilly, *Measurement of dimeric inhibin B throughout the human menstrual cycle*, J. Clin. Endocrinol. Metab. 81 (1996), pp. 1401–1405.

49. F.J. Hayes, J.E. Hall, P.A. Boepple, and W.F. Crowley, Jr., *Differential control of gonadotropin secretion in the human: Endocrine role of inhibin*, J. Clin. Endocrinol. Metab. 83 (1998), pp. 1835–1841.

50. A. Sehested, A. Juul, A.M. Anderson, J. Peterson, T. Jensen, J. Muller, and N. Skakkebaek, *Serum inhibin A and inhibin B in healthy prepubertal, pubertal, and adolescent girls and adult women: Relation to age, stage of puberty, menstrual cycle, follicle stimulatin hormone, luteinizing hormone, and estradiol levels*, J. Clin. Endocrinol. Metab. 85 (2000), pp. 1634–1640.

51. J. Keener and J. Sneyd, *Mathematical Physiology I: Cellular Physiology*, 2nd ed., Springer-Verlag, New York, 2009.

52. J.C. Marshall, C.A. Eagleson, and C.R. McCartney, *Hypothalamic dysfunction*, Mol. Cell. Endocrinol. 183 (2001), pp. 29–32.

53. C.K. Welt, D.J. McNicholl, A.E. Taylor, and J.E. Hall, *Female reproductive aging is marked by decreased secretion of dimeric inhibin*, J. Clin. Endocrinol. Metab. 84 (1999), pp. 105–111.

11 Molecular Structural Characteristics That Influence Partitioning of Xenobiotics into Human Breast Milk

Snezana Agatonovic-Kustrin and David W. Morton

CONTENTS

ABSTRACT

Breast milk is the most complete infant nutrition, which is why breastfeeding is recommended as the optimal feeding choice for most infants. However, humans are also constantly exposed to environmental pollutants, often with potentially synergistic effects and at levels that can cause side effects. Because milk is the only nutrient source for the infant, a newborn will be exposed to all the xenobiotics present in the milk. The milk-to-plasma-concentration ratio is a key parameter used to estimate an infant's exposure to different xenobiotics. Due to the countless number of chemicals released into environment, computational *in silico* methods and quantitative structure–activity relationships (QSARs) are gaining more and more attention in assessing this risk. The ability to predict the approximate amount of a chemical that might be present in milk from its structure can be very useful in the clinical setting. Molecular descriptors are numerical values that characterize properties of molecules, i.e., experimentally measured physicochemical properties (empirical) or calculated values from algorithms, such as two-dimensional fingerprints or three-dimensional structure. *In silico* QSAR models enable us to identify the essential structural characteristics that are responsible for secretion of a xenobiotic into milk. These models can be used to screen the milk/plasma partitioning potential for a huge number of compounds using data in existing xenobiotics/drugs databases.

KEYWORDS

In silico modelling, milk-to-plasma-concentration ratio, molecular descriptor, quantitative structure activity relationship, xenobiotic

11.1   INTRODUCTION

Despite considerable research over the past few decades, the extent to which many drugs taken by a breastfeeding mother pass into breast milk is not well known [1–3]. There is also increasing concern over the extent to which many environmental pollutants [4–6] and infection agents like human immunodeficiency virus (HIV) pass into breast milk. HIV breastfeeding is contraindicated in the developed world, but breast milk transmission of HIV-1 is a major concern in the underdeveloped world where bottle feeding is not a viable option [7].

The milk-to-plasma (M/P) ratio is commonly used to express the relative concentration of a chemical in milk compared to its concentration in the maternal plasma [8]. Xenobiotics with a low M/P ratio are likely to cause fewer dose-related adverse effects in the infant than those with a high M/P ratio. However, an accurate determination of the M/P ratio requires a carefully planned and executed program of breast milk sampling and analysis. The most straightforward way to assess an infant's exposure to various chemicals and drugs through breastfeeding is to directly measure their concentrations in the breast milk. Thus, various methods have been developed to assess the extent of drug binding to plasma milk proteins and lipids [9], including in vitro experiments using mammary cell monolayers [10,11] and in vivo experiments in animals [12]. Although these methods allow a relatively quick and inexpensive way for detecting the presence and concentration of these chemicals in milk, testing for a large number of different compounds is still far too expensive and time-consuming. Given these difficulties, it is not surprising that the M/P ratios for many compounds have not been determined in humans. Theoretical predictive methods have therefore become important tools in estimating the M/P ratio for many compounds that have not been experimentally determined. This generally involves performing a computer simulation of the chemical partitioning from plasma into milk using theoretical regression models, based on the compound's physicochemical characteristics.

Registration, Evaluation, Authorisation and Restriction of Chemicals (REACH), the European Community Regulation on chemicals and their safe use, have nominated quantitative structure–activity relationships (QSARs) for the toxicological and ecotoxicological preevaluation of chemicals [11]. The limiting factor in the development of a good QSAR is the availability of high-quality experimental data, appropriate descriptor selection, and statistical methods. In QSAR analysis, it is imperative that the input data be both accurate and precise in order to develop a meaningful model. Data used in QSAR modeling can be obtained from the literature or can be experimentally measured in the laboratory. These data can consist of a homogenous series of chemicals with similar chemical structure or have structural diversity even within a chemical class. The diversity of a data set allows for the development of more robust QSARs that are able to provide good predictability for molecules that differ from those used in the initial development of the model.

Molecular descriptors are numerical values that characterize properties of molecules, that is, experimentally measured physicochemical properties (empirical) or calculated values from algorithms, such as two-dimensional (2D) fingerprints. They vary in the complexity of encoded information they contain. Different descriptors represent different ways to view and encode a molecular structure and its respective properties, taking into account the various features of its chemical structure. These include not only monodimensional descriptors such as the simple counts of atoms and groups but also topological descriptors calculated from a 2D graph or three-dimensional (3D) descriptors derived from the most stable conformation with an optimized energy that usually must take into account conformational flexibility. The development of robust QSAR models requires the use of calculated structure-based descriptors, taking into consideration both geometrical and topological approaches without discrimination, and to use them in a structure–activity relationship procedure, strengthened with a natural selection algorithm for obtaining the best relationship model for given sets of compounds, for the given properties or activities being investigated. The success of any QSAR model depends on the accuracy of the input data, selection of appropriate descriptors and statistical tools, and, most importantly, validation of the developed model.

A big problem regarding the use of molecular descriptors is their reproducibility. For example, there can be a significant variation in experimentally measured values for log P, reported by different workers for a particular compound. For that reason, several approaches have been developed for the theoretical calculation of log P [13]. However, it is not uncommon to have differences of several orders of magnitude between these approaches [14]. In the development of modern QSARs, large sets of theoretical molecular descriptors are often used. The advantage of theoretical descriptors is that they are reproducible (provided the same calculation method is used) and can be calculated for all chemicals, even those that are not yet synthesized. Choosing the most appropriate molecular descriptors for use in a QSAR model can be an issue as there are vast numbers of descriptors available to choose from [15,16].

11.2 MOLECULAR DESCRIPTORS AND THEIR CLASSIFICATION

Molecular descriptors can be defined as the final result of a logical and mathematical procedure that transforms chemical information encoded within a graphical representation of a molecule into a unique and useful number (theoretical descriptor), or as the result of some standardized experiment (experimental descriptor). The numerical value is unique as it describes a variable that characterizes each particular chemical structure. Thousands of different descriptors are currently available, and the number is still growing [17]. They are traditionally divided into several subclasses according to the type of information they capture.

Theoretical chemical descriptors can be categorized in a number of ways. For example, they can be categorized as conformational, electronic, quantum mechanical, topological, spatial, or structural descriptors, or be based on the dimensionality of the molecular representation (i.e., zero-dimensional [0D], one-dimensional [1D], 2D, and 3D descriptors).

Constitutional descriptors capture information about the chemical composition of compounds such as counts of atoms, bonds, functional groups, and so on. Constitutional descriptors characterize the 1D properties of molecules, where the chemical formula is sufficient to calculate the descriptors.

Topological descriptors are 2D conformationally independent descriptors, which require information about the connectivity in molecules. This is usually derived from the 2D structure/graph of a molecule. A topological descriptor reduces the molecular graph into a number that characterizes the structure and the branching pattern of the molecule.

Connectivity indices are the most important topological descriptors that provide quantitative characterization of skeletal variation in a molecule. These descriptors are based on substructure features in the molecular graph, such as bonds, clusters, and rings. They can also include information about the nature of atoms, bond multiplicity, stereo-chemical features, and electronic parameters associated with various atoms.

Geometrical descriptors are derived from information that defines the orientation and position of atoms in space. They are calculated from the 3D molecular graph of a molecule. Typical examples are molecular surface areas, solvent-accessible molecular surface areas, moments of inertia of a molecule, and so on.

Charge distribution–related descriptors combine the 3D coordinates and the information about the electronic properties of molecules. The electronic structure is particularly important because the electron densities and charges in molecules determine the physicochemical properties (polar interactions) and reactivity of chemicals (covalent interactions).

Quantum-chemical descriptors are based on the molecular quantum mechanical calculations that solve the time-independent Schrödinger equation for the stationary states of a molecule. Quantum mechanical calculations can range from various semiempirical approximations to the Schrödinger equation and a wide range of *ab initio* methods. From the vast number of quantum chemical descriptors, the most commonly used ones are the energies of the highest occupied molecular orbital (HOMO) and the lowest unoccupied molecular orbital (LUMO).

The classification of theoretical molecular descriptors can also be based on the dimensionality of the molecular information. The so-called 0D descriptors are derived from the chemical formula of the molecule. Descriptors in this group are the number and type of atoms in the molecule, its molar mass, and any function of atomic properties (e.g., the sum of atomic van der Waals volumes). A substructure list representation of a molecule can be considered as a 1D molecular representation and consists of a list of molecular fragments (e.g., count descriptors of functional groups, rings, bonds, substituents, etc.). A molecular graph contains topological or 2D information. It describes how the atoms are bonded in a molecule, both the type of bonding, and the interaction of particular atoms. The derived molecular properties (such as total path count) from the molecular graph are called 2D descriptors. Another group of theoretical descriptors consists of 3D descriptors, which are determined from the geometrical or 3D representation of a molecule. Finally the descriptors, which are derived from a stereo-electronic or lattice representation, are called four-dimensional (4D) descriptors.

When dealing with a large number of molecular descriptors, experimental or theoretical, the careful selection of relevant and meaningful descriptors (variables) for use in the QSAR models becomes important. This is a particularly challenging problem in the development of artificial neural network (ANN) models. Sometimes two or more different molecular descriptors provide different views of the same molecular property and thus are highly correlated. Therefore, when dealing with a large number of highly correlated descriptors, a selection process for descriptors is necessary in order to find a simple and predictive QSAR model. The final model(s) should be based on a minimum possible number of descriptors to give good predictability. The descriptors in the final model(s) should theoretically provide the best predictability if they are the least correlated among each other. The descriptors for lipid solubility (log K_{ow}, Clog P, and Mlog P), molecular size (MW and MgVol), and polarizability (CMR, AMR, and PolarizG) are collinear, and each correlates well with the milk/serum partition coefficient (log P).

11.3 PHYSICOCHEMICAL FACTORS AFFECTING THE TRANSFER OF XENOBIOTICS INTO BREAST MILK

Understanding the mechanisms of partitioning of xenobiotic substances from plasma into breast milk and quantifying the process by the use of molecular descriptors is essential in order to assess the risk posed by xenobiotics to breastfed neonates. The transfer of a xenobiotic molecule from plasma into breast milk is affected by factors such as the extent of molecular ionization, plasma protein binding, molecular weight, the lipophilicity of the molecule, molecule–membrane interactions [18], the pharmacology of the xenobiotic in the mother, the composition of the breast milk, and the feeding habits of the neonate. In general, low plasma protein binding, low molecular weight, and lipophilic cations will favor partitioning of molecules into breast milk via a simple passive diffusion process [19]. They exhibit a nonuniform distribution throughout the body that is largely determined by their ability to pass through particular membranes at different points in the body. Breast milk can be considered as a compartment with bidirectional transfer across the plasma–breast milk membrane barrier, rather than simply a reservoir, with milk accumulation between consecutive collections. Drugs appear to equilibrate rapidly between plasma and milk, resulting in the drug concentration in milk being related directly to the drug plasma concentration.

It is generally assumed that passive diffusion is the process that controls the transfer of most chemicals between milk and plasma, and this has been confirmed to be the case for many drugs [20]. However, the partitioning of drugs and chemicals between plasma and milk is not always just by simple diffusion and may often involve various transport mechanisms analogous to those that operate across other membranes in the body [21].

Several unexplained but significant differences between the values expected and obtained for M/P ratios suggest that active transport processes also operate across mammary gland membranes. For example, the excretion of drug cations such as cimetidine [22], ranitidine [23], and nitrofurantoin [24,25] is higher than expected from passive diffusion alone, which suggests carrier-mediated transport system(s) for organic cations to be also involved. In the case of nitrofurantoin, the transfer from plasma to breast milk is nearly 20-fold higher than predicted by passive diffusion alone. Gerk et al. have suggested the involvement of two organic cation transporters in the active transfer process [26]. Benzylpenicillin was a significant "outlier" among the acidic drugs investigated by the passive diffusion model and its excretion into milk was substantially higher than expected. An in vivo animal study showed that the transport of benzylpenicillin across the mammary gland of goats and cows is reduced by concomitantly given probenecid, suggesting a presence of a carrier-mediated system for the drug [27]. Active transport requires specific transmembrane carrier proteins. The substrate specificity of transporters is generally broad, and it has been proposed that the only requirement is a degree of hydrogen bonding [28]. The ability of transmembrane carrier proteins to bind a wide variety of drugs of varying molecular weight and physicochemical properties is attributed to the relatively large active site that facilitates weak hydrophobic interactions with substrates.

11.3.1   Acid–Base Characteristics and Ionization (pKa Values)

11.3.1.1   Degree of Ionization

Milk is primarily buffered by citrates and phosphates; however, its buffer capacity is rather low [29]. Human milk is generally more acidic than plasma (values of 7.0, 7.1, and 7.25 have been reported) [30]; thus, its pH tends to be lower than that of plasma (pH 7.2). The degree of ionization and hence the percentage of unionized drug available to cross the membrane may be calculated using the Henderson–Hasselbalch equation.

The pKa^0 is the ionization constant of the unsubstituted parent compound. Since pKa and pKa^0 are descriptors for ionization of a chemical compound, as a function of pH, it is expected that these would be important descriptors in ANN models. Given the difference between the pH of plasma and pH of milk, there will be differences in the equilibrium concentrations of ionized and unionized drug in milk and plasma, which will depend on the pKa of the drug. Ionization of the drug in plasma will also affect the concentration of the unionized diffusible species and thus will affect the kinetics of transfer from plasma to milk. This could be an important factor in modeling the extent of transfer of molecules from plasma to breast milk, but it is unclear whether much of the data reported in the literature are true equilibrium milk plasma ratios or whether they relate to nonequilibrium conditions where the kinetics of transfer is significant.

A rate-limiting condition for many biological responses involves the movement of the drug through a large number of cellular compartments essentially made up of aqueous or organic phases. The ratio between K and K^0 is a measure of the substituent's contribution to solubility behavior in such a series of partitioning steps. As expected, higher pKa^0 values promote drug transfer into milk. Hammett [31] has quantified these effects by correlating the electronic properties of organic acids and bases with their equilibrium constants and reactivity. It was shown that the ratio between K and K^0 is linear and relates the effect of substituents to the dissociation equilibrium constant. The magnitude of the slope gives the relative strength of the electron-withdrawing or electron-donating properties of the substituents. If the slope is positive, the substituent is electron-withdrawing, and if it is negative, it is electron-donating. Drugs with a higher pKa generally have a higher M/P ratio. Drugs that typically pass readily into milk are weak bases (higher pKa) that are lipid soluble and do not bind significantly to plasma proteins.

11.3.1.2 Ion Trapping and pH

A drug in plasma or breast milk may exist as an unionized, ionized, or protein-bound species, depending on the pKa of the drug and the pH of the medium (plasma or breast milk). Unbound drugs may also (a) bind to proteins in milk and plasma or (b) partition into fat globules present in milk [32]. A neutral molecule or a weak base that passes from plasma to milk may protonate in the milk and, once ionized, become trapped in the milk since in its ionized form perfusion back into the maternal circulation is prevented (or minimized). Thus, weak bases may reach higher concentrations in milk, while acidic compounds present in the milk may be reabsorbed back into the blood, resulting in lower concentrations in the milk. It is also important to note that drugs that are more lipophilic will partition more effectively into the lipids that are present in much higher concentration in breast milk [33]. By these mechanisms, a drug could partition into the aqueous phase of milk or could partition into the separate fat phase of milk. Therefore, pH differences between plasma and milk may lead to a higher or lower M/P ratio because of the acid/base properties of the drug [34]. For instance, weakly basic drugs, such as β-blockers [35], may concentrate in milk, while weakly acidic drugs usually behave in the opposite fashion.

Unbound, unionized drug enters the milk until equilibrium is reached between the milk and the plasma. Drug ionization is a function of drug pKa and the pH of the milk and serum, milk and serum protein binding, and milk fat partitioning [9]. Assuming that only unbound, unionized drug freely equilibrates across the mammary epithelium, the extent of entry into breast milk can be predicted using the Henderson–Hasselbalch equation [36]. Therefore, partitioning of a drug into breast milk can be seen as a partitioning through a lipid membrane that the drug crosses by simple diffusion. However, the determination of the pKa of drugs in blood plasma has not been reported; hence, there are no available experimental data to support the assumption that the pKa of a drug in plasma will be the same as it is in water.

It is important to note that the amount of a xenobiotic available to cross into the milk depends not only on the fraction of unionized molecules present (as only the unbound fraction that is unionized is available to diffuse into milk) but also on the degree of protein binding, extent of lipophilicity, and molecular weight. Furthermore, some xenobiotics may be actively transported, resulting in a greater M/P ratio than is predicted by passive diffusion alone. Thus, for a particular drug, we can conclude that the most important factors that govern the partitioning into milk are the primary molecular descriptors: molecular size, lipid solubility, molecular charge or acid–base characteristics attributed to the difference in pH between milk and plasma [34,37], and its affinity to bind to proteins in plasma and milk [38].

11.3.2 SOLUBILITY AND SOLUTE DESCRIPTORS

Solubility is probably the most simple parameter to consider when modeling simple diffusion but, at the same time, the most difficult to accurately predict. Solubility can be modeled using (a) the linear solvation energy relationship (LSER) [39,40], (b) the mobile order theory [41], and (c) a quantitative structure–property relationship (QSPR) [42]. The LSER method is based on a multilinear regression analysis of the solubilities of solutes in different solvents. This method was developed by Kamlet and Taft [43], who introduced the solvatochromic relationship, which measures separately the hydrogen bond donor, hydrogen bond acceptor, and dipolarity/polarizability properties of solvents, each of which contribute to the overall solvent polarity. The method was further refined and developed by Abraham [44] who has applied it to numerous solutes.

Abraham et al. used the general linear free energy equation to correlate equilibrium plasma–milk solute distributions of drugs and environmental pollutants with five solute descriptors [45]. In addition to the five Abraham descriptors they developed, Abraham et al. also used a number of other descriptors with their method but were unable to obtain a useful linear equation that would accurately correlate equilibrium plasma–breast milk solute distributions of drugs and environmental pollutants. The Abraham method assumes that the partition coefficient (SP) between water and a solvent is given by the ratio of solubilities of a solute in the solvent (Ss), and in water (Sw), where

$$SP = Ss/Sw. \tag{11.1}$$

Making solubility estimations for most organic compounds in a wide range of solvents has always been a difficult task. Abraham et al. attempted to address this problem by developing a general solvation equation referred to as the *Abraham solvation equation*:

$$SP = c + eE + sS + aA + bB + vV, \tag{11.2}$$

where the five parameters, E, S, A, B, and V, referred to earlier, are the Abraham descriptors. The regression coefficients, c, e, s, a, b, and v, encode chemical information about the barrier and the contributing equilibria and kinetic processes. Furthermore, the coefficients for different putative biological mimetic systems can be compared, in order to uncover significant mechanistic differences. E is the excess molar refractivity (MR) that is obtained from the refractive index of the solution. S represents the dipolarity/polarizability solute–solvent interactions that can be obtained from gas–liquid chromatographic measurements on polar stationary phases or more generally from water/solvent partitions. The parameters A and B are overall hydrogen bond acidity and basicity, respectively, and represent the strength and number of hydrogen bonds formed by donor and acceptor groups in solute–solvent interactions. V is the McGowan characteristic volume that can be calculated from bond and atom contributions in a molecule. The solute descriptors A and B are based on the theoretical cavity model of solute–solvent interactions and these have been widely applied in the prediction of a variety of properties, such as solubility [46], blood–brain partitioning [47], and skin permeability [48]. The use of the Abraham descriptors allows a more detailed understanding of possible hydrogen-bonding patterns in a solute–solvent system.

11.3.3 LIPOPHILICITY AND LOG P

In general, lipid-soluble xenobiotics tend to easily penetrate lipid membranes, whereas water-soluble xenobiotics have to move through the narrow channels between cells. In general, most molecules partition into breast milk in accordance with their lipophilicity. The extent of partitioning can be approximately correlated to their passive transport across cell membranes and their ability to partition through the cell membranes [49]. Therefore, lipophilicity has also been described as being an appropriate measure for the distribution of a drug between milk fat and milk water [33].

The octanol–water partition coefficient (log P) is a key parameter used to quantify the uptake and distribution of xenobiotic molecules into biological systems. For a two-phase system at equilibrium, this is the ratio of a chemical's concentration in the n-octanol phase to its concentration in the aqueous phase [50] and is a parameter frequently used in QSARs [51,52] as a measure of the lipophilic character of a substance. However, log P is not sufficient on its own to establish a reliable QSAR, as it does not account for intramolecular interactions, and other important effects, such as the steric and electronic effects of the molecule. For example, intramolecular hydrogen bonding [53] can dramatically influence the extent of membrane penetration of a molecule. Correlations of lipophilicity and membrane penetration of molecules have been extensively reviewed by Seydel and Schaper [54], and the role of the importance of lipophilicity has also been the subject of some recent work by Escribano et al. [55]. We should also keep in mind that log P is a ratio [56] of solubility between two different phases, and a compound with low solubility in both octanol and water could have the same log P as a compound with 100 times higher solubility in both solvents. Therefore, many other descriptors need to be incorporated into a QSAR in addition to log P in order to account for these additional factors.

The acidity of drugs has an opposite effect to molecular size (or molecular weight) and log P. If the data in the literature for milk plasma ratios are truly equilibrium data, it might be expected that increasing lipophilicity (increasing log P value) would increase the milk plasma ratio given the higher lipid content of milk compared with plasma. However, high lipid solubility favors protein binding, reducing the amount of xenobiotic available for diffusion into milk. Therefore, an increase in lipophilicity or log P of a substance results in a decrease in its M/P ratio. Protein binding must be considered as an input variable and is required to be estimated from the physico-chemical properties of the molecule. An increase in log P correlates with an increase in protein binding for a xenobiotic molecule, since greater hydrophobic interactions permit greater binding affinity.

Breast milk has higher lipid content and is more acidic than plasma (i.e., has a lower pH); thus, alkaline fat-soluble substances will become trapped in milk. Nevertheless, the predictions using a model based on molecular weight and log P [57] do not correlate well with in vivo observations, because of a lack of consider-ation for differences in protein content between milk and plasma and also the deriva-tion of the partition coefficients employed. However, researchers [58] have observed a higher solubility of drugs in milk relative to water (phosphate buffer pH 6.5) than was expected. Thus, the sole use of log P values may be a significant source of error.

Syversen and Ratkje [59], using in vitro techniques, observed that the lipid solu-bility of a drug appeared to be the most important property responsible for variations in the drug concentration in milk. In a number of studies, lipophilic descriptors such as log P have also been identified as important factors for human serum albumin (HSA) binding [60,61]. However, log P has not been included in the recent general regression neural network [62] and ANN models [63] of drug partitioning into milk. It is possible that descriptors such as log P do not contain as much information as the current descriptor subset for describing molecule–protein interactions.

Morris and Bruneau [64] developed a linear model relating protein binding to either experimentally determined log P values for neutral and acidic compounds or log D (distribution coefficient) values for basic compounds. For bases, however, no correlation between binding to HSA and lipophilicity could be observed. Recently, it has been emphasized that a sigmoidal relationship between the percentage plasma protein binding and log D values (pH 7.4) within a large class of different compounds (acids, bases, neutrals) exists [65]. Therefore, essential physicochemical properties such as log D values or log P values alone are not sufficient to explain plasma protein binding for a diverse set of compounds. Binding to albumin is also dependent on spe-cific molecular recognition, such as directed hydrogen bonds, charge interactions, and space-filling of binding pockets [66–68].

Recent efforts over several years to develop QSAR models for serum protein or HSA binding of drug or drug-like compounds have focused on moderately sized data sets [60,61,69–72] with the exception of one large chemometric analysis [73]. In all of these studies, calculated log P was found to be the most important descrip-tor in QSAR models involving a number of other physicochemical parameters. By contrast, log D was found to have little or no correlation to HSA binding [74,75]. Topological descriptors in combination with log P could be used to tackle the dif-ficult problem of human serum protein binding prediction. The roles of nitrogen-containing compounds, acids, aromatic entities (atom-type and bond-type E-State descriptors), and skeletal ramification (molecular connectivity and chi indices) are all included in the structure information [76].

It is important to note that lipophilicity is not sufficient to explain plasma protein binding alone for a structurally diverse set of compounds. Kratochwil et al. [74] analyzed the effect of different molecular properties to protein binding in order to design a predictive model for the albumin binding affinities of drugs. Lipophilicity was rather poorly correlated to HSA binding for a diverse set of molecules, in contrast to a congeneric set, where lipophilicity is often found to be the dominant factor, suggesting that other specific molecular recognition elements beside physicochemical parameters are essential.

The composition of milk is important in determining the M/P ratio of a xenobiotic. In theory, increasing the concentration of milk fat would aid in the sequestration of a lipid-soluble drug into milk and increasing the concentration of milk protein would enhance the proportion of acidic drug binding in milk. It is likely that access to milk fat is partially controlled by protein binding in the plasma. The lipids in milk are present in the form of the milk fat globules whose membranes contain high-molecular-weight glycoproteins and a proteinous coat on each side of the membrane. Additionally, each milk fat globule is covered in a large surface casein micelle [77]. Thus, a drug that is bound to a protein is likely to be brought into proximity with, and may later be sequestered, into the fat globules. Unbound drug molecules may have only limited access to the fat globules regardless of their lipid solubility.

Increasing the protein component in milk increases the amount of drug exposed to fat and thus encourages its uptake for low protein bound drugs. Lipid-soluble drugs will dissolve preferentially in the lipid component of milk and, therefore, may not be available for diffusion back into the plasma. Because of this higher preference for the lipid component of milk, and because milk is the only fat reservoir that is periodically emptied, milk can be a very effective sink for fat-soluble drugs. Highly plasma protein bound drugs are not affected by changes in milk protein levels, and for these drugs, lipid solubility is the primary controller of drug entry into milk. Unlike plasma, breast milk contains emulsified fat, ranging on average from 3.2% to 3.6% in humans [78]. Milk fat can concentrate lipid-soluble drugs, in some instances producing higher levels in milk than in plasma [79]. It is important to note that the lipid content of milk varies considerably within a single feeding of an infant, between feedings, and among breastfeeding mothers. Consequently, drugs that are lipid soluble may be ingested by the infant in different quantities over the length of each feeding [80,81].

11.3.4 Size of the Molecule

Molecular size limits the absorption of molecules through membranes. Small lipid-insoluble substances penetrate cell membranes via the pores between aqueous phases on both sides of the membrane. The rate of such passive diffusion depends on the size of the pores, the molecular volume of the solute, and the solute concentration gradient. Medications with a low molecular mass [82] that are unionized and lipophilic will tend to be excreted into breast milk to a greater extent simply because diffusion through the alveolar epithelial cells is much easier.

Molecular weight is related to molecular size, and chemical compounds with a molecular weight greater than 200 have difficulty in crossing cell membranes [83–85].

However, under certain circumstances, large molecules will be transferred from plasma into breast milk [86,87]. In general, the smaller the molecule, the easier its transfer through the water-filled pores and channels of the lipid barrier of the alveolar cell [19].

The passive diffusion of a chemical compound is mainly affected by its plasma concentration, its physicochemical properties, and the protein and lipid content of breast milk. Passive transport can occur by simple diffusion owing to the concentration gradient present across the membrane or by passage of drugs through the aqueous pores (filtration) in the membrane. The passage of chemicals across a biological membrane involves (a) partitioning into, and then out of, the membrane, determined mainly by its lipid solubility, and (b) diffusion within the membrane, which is mainly dependent on its molecular size or molecular weight. Molecules pass either directly through the lipid bilayer and thus their diffusion will depend on their lipid solubility or through aqueous pores created by certain integral membrane proteins, in which case it will depend on both their water solubility and molecular size (weight). Therefore, an uncharged molecule, present in its nonionized form, may diffuse across a concentration gradient until equilibrium is reached.

Since lipophilic compounds partition into fat, the fat content of breast milk is a major determinant of the chemical level in whole breast milk. In general, lipophilic chemicals penetrate membrane barriers easily and are preferentially concentrated in the milk fat globules, which can lead to a high concentration ratio of the chemical between breast milk and plasma. Therefore, there is no substrate specificity and very small molecules (molecular weight <200) can easily move through aqueous pores between cells to reach the breast milk, while lipophilic compounds need to diffuse through the lipid bilayer. Meskin and Lien [88] developed a relationship that correlated transfer into breast milk as a function of molecular weight (MW) and the lipophilicity of the drug (expressed as the lipid/water coefficient or log P value). The concept of the aqueous pore has been quite useful for explaining the passive transport of hydrophilic solutes across cellular membranes. However, the effects of molecular geometry and the flexibility of the molecule on the permeability of the molecule are not straightforward. For instance, the simple molecular diameter that describes molecular size is not directly correlated to the variation of partitioning for highly lipophilic compounds. The hypothesis that the effective diameter controls permeability of chemicals, assumes a strict spatial orientation of the molecules toward the cell membrane surface where the molecular projection over the membrane pore does not exceed a certain threshold value around 0.95 nm [89]. The geometric relationship between the diffusing molecule and the channel is such that diffusion is severely restricted when the molecular dimension approaches half the channel size. The required orientation, however, is prevented by the chaotic movement of the molecules, which may explain the unsuccessful use of the effective cross section of molecules to predict their permeability across the membrane. Conversely, the chaotic collision of molecules with the cell membrane surface at different angles could help explain why geometric properties, instead of the generally accepted effective diameter, can be better correlated with the diffusion of a molecule across the cell membrane. The drop in partitioning of chemicals at a maximum cross-sectional diameter of approximately 1.5 nm is an indication that, above this threshold, there is a change in the mechanism involved in the partitioning of chemicals into breast milk. Therefore, the value of this transition point could be used as an additional parameter in addition to lipophilicity for regression modeling of the variation in the M/P ratio. Conformational flexibility tends to further increase the significance of chaotic movement in relation to membrane permeability, which leads to an additional decrease in the observed log P. The effect of this structural characteristic, however, needs to be further evaluated in order for it to be used successfully to quantify the extent of partitioning into breast milk.

MR is used as a descriptor and is a composite of refractive index, density, and molecular weight and is also related to molecular size and steric effects. MR is the molar volume corrected by the refractive index and represents the size and polarizability of a fragment or the whole molecule. MR is an additive value and can be considered as the sum of either atom or bond refractivities [90,91]. It is determined using the Lorentz–Lorenz formula [92]:

$$MR = (n^2 - 1)/(n^2 + 2) \, M/\rho, \tag{11.3}$$

where n is the refractive index, M is the molar mass, and ρ is the density of the substance. The density of a substance is the ratio of its mass to its volume. Molar density has been proposed to estimate diffusion coefficients for hydrocarbon systems [93]. The increase in density increases drug transfer to milk perhaps because of the decrease in molar volume and therefore molecular size.

Note that the refractivity of molecules containing double bonds is not the same as those with the same atoms singly bonded. Bonds of higher order increase refractivity, but for bonds with heteroatoms, the effect may be negative. Conjugation within rings has a negligible effect on MR, while, in chains, it is strongly positive. MR can also provide information on nucleophilic susceptibility, be used as a measure of the presence of olefinic functional groups, and serve as a measure of the binding force between the polar portions of an enzyme and its substrate [94]. As expected, the increase in molecular mass and MR decreases drug transfer from plasma into breast milk.

Solvent-accessible surface area (SASA) and molar volume are highly correlated geometrical descriptors that can provide information about contact surface, surface diffusion, absorption, and information on the size of the molecules. The contact surface area can be viewed as an indicator of the extent to which the solute is exposed to intermolecular interactions with the solvent [95] and is shown to be a remarkably accurate predictor of its water solubility [96]. Solute–solvent interactions influence both the transport of the drug molecule and the strength of the drug–protein interaction. The increase in SASA and the corresponding decrease in the M/P ratio are perhaps attributed to an increase in solubility and molecular size.

11.4 HYDROGEN-BONDING DESCRIPTORS

Hydrogen bonds important in biochemistry and molecular pharmacology are an essential component of intermolecular interactions. The fact that hydrogen bonding and log P have positive sensitivity correlates with the expectation that membrane permeation will be easier for compounds that form hydrogen bonds with the solvent and with the observation that milk is more lipophilic than blood, making lipophilic compounds more likely to penetrate into breast milk.

The polar surface area (PSA) of a molecule is also an indication of a compound's capacity to form hydrogen bonds. Higher M/P values are observed for compounds with high calculated log P and low PSA values. Calculated surface characteristics correlate with a number of physicochemical properties of drug molecules including lipophilicity, the energy of hydration, and the hydrogen bond formation capacity [97,98]. The surface properties of a molecule involved in intramolecular hydrogen bonding may be less polar, resulting in enhanced membrane permeability in comparison to a homologous molecule that exposes the (polar) hydrogen-bonding group(s) on its surface. It can, therefore, be hypothesized that the relative importance of each of the physicochemical factors will be reflected in a single measure such as the polar molecular surface area.

PSA and hydrogen-bonding capacity can be used to predict partitioning through biological membranes [99]. The utility of PSA as a predictor of absorption has even been identified by Palm et al. [100]. PSA is usually defined as those parts of the van der Waals or solvent-accessible surface of a molecule that are associated with hydrogen bond accepting capability (e.g., the number of N or O atoms) and hydrogen bond donating capability (e.g., NH_2 or OH groups). The simplest way of calculating the hydrogen-bonding capacity of a molecule is to count the number of hydrogen bond donor and acceptor atoms or to count the number of lone pairs of electrons on certain kinds of atoms. PSA is defined as a sum of surfaces of polar atoms (usually oxygen and nitrogen atoms) and their attached hydrogen atoms. It can be used to describe the hydrogen-bonding potential of a drug.

Certainly, these simplified models are not highly accurate descriptions of the hydrogen-bonding properties of the molecules, but they have in some cases provided reasonable predictions of membrane permeability. Österberg and Norinder analyzed the relationship between PSA and three H-bonding descriptors (the number of H-bond nitrogen atoms, the number of H-bond oxygen atoms, and the number of H-bond donor atoms on nitrogen and oxygen) and found high linear correlations between the hydrogen-bonding descriptors and PSA of five chemically diverse sets of drugs ($r^2 > 0.93$, $q^2 > 0.69$) [101].

11.5 PLASMA PROTEIN–BINDING DESCRIPTORS

Drug binding to plasma proteins is of significant practical therapeutic importance as this property influences the pharmacological efficacy of a drug, the pharmacokinetic profile of a drug, and drug–drug interactions. Drugs in the circulation exist in equilibrium between the free drug, plasma protein–bound and tissue-bound forms. Low protein-bound drugs (<20% binding) generally have higher partitioning into milk fat as the protein content in milk increases [10]. Drugs with intermediate to high protein binding (>36%) do not show increased fat partitioning with increasing milk fat. Protein binding and partitioning into fat are not additive factors for transfer into milk but act synergistically as a function of each other.

The most important carrier proteins responsible for binding xenobiotics in human plasma are HSA, α1-acid glycoprotein, and lipoproteins, which mainly interact with acidic, basic, and neutral compounds, respectively. The binding of drugs and other chemicals to plasma and milk proteins can influence the rate of passage of drugs into the alveolar lumen. Drugs that bind strongly to plasma protein tend to have lower volumes of distribution. They are too large to pass through the cellular membranes and cannot permeate through the aqueous pores. Extensive plasma protein binding will reduce partitioning into milk and will cause a drug to stay in the central blood compartment. Essentially, only the free, unbound drug molecules diffuse into breast milk regardless of their other physicochemical characteristics and are available for transfer across this barrier. Some protein binding also occurs in milk, but generally the extent of protein binding here is only 20%–60% compared to plasma [102,103]. The total plasma protein concentration is approximately 75 g/L of protein, whereas milk contains approximately 9 g/L of protein [79,104]. Of these plasma proteins, albumin, which comprises 50% of the total proteins present, is a major drug-binding protein and binds the widest range of drugs. However, its concentration in milk is only 0.4 g/L. Acidic drugs commonly bind to albumin, while basic drugs often bind to α1-acid glycoproteins and lipoproteins. Drug binding to plasma protein may be attributed to ionic, van der Waals, and hydrogen bonding.

Despite the large amount of data available on plasma protein binding of drugs, empirical rules for protein binding have not yet been established and it is still not clear how to go about modeling this parameter in multidimensional lead optimization strategies in order to develop a successful predictive QSAR [74]. The lipophilicity of drugs, traditionally felt to dominate binding to HSA, is not the only relevant descriptor. The studies show that the plasma protein binding of drugs is difficult to classify and no general implications for the design of lead-like compounds are able to be derived from the examination of the properties of existing drugs, except for chemotherapeutic drugs where >90% binding is observed. Lemont B. Kier developed a predictive topological QSAR based on a combination of general and specific structure features that are important for estimating protein binding of β-lactams [105]. For protein binding by β-lactams, the model indicates that binding is increased by the presence of aromatic groups, branching in substituents, and the presence of =N-, -F, and -Cl atoms, but is decreased by the presence of amino groups and carbonyl oxygen atoms. Similar conclusions were drawn for the albumin binding affinity study, except that, in this case, amino groups and carbonyl atoms are not explicitly implicated in the binding process.

A model based on the pharmacophoric similarity concept and the partial least square analysis for the prediction of drug association constants to HSA was able to single out the submicromolar to nanomolar binders, that is, to differentiate between 99.0% and 99.99% plasma protein binding. A model based on topological descriptors indicated descriptors that increase (positive) or decrease (negative factors) protein binding [72]. Positive factors for binding affinity include electron accessibility and the number of aromatic rings, aliphatic CH groups ($-CH_3$, $-CH_2-$, $>CH-$), halogens (fluorine and chlorine), and -OH groups. Five-membered heteroatomic rings present a negative factor, whereas six-membered heteroatomic rings present a positive factor. For environmental pollutants, such as polybrominated diphenyl ethers, the oxygen atom bridging two halogenated aryl groups, which functions as a hydrogen bond acceptor, appeared to reduce the transfer of the molecule from plasma to milk.

A relatively new approach is to use molecular similarity calculations, and to describe a molecular structure using, for example, topological pharmacophore description [106] or daylight fingerprint description (DFP) [107]. The use of pharmacophore type descriptors is critical for the derivation of 3D-QSAR or 4D-QSAR models [108,109]. According to the similarity principle [110], similar molecules, that is, compounds with a similar distribution of their pharmacophore units (hydrogen acceptor, hydrogen donor, hydrophobic parts), would experience comparable interactions with a protein and, thus, would also have similar binding constants.

The characterization of the pharmacophoric properties of chemical structures consists of two components: (a) a description in terms of a relevant scheme of any single structure and (b) instruction on how to compare any pair of structures. The description attributes to each structure a set of values, such as counts of substructure elements, physical parameters, geometrical quantities, and so on. The similarity measure combines the two value sets of a pair of structures to yield a value, between zero and one, which describes the similarity of the pair. A value of zero indicates that the two structures have nothing in common, while a value of one tells us that the two structures are identical with respect to the chosen description.

A second example is the pharmacophoric similarity concept [111] as implemented in the software Moloc [112]. Molecules showing biological activity are often called pharmacophores. They are characterized by their constituent pharmacophoric units, a set of centers with pharmacophoric properties that are called agons, and a topological distance matrix between the agons. Agons are classified into hydrogen binders and hydrophobics. In the most detailed case, each atom is an agon by itself. An atom is a hydrogen binder, when it has a hydrogen bond donor or acceptor strength above a corresponding donor or acceptor strength threshold. Otherwise, it is classified as hydrophobic.

Topological distances are derived by taking the reference bond distance between bonded atoms and the sum of these for the shortest path between two atoms that are not directly connected. This single atom description results in long computation times and is in general much too detailed for a fast comparison of molecular structures.

The hydrogen acceptor strength values have been derived from the force field [113,114], and the hydrophobic pharmacophore units are characterized by their size. The pharmacophore is further characterized by a distance matrix between pharmacophore units, which is derived from topological bond distances between all the atoms in each pharmacophore unit. The number of pharmacophore units, which are needed to represent a given molecule, is dependent on its size. Large molecules, therefore, have more pharmacophore units.

Fingerprints essentially monitor the occurrence of linear structural elements of different types. This approach can handle large data sets at high speed and is neither restricted to nor optimally suited for pharmacophoric applications.

11.6   CONSTITUTIONAL DESCRIPTORS/FUNCTIONAL GROUP COUNTS

The most important constitutional descriptors that have been used in modeling the milk/plasma partitioning are methyl; methylene; hydroxy; ether; nitro and amino group count; weight percent of C, H, and O in molecular mass; and molecular mass. Functional groups exhibit a characteristic reactivity and characteristic chemical behavior when present in a compound. Particular functional groups will be the most important for the specific interactions between a drug and a transporter receptor. Presumably, the functional group accounts for many of the dipole–dipole, dipole-induced dipole, and hydrogen bond interactions. Polar functional group counts accommodate additional interaction in polar and in hydrogen-bonding compounds. Hydrogen bonding can be facilitated by the presence of hydroxy or amino groups. Amino groups may account for higher solubility arising from the presence of protonated amine. It was found that the presence of amino and hydroxyl groups decreases drug transfer into milk, while the presence of an ether functional group increases drug transfer, perhaps because of an increase in liposolubility. Dipole interactions are related to the dipole moment of a whole molecule or a part of a molecule, such as a functional group (e.g., nitro groups). High charge-transfer properties (dipole, nitro group) decrease drug transfer into milk. The presence of a methyl group as an electron-donating group enhances nonpolar solubility. It is shown that the methyl substitutions have substantial effects on the proton affinity and increase the M/P ratio. The effect is opposite for the methylene group as it shows electron-withdrawing tendencies.

11.7 TOPOLOGICAL OR 2D SHAPE DESCRIPTORS

A current trend in quantitative structure–property/activity relationship (QSPR/QSAR) studies is the use of theoretical molecular descriptors that can be calculated directly from molecular structure. A variety of topological or shape descriptors have been developed as alternative descriptors in quantitative structure–activity studies for the characterization of molecular structure in combination with molecular dynamic analysis [115,116]. The advantage of these descriptors is that they can be calculated for any chemical structure, real or hypothetical. Topological shape descriptors (Kappa 1–Kappa 3) and connectivity indices (Chi 0–Chi 2) were found to be useful in effective quantification of molecular shape and bulk properties and therefore are important descriptors in modeling milk to plasma partitioning [117]. They are derived from different classes of weighted graphs, representing various levels of chemical structural information. Topological descriptors describe the atomic connectivity in the molecule, including valence and nonvalence molecular connectivity indices, calculated from the hydrogen-suppressed formula of the molecule, encoding information about the size, composition, and the degree of branching of a molecule. Topological indices are 2D theoretical descriptors that take into account the internal atomic arrangement of compounds and encode in numerical form information about molecular size, shape, branching, unsaturation, the number of heteroatoms, and cyclicity. They are suitable for describing similarity or dissimilarity of molecules. If two compounds have close values for a number of indices, they can be regarded as similar.

Topology refers to properties of the shape that do not change. The connectivity method directly correlates structural information to molecular activity, rather than indirectly through an intermediate physical property. The structure base of the molecular connectivity chi indices has created the potential to extract from QSAR equations sufficient information so that molecules may be designed directly from those equations. This objective has been a goal of those who use topological indices, because the purpose of those indices is to encode molecular structure information. The shape may not be entirely reconstructable from the descriptors, but the descriptors for different shapes should be different enough that the shapes are able to be discriminated.

Molecular connectivity [118] is a method of molecular structure quantification in which molecular size and shape are encoded with weighted counts of substructure fragments, weighted by the degree of skeletal branching. Low-order molecular connectivity indices are converted into path counts and later into counts of atom types (vertex degrees) and assembled directly into molecular structures [119,120]. Connectivity indices up to the fourth encode molecular density, branching, and aromatic ring substitutions.

Each carbon atom in a molecular skeleton is assigned a number according to the number of neighboring carbon atoms. The molecular skeleton is then fragmented into all its two carbon atom bonds. The sum of these values over the structure forms the Chi index. Molecules can be further dissected into two bond fragments, three bond fragments, and so on. Molecular structure is quantified so that weighted counts of substructure fragments are incorporated into numerical indices and an index is derived from a consideration of pairs of atoms forming bonds. Chi 0, zeroth order (atomic) connectivity index, conveys information about the number of atoms in a molecule. It is shown that the increase in Chi 0 decreases drug transfer to milk owing to the increase in the size of the skeleton (i.e., molecular size). The molecular connectivity index of the first order, Chi 1, encodes size and branching information. It is a weighted count of bonds, related to the types and position of branching in the molecule. The Chi 2 index encodes more specific information about skeletal branching. Chi 2 (path) is derived from fragments of two-bond length. It also provides information about types and position of branching and may be an indication of the amount of structural flexibility. Chi 1 increases with the number of atoms present but decreases with skeletal branching, whereas Chi 2 increases with both atom count and skeletal branching. An increase in Chi 1 (the first-order [bond] connectivity indices) and Chi 2 results in an increase in the M/P ratio.

An increase in branching increases surface area and molecular volume [121], resulting in an increase in solubility and a lower partition coefficient. A statistical analysis has shown that Chi 1 and Chi 2 are covariant to an extent. However, there is enough difference between the information in Chi 1 and Chi 2 to reflect structural features contributing in a different way to the numerical value. Chi 2 can differentiate between structural isomers, while Chi 1 values are identical. Low values of Chi 1 and Chi 2 are found for more elongated molecules or those with only one branching atom. An increase in the length of the carbon chain or the nonpolar portion of the molecule results in an increase in lipid solubility (log P) and an increase in molecular size.

Chi 3 is derived from three bonds' subgraphs with one atom common to all three. Among the molecules with tertiary carbons, those with terminal branching yield larger Chi 3 values than those with midchain branching. Thus, this index is quite specific for deriving relative numeric structural information about branching in the molecule. The valence connectivity index [122] uses the same invariant but modifies vertex degrees to account for heteroatoms by using the number of valence electrons in the corresponding atoms.

Topological Kappa indices [123] are the basis of a method of molecular structure quantification in which attributes of molecular shape and size are encoded into three indices (Kappa values 1–3). They model molecular shape and flexibility [124]. They are always used to predict if molecular cavities can be filled up with a candidate molecule. Steric shape relates to the ability of a drug molecule to be able to penetrate cell membranes and to bind in a receptor site and also influences molecular transport via protein binding that may act as a transport system for the drug and also limit access to the milk compartment. These shape indices (Kappa values) are derived from counts of one-bond, two-bond, and three-bond fragments, each count being made relative to fragment counts in reference structures that possess a maximum and minimum value for that number of atoms. The shape index of the first-order Kappa 1 encodes molecular cycles, Kappa 2 encodes linearity, and Kappa 3 encodes branching. It was shown that an increase in Kappa 1 and Kappa 2 decreases the M/P ratio because of an increase in molecular size and lipid solubility, while an increase in Kappa 3 (branching) promotes the transfer of the drug molecule into breast milk. Molecular branching decreases molecular size, decreases molecular length, increases molecular complexity, and perhaps decreases protein binding.

11.8   3D MOLECULAR DESCRIPTORS

11.8.1   ELECTRONIC EFFECTS

3D descriptors require geometry optimization, whereas 2D and 1D descriptors can be calculated directly from connectivity tables of molecules. Although solubility parameters (1D), topological shape, and connectivity indices (2D) are often successful in predicting solubilities and partition coefficients, they cannot account for conformational changes and they do not provide information about electronic influence through bonds or across space. Once the 3D molecular structure is optimized, in the minimum energy state, it is possible to calculate a number of molecular descriptors that mathematically characterize the molecule.

Electrostatic descriptors reflect the characteristics of the charge distribution of the molecule (e.g., max/min partial charge), the count of H acceptor sites, and the topographic electronic index. The quantum chemical descriptors provide information about binding and formation energies, partial atom charge, dipole moment, and molecular orbital energy levels (e.g., HOMO/LUMO energy), maximum/minimum electron–electron repulsion, and maximum/minimum electron–nuclear attraction. These electronic influences may play a role in the magnitude of biological activity, along with structural features encoded in indexes. Electronic effects are quantified explicitly by the use of molecular orbital calculations to estimate total energy, LUMO energy, electron affinity, and steric energy. These quantum chemical descriptors can give great insight into structure and reactivity and can be used to establish and compare the conformational stability, chemical reactivity, and intermolecular interactions.

The LUMO energy represents the electron affinity of a molecule or its reactivity as an electrophile. Good electrophiles are those where the electrons reside in low energy orbitals. Electrophiles are often reducing agents. Electron affinity [125] also incorporates electron correlation and relaxation, whereas LUMO does not, and is also a measure of the reduction capacity of the molecule. Since living organisms function at an optimum redox potential range, it is assumed that the redox potential of particular compounds may also be correlated with a particular biological effect. Note that an increase in molecular reactivity also results in an increase in the rate of metabolic processes.

The preferred structure of a molecule is obtained by minimizing the steric energy. For a given molecule, the atoms will adjust their positions by stretching and bending bonds away from standard values so as to produce a minimum energy configuration. Deviation from those standard values results in an increase in steric energy in the molecule. Molecular modeling software with molecular mechanics capability computes and displays an overall steric energy for an optimized structure as well as six or eight components of this steric energy. All of these components arise from essentially five phenomena: bond stretching, bond angle bending, bond rotation (torsion), van der Waals interactions, and electrostatic interactions. Calculation of the steric energy for different molecules allows for an assessment of the relative stability of those molecules. As expected, molecules with lower values of total and steric energies, lower LUMO energy, and higher electron affinity have a higher M/P ratio [63].

Electronic properties are more important factors in determining the binding affinity to HSA than the shape of the molecule. This is consistent with findings that HSA can bind to a large variety of compounds with different shapes and sizes [61]. Several QSAR models have identified charge distribution in a molecule [126,74], electrostatic interactions, and the presence and electron accessibility of certain molecular substituents [127] as important elements for HSA binding. These studies consistently suggested the importance of electronic descriptors such as electronegativity in the prediction of HSA binding.

11.8.2 ELECTROSTATIC DESCRIPTORS

Once the 3D molecular structure is optimized, in the minimum energy state, it is possible to calculate a number of molecular descriptors that mathematically characterize the molecule. It has long been agreed, especially in the modeling of biological systems, that the most relevant molecular descriptors are shape dependent. However, further studies have shown that molecular interactions between a protein (receptors or enzymes) and a drug, or a catalyst and its substrate, which produces an observed biological effect, are dominated by noncovalent (nonbonded) interactions. The two main components of such noncovalent interactions are steric or shape-dependent and electrostatic interactions. Therefore, noncovalent interactions that are responsible for the 3D configuration and that describe steric and electrostatic forces could describe more precisely a great variety of molecular properties. It seems reasonable that a suitable sampling of the steric and electrostatic fields surrounding a set of ligand (drug) molecules might provide all the information necessary for understanding their observed biological properties. In contrast to atom-based descriptors that only describe the magnitude of particular physical properties but no directional preferences that these properties may have, field-based descriptors describe the microenvironment surrounding the molecules. This approach to structure/activity correlation, also called Comparative Molecular Field Analysis (CoMFA), has pioneered a new paradigm of 3D QSAR studies, where properties of molecules are related to their specific structural and electronic features and their spatial arrangement. CoMFA looks at the molecule in 3D and describes the magnitude and directions of electronic and steric interactions [127].

The CoMFA approach for ligand QSAR is based on the assumption that non-covalent interactions affect the catalytic activity and therefore should correlate with the steric and electronic fields of these molecules. To develop the numerical representation of those fields, all the molecules under investigation are first structurally aligned and the steric and electrostatic fields around them are sampled with probe atoms. This is done by moving a positively charged sp^3 carbon atom on a rectangular grid that encompasses the aligned molecules. In most cases, the molecular field is developed from the quantum-chemically calculated atomic partial charges of the investigated molecule. The result of the molecular alignment is a schematic representation similar to that of the pharmacophore in drug design. Results can be mapped into 3D space and special distributions of properties that are related to the activity can be localized. Calculated descriptors, derived solely from 3D structures of the molecules, reflect the characteristics of charge distribution in the molecule. As expected, they correlate with the octanol–water partition coefficient [128]. However, CoMFA models are not successful in studying the highly heterogeneous data sets because of the problem of the compounds' alignment. A model developed by Zhao et al. [129] identified the electronegativity-based maximum partial charge for a C atom $[PC_{max}(C)]$ and the minimum coulombic interaction for a C–C bond $[CI_{min}(CC)]$ as important descriptors in modeling the passage of drugs into breast milk. Maximum partial charge on the C atom in a molecule reflects the characteristics of the charge distribution of the molecule. The empirical partial charges in the molecule are calculated using the approach proposed by Zefirov et al. [130]. This method is based on the Sanderson electronegativity scale and uses the concept that represents molecular electronegativity as a geometric mean of atomic electronegativities. These electrostatic descriptors denote that the affinity of the molecules in the electrostatic interaction minimum electron–electron repulsion for a C–C bond is one of the quantum chemical descriptors used to establish conformational stability, chemical reactivity, and intermolecular interactions. The descriptor characterizes the nuclear repulsion-driven processes in the molecule and may be related to the conformational (rotational, inversional) changes or atomic reactivity in the molecule. The energy of the electrostatic interactions between the chemically bonded atoms is important in determining conformational change within the molecule, which, in turn, may affect protein bonding and therefore diffusion of the drug in human breast milk. Consequently, this descriptor may also be related to log M/P values.

TABLE 11.1
Molecular Descriptors for M/P Ratio Prediction

Molecular Property	Descriptors
Acid–base characteristics/ionization prediction	Degree of ionization (pKa)
Lipophilicity and hydrophilicity	Octanol/water partitioning (log P)
Solubility and solvate descriptors	Abraham descriptors
Molecular size and shape	Molecular weight (MW), molecular volume, molecular surface (total surface area [TSA], solvent-accessible surface area [SASA]), molar refractivity (MR), effective molecular diameter
Hydrogen bonding	Polar surface area (PSA), hydrogen-bonding donor capacity, hydrogen-bonding acceptor capacity
Plasma protein binding	Constitutional descriptors/functional group counts, Descriptors based on molecular similarity (topological pharmacophore [TPR] description, daylight fingerprint description [DFP])
Topology or 2D molecular shape	Topological shape descriptors (Kappa 1–Kappa 3), Connectivity indices (Chi 0–Chi 2)
3D molecular shape	Electronic properties (electron affinity and lowest unoccupied molecular orbital [LUMO] energy), Electrostatic descriptors (comparative molecular field analysis [CoMFA] approach)

11.9   CONCLUSION

The use of computational methods in the QSAR analysis is gaining more and more attention because of two main reasons. First, the number of chemicals in use is growing rapidly, followed by the multitude of information required in the research and development and by the legislations and regulatory bodies for the risk assessment. Second, the number of chemicals in use is countless (enormous), and although the human experience is invaluable, the amount of required information for the safe use of drugs and chemicals is growing rapidly. Thus, computational methods have been investigated to get insight into many biological phenomena related to drug design, development, and risk assessment. A number of recent studies have confirmed the utility of the theoretical molecular descriptors and the QSAR models derived on their basis for the effective prediction of complex biomedical molecular properties such as partitioning into breast milk (Table 11.1). Although the classification for M/P ratios is only one component of the complex process for a drug transfer to human milk, it probably forms the most selective filter in drug screening. Computational methods are expected to have a similar effect on the search for drugs and chemicals whose M/P ratios are not currently available.

REFERENCES

1. P.J. McNamara and S. Ito, *Drug excretion in breast milk: Mechanisms, models and drug delivery implications for the infant*, Adv. Drug Deliv. Rev. 55 (2003), pp. 615–616.

2. S. Ito and A. Lee, *Drug excretion into breast milk*, Adv. Drug Deliv. Rev. 55 (2003) pp. 617–627.
3. J.L. McManaman and M.C. Neville, *Mammary physiology and milk secretion*, Adv. Drug Deliv. Rev. 55 (2003), pp. 629–641.
4. J.S. LaKind, A. Wilkins, and C.M. Berlin Jr., *Environmental chemicals in human milk: A review of levels, infant exposures and health, and guidance for future research*, Toxicol. Appl. Pharmacol. 198 (2004), pp. 184–208.
5. D. Costopoulou, I. Vassiliadou, A. Papadopoulos, V. Makropoulos, and L. Leondiadis, *Levels of dioxins, furans and PCBs in human serum and milk of people living in Greece*, Chemosphere, 65 (2006), pp. 1462–1469.
6. WHO (World Health Organization), *Levels of PCBs, PCDDs and PCDFs in Human Milk. Protocol for Third Round of Exposure Studies*, World Health Organization European Centre for Environment and Health, Bilthoven, 2000.
7. M. de Martino, P.A. Tovo, A.E. Tozzi, P. Pezzotti, L. Galli, S. Livadiotti, D. Caselli, E. Massironi, E. Ruga, F. Fioredda, A. Plebani, C. Gabiano, G. Zuccotti, and Y. Gian, *HIV-1 transmission through breast-milk: Appraisal of risk according to duration of feeding*, AIDS 6 (1992), pp. 991–997.
8. S. Ito and G. Koren, *A novel index for expressing exposure of infants to drugs in breast milk*, Rev. J. Clin. Pharmacol. 38 (1994), pp. 99–102.
9. J.C. Fleishaker, N. Desai, and P.J. McNamara, *Factors affecting the milk-to-plasma drug concentration ratio in lactating women: Physical interactions with protein and fat*, J. Pharm. Sci. 76 (1987), pp. 189–193.
10. L.J. Notarianni, D. Belk, S.A. Aird, and P.N. Bennett, *An in vitro technique for rapid determination of drug entry into breast milk*, Br. J. Clin. Pharmacol. 40 (1995), pp. 333–337.
11. L. Beaulac-Baillargeon, A. Auclair, L. Matte, R.C. Gaudreault, and J.-C. Forest, *A novel approach for the determination of human milk/plasma ratios: Correlation of in vivo and in vitro paracetamol milk/plasma ratios*, Drug Invest. 7 (1994), pp. 57–62.
12. P.J. McNamara, D. Burgio, and S.D. Yoo, *Pharmacokinetics of caffeine and its demethylated metabolites in lactating rabbits and neonatal offspring. Predictions of breast milk to serum concentration ratios*, Drug Metab. Dispos. 20 (1992), pp. 302–308.
13. R. Mannhold and A. Petrauskas, *Substructure versus whole-molecule approaches for calculating log P*, QSAR Comb. Sci. 22 (2003), pp. 466–475.
14. E. Benfenati, G. Gini, N. Piclin, A. Roncaglioni, and M.R. Varì, *Predicting log P of pesticides using different software*, Chemosphere 53 (2003), pp. 1155–1164.
15. J. Devillers and A.T. Balaban (eds.), *Topological Indices and Related Descriptors in QSAR and QSPR*, Gordon Breach Sci. Pub., Amsterdam, 1999.
16. M. Karelson, *Molecular Descriptors in QSAR/QSPR*, Wiley-InterScience, New York, 2000.
17. R. Todeschini and V. Consonni, *Handbook of Molecular Descriptors*, Wiley–VCH, Weinheim, Germany, 2000.
18. D.R. Hennessy and M.R. Alvinerie, *Pharmacokinetics of the macrocyclic lactones: Conventional wisdom and new paradigms*, in *Macrocyclic Lactones in Antiparasitic Therapy*, J. Vercruysse and R.S. Rew, eds., CAB International Publishing, Wallingford, UK, 2002, pp. 97–123.
19. F. Rasmussen, *Excretion of drugs by milk*, in *Concepts in Biochemical Pharmacology, Part 1, Handbook of Experimental Pharmacology*, Vol. 28, B.B. Brodie and J.R. Gillette, eds., Springer-Verlag, Berlin, New York, 1971, pp. 390–402.
20. D.C. Pang, G.L. Amidon, P.C. Preusch, and W. Sadee, *Meeting report. Second AAPS-NIH frontier symposium 1999: Membrane transporters and drug therapy*, AAPS Pharm. Sci. 2 (1999), p. 5.

21. J.Z. Byczkowski, J.M. Gearhart, and J.W. Fisher, *"Occupational" exposure of infants to toxic chemicals via breast milk*, Nutrition 10 (1994), pp. 43–48.

22. C.Y. Oo, R.J. Kuhn, N. Desai, and P.J. McNamara, *Active transport of cimetidine into human milk*, Clin. Pharmacol. Ther. 58 (1995), pp. 548–555.

23. P.J. McNamara, J.A. Meece, and E. Paxton, *Active transport of cimetidine and ranitidine into the milk of Sprague Dawley rats*, J. Pharmacol. Exp. Ther. 277 (1996), pp. 1615–1621.

24. P.M. Gerk, R.J. Kuhn, N.S. Desai, and P.J. McNamara, *Active transport of nitrofurantoin into human milk*, Pharmacotherapy 21 (2001), pp. 669–675.

25. F.W. Kari, R. Weaver, and M.C. Neville, *Active transport of nitrofurantoin across the mammary epithelium in vivo*, J. Pharmacol. Exp. Ther. 280 (1997), pp. 664–668.

26. P.M. Gerk, C.Y. Oo, E.W. Paxton, J.A. Moscow, and P.J. McNamara, *Interactions between cimetidine, nitrofurantoin, and probenecid active transport into rat milk*, J. Pharmacol. Exp. Ther. 296 (2001), pp. 175–180.

27. A.M. Schadewinkel-Scherkl, F. Rasmussen, C.C. Merck, P. Nielsen, and H.H. Frey, *Active transport of benzylpenicillin across the blood-milk barrier*, Pharmacol. Toxicol. 73 (1993), pp. 14–19.

28. A. Seelig, *A general pattern for substrate recognition by P-glycoprotein*, Eur. J. Biochem. 251 (1998), pp. 252–261.

29. H. Vorherr, *The Breast: Morphology, Physiology, and Lactation*, Academic Press, New York, 1974, pp. 1–124.

30. J.C. Allen and M.C. Neville, *Ionized calcium in human milk determined with a calcium-sensitive electrode*, Clin. Chem. 29 (1983), pp. 858–861.

31. C. Hansch, A. Leo, and R.W. Taft, *A survey of Hammett substituent constants and resonance and field parameters*, Chem. Rev. 91 (1991), pp. 165–195.

32. J.C. Fleishaker, *Models and methods for predicting drug transfer into human milk*, Adv. Drug Deliv. Rev. 55 (2003), pp. 643–652.

33. H.C. Atkinson and E.J. Begg, *Relationship between human milk lipid–ultrafiltrate and octanol–water partition coefficients*, J. Pharm. Sci. 77 (1988), pp. 796–798.

34. J.T. Wilson, R.D. Brown, D.R. Cherek, J.W. Dailey, B. Hilman, P.C. Jobe, B.R. Manno, J.E. Manno, H.M. Redetzki, and J.J. Stewart, *Drug excretion in human breast milk: Principles, pharmacokinetics and projected consequences*, Clin. Pharmacokinet. 5 (1980), pp. 1–66.

35. H. Liedholm, A. Melander, P.O. Bitzén, G. Helm, G. Lönnerholm, I. Mattiason, B. Nilsson, and E. Wåhlin-Boll, *Accumulation of atenolol and metoprolol in human breast milk*, Eur. J. Clin. Pharmacol. 20 (1981), pp. 229–231.

36. F. Rasmussen, *Mammary excretion of sulphonamides*, Acta Pharmcol. Toxicol. 15 (1958), pp. 139–148.

37. J.T. Wilson, R.D. Brown, J.L. Hinson, and J.W. Dailey, *Pharmacokinetic pitfalls in the estimation of the breast milk/plasma ratio for drugs*, Ann. Rev. Pharmacol. Toxicol. 25 (1985), pp. 667–689.

38. D.I. George and T.J. O'Toole, *A review of drug transfer to the infant by breast-feeding: Concerns for the dentist*, JADA 106 (1983), pp. 204–208.

39. J.L. Abboud and R.W. Taft, *Regarding a generalized scale of solvent polarities*, J. Am. Chem. Soc. 99 (1977), pp. 8325–8327.

40. R.W. Taft, M.H. Abraham, G.R. Famini, R.M. Doherty, J.L.M. Abboud, and M.J. Kamlet, *Solubility properties in polymers and biological media. 5. An analysis of the physicochemical properties which influence octanol water partition-coefficients of aliphatic and aromatic solutes*, J. Pharm. Sci. 74 (1985), pp. 807–814.

41. W.E. Acree Jr., J.R. Powell, M.E.R. McHale, S. Pandey, T.L. Borders, and S.W. Campbell, *Thermodynamics of mobile order theory*, in *Research Trends in Physical Chemistry*, Vol. 6, Council of Scientific Research Integration, TriVandrum, India, 1997, pp. 197–233.

42. A.R. Katritzky, A.A. Oliferenko, P.V. Oliferenko, R. Petrukhin, D.B. Tatham, U. Maran, A. Lomaka, and W.E. Acree, *General treatment of solubility. 2. QSPR prediction of free energies of solvation of specified solutes in ranges of solvents*, J. Chem. Inf. Comput. Sci. 43 (2003), pp. 1806–1814.

43. M.J. Kamlet and R.W. Taft, *Solvatochromic comparison method. 1. The β-scale of solvent hydrogen-bond acceptor (HBA) basicities*, J. Am. Chem. Soc. 98 (1976), pp. 377–383.

44. M.H. Abraham, *Physicochemical and biological processes*, Chem. Soc. Rev. 22 (1993), pp. 73–83.

45. M.H. Abraham, J. Gil-Lostes, and M. Fatemi, *Prediction of milk/plasma concentration ratios of drugs and environmental pollutants*, Eur. J. Med. Chem. 44 (2009), pp. 2452–2458.

46. M.H. Abraham and J. Le, *The correlation and prediction of the solubility of compounds in water using an amended solvation energy relationship*, J. Pharm. Sci. 88 (1999), pp. 868–880.

47. M.H. Abraham and P.K. Weathersby, *Hydrogen bonding. XXX. Solubility of gases and vapors in biological liquids and tissues*, J. Pharm. Sci. 83 (1994), pp. 1450–1456.

48. M.H. Abraham, F. Martins, and R.C. Mitchell, *Algorithms for skin permeability using hydrogen bond descriptors: The problem of steroids*, J. Pharm. Pharmacol. 49 (1997), pp. 858–865.

49. B.B. Brodie and C.A.M. Hogben, *Some physico-chemical factors in drug action*, J. Pharm. Pharmacol. 9 (1957), pp. 345–380.

50. J.C. Dearden, *Partitioning and lipophilicity in quantitative structure–activity relationships*, Environ. Health Perspect. 61 (1985), pp. 203–228.

51. G.R. Famini and C.A. Penski, *Using theoretical descriptors in quantitative structure-activity relationships-some physicochemical properties*, J. Phys. Org. Chem. 5 (1992), pp. 395–408.

52. M. Heaberlein and T. Brinck, *Prediction of water-octanol partition coefficients using theoretical descriptors from the molecular surface area and the electrostatic potential*, J. Chem. Soc. Perkin Trans. 2 (1997), pp. 289–294.

53. U. Norinder, T. Osterberg, and P. Artursson, *Theoretical calculation and prediction of intestinal absorption of drugs in humans using MolSurf parametrization and PLS statistics*, Eur. J. Pharm. Sci. 8 (1999), pp. 49–56.

54. J.K. Seydel and K.J. Schaper, *Quantitative structure-pharmacokinetic relationships and drug design*, Pharmacol. Ther. 15 (1981), pp. 131–182.

55. E. Escribano, A.C. Calpena, T.M. Garrigues, J. Freixas, J. Domenech, and J. Moreno, *Structure-absorption relationships of a series of 6-flouroquinolones*, Antimicrob. Agents Chemother. 41 (1997), pp. 1996–2000.

56. W.J. Egan, K.M. Merz, and J.J. Baldwin, *Prediction of drug absorption using multivariate statistics*, J. Med. Chem. 43 (2000), pp. 3867–3877.

57. E.J. Begg and H.C. Atkinson, *Modelling of the passage of drugs into milk*, Pharm. Ther. 59 (1993), pp. 301–310.

58. P.E. Macheras, M.A. Koupparis, and S.G. Antimisaris, *Drug binding and solubility in milk*, Pharm. Res. 7 (1990), pp. 537–541.

59. G.B. Syversen and S.K. Ratkje, *Drug distribution within human milk phases*, J. Pharm. Sci. 74 (1988), pp. 1071–1074.

60. G. Colmenarejo, A. Alvarez-Pedraglio, and J.L. Lavandera, *Cheminformatic models to predict binding affinities to human serum albumin*, J. Med. Chem. 44 (2001), pp. 4370–4378.

61. G. Colmenarejo, *In silico prediction of drug-binding strengths to human serum albumin*, Med. Res. Rev. 23 (2003), pp. 275–301.

62. C.W. Yap and Y.Z. Chen, *Quantitative structure–pharmacokinetic relationships for drug distribution properties by using general regression neural network*, J. Pharm. Sci. 94 (2005), pp. 153–168.

63. S. Agatonovic-Kustrin, L.H. Ling, S.Y. Tham, and R.G. Alany, *Molecular descriptors that influence the amount of drugs transfer into human breast milk*, J. Pharm. Biomed. Anal. 29 (2002), pp. 103–119.

64. J.J. Morris and P.P. Bruneau, *Prediction of physicochemical properties*, in *Virtual Screening for Bioactive Molecules*, H.-J. Böhm and G. Schneider, eds., Wiley-VCH Verlag, Weinheim, GmbH, 2000, pp. 33–56.

65. H. van de Waterbeemd, D.A. Smith, and B.C. Jones, *Lipophilicity in PK design: Methyl, ethyl, futile*, J. Comput. Aided Mol. Des. 15 (2001), pp. 273–286.

66. W. Scholtan, *Hydrophobic binding of drugs to human albumin and ribonucleic acids*, Arzneimittelforschung 18 (1968), pp. 505–517.

67. W. Scholtan, *Methods of determination and theoretical principles of the serum protein binding of drugs*, Arzneimittelforschung 28 (1978), pp. 1037–1047.

68. J.K. Seydel and K.-J. Schaper, *Quantitative Struktur-Pharmakokinetik Beziehungen*, in *Chemische struktur und biologische aktivitaet von wirkstoffen, methoden der quantitativen struktur-wirkungs-analyse*, J.K. Seydel and K.-J. Schaper, eds., Wiley-VCH Verlag, Weinheim, GmbH, 1979, p. 223.

69. K. Yamazaki and M. Kanaoka, *Computational prediction of the plasma protein-binding percent of diverse pharmaceutical compounds*, J. Pharm. Sci. 93 (2004), pp. 1480–1494.

70. R. Saiakhov, L.R. Stefan, and G. Klopman, *Multiple computer-automated structure evaluation model of the plasma protein binding affinity of diverse drugs*, Perspect. Drug Discov. Des. 19 (2000), pp. 133–155.

71. F. Beaudry, M. Coutu, and N.K. Brown, *Determination of drug-plasma protein binding using human serum albumin chromatographic column and multiple linear regression model*, Biomed. Chromatogr. 13 (1999), pp. 401–406.

72. L.M. Hall, L.H. Hall, and L.B. Kier, *Modeling drug albumin binding affinity with E-State topological structure representation*, J. Chem. Inf. Comput. Sci. 43 (2003), pp. 2120–2128.

73. P.J. Hajduk, R. Mendoza, A.M. Petros, J.R. Huth, M. Bures, S.W. Fesik, and Y.C. Martin, *Ligand binding to domain-3 of human serum albumin: A chemometric analysis*, J. Comput.-Aided Mol. Des. 17 (2003), pp. 93–102.

74. N.A. Kratochwil, W. Huber, F. Muller, M. Kansy, and P.R. Gerber, *Predicting plasma protein binding of drugs: A new approach*, Biochem. Pharmacol. 64 (2002), pp. 1355–1374.

75. K. Valko, S. Nunhuck, C. Bevan, M.H. Abraham, and D.P. Reynolds, *Fast gradient HPLC method to determine compounds binding to human serum albumin. Relationships with octanol/water and immobilized artificial membrane lipophilicity*, J. Pharm. Sci. 92 (2003), pp. 2236–2248.

76. R.J. Votano, M. Parham, L.M. Hall, L.H. Hall, L.B. Kier, S. Oloff, and A. Tropsha, *QSAR modeling of human serum protein binding with several modeling techniques utilizing structure-information representation*, J. Med. Chem. 49 (2006), pp. 7169–7181.

77. U. Welsch, W. Buchheim, U. Schumacher, I. Schinko, and S. Patton. *Structural, histochemical and biochemical observations on horse milk-fat-globule membranes and casein micelles*, Histochemistry 88 (1988), pp. 357–365.

78. U.S. Environmental Protection Agency, *Exposure Factors Handbook, Vol. I, General Factors*, EPA Publication No. EPA/600/P-95/002Fa, United States Environmental Protection Agency, Office of Environmental Information, Washington, DC, 1997. Available at http://www.epa.gov/ncea/pdfs/efh/front.pdf.

79. P.O. Anderson, *Therapy review: Drug use during breast feeding*, Clin. Pharmacokinet. 10 (1991), pp. 594–624.

80. J.T. Wilson, *Pharmacokinetics of drug excretion*, in *Drugs in Breast Milk*, J.T. Wilson, ed., ADIS Press, Australia, 1981.

81. J.T. Wilson, J.L. Hinson, R.D. Brown, and I.J. Smith. *A comprehensive assessment of drugs ad chemical toxins excreted in breast milk*, in *Human Lactation 2: Maternal and Environmental Factors*, M. Hamosh and A.S. Goldman, eds., Plenum Press, New York, 1986.

82. J.A. Mennella and G.K. Beauchamp, *The transfer of alcohol to human milk. Effects on flavor and the infant's behaviour*, N. Engl. J. Med. 325 (1991), pp. 981–985.

83. C.M. Berlin, *Pharmacologic considerations of drug use in the lactating mother*, Obstet. Gynecol. 58 (1981), pp. 17S–23S.

84. C.M. Berlin, *Drug excretion into human milk*, in *Human Lactation 2: Maternal and Environmental Factors*, M. Hamosh and A.S. Goldman, eds., Plenum Press, New York, 1986.

85. G. Pons, E. Rey, and I. Matheson, *Excretion of psychoactive drugs into breast milk: Pharmacokinetic principles and recommendations*, Clin. Pharmacokinet. 27 (1994), pp. 270–289.

86. I.B. Frederick, R.J. White, and S.W. Braddock, *Excretion of varicella-herpes zoster virus in breast milk*, Am. J. Obstet. Gynecol. 154 (1986), pp. 1116–1117.

87. R.F. Black, *Transmission of HIV-1 in the breast-feeding process*, J. Am. Diet. Assoc. 96 (1996), pp. 267–274.

88. M.S. Meskin and E.J. Lien, *QSAR analysis of drug excretion into human breast milk*, J. Clin. Hosp. Pharm. 10 (1985) pp. 269–278.

89. S.D. Dimitrov, N.C. Dimitrova, J.D. Walker, G.D. Veith, and O.G. Mekenyan, *Predicting bioconcentration factors of highly hydrophobic chemicals. Effects of molecular size*, Pure Appl. Chem. 74 (2002), pp. 1823–1830.

90. S. Glasstone, *Textbook of Physical Chemistry*, 2nd Ed., D. VanNostrand, New York, 1946, p. 529.

91. J.A. Padrón, R. Carrasco, and R.F. Pellón, *Molecular descriptor based on a molar refractivity partition using Randic-type graph-theoretical invariant*, J. Pharm. Pharm. Sci. 5 (2002), pp. 258–266.

92. M.I. Bykov, *Calculation of specific and molar refraction of hydrocarbons*, Chem. Technol. Fuels. Oils 20 (1984), pp. 310–312.

93. M.R. Riazi and C.H. Whitson, *Estimating diffusion-coefficients of dense fluids*, Ind. Eng. Chem. Res. 32 (1993), pp. 3081–3088.

94. A. Carotti, R.N. Smith, S. Wong, C. Hansch, J.M. Blaney, and R. Langridge, *Papain hydrolysis of X-phenyl-N-methanesulfonyl glycinates: A quantitative structure-activity relationship and molecular graphics analysis*, Arch. Biochem. Biophys. 229 (1984), pp. 112–125.

95. R.B. Hermann, *Modeling hydrophobic solvation of nonspherical systems: Comparison of use of molecular surface area with accessible surface area*, J. Comput. Chem. 18 (1997), pp. 115–125.

96. N. Bodor and M.J. Huang, *A new method for the estimating of the aqueous solubility of organic compounds*, J. Pharm. Sci. 81 (1992), pp. 954–960.

97. T. Ooi, M. Oobatake, G. Némethy, and H.A. Scheraga, *Accessible surface areas as a measure of the thermodynamic parameters of hydration of peptides*, Proc. Natl. Acad. Sci. U.S.A. 84 (1987), pp. 3086–3090.

98. W.J. Dunn, M.G. Koehler, and S. Grigoras, *The role of solvent-accessible surface area in determining partition coefficients*, J. Med. Chem. 30 (1987), pp. 1121–1126.

99. S. Winiwarter, N.M. Bonham, F. Ax, A. Hallberg, H. Lennernaes, and A. Karlen, *Correlation of human jejunal permeability (in vivo) of drugs with experimentally and theoretically derived parameters. A multivariate data analysis approach*, J. Med. Chem. 41 (1998), pp. 4939–4949.

100. K. Palm, P. Stenberg, K. Luthman, and P. Artursson, *Polar molecular surface properties predict the intestinal absorption of drugs in humans*, Pharm. Res. 14 (1997), pp. 568–571.

101. O.A. Raevsky, V.I. Fetisov, E.P. Trepalina, J.W. McFarland, and K.J. Schaper, *Quantitative estimation of drug absorption in humans for passively transported compounds on the basis of their physico-chemical parameters*, Quant. Struct.-Act. Rel. 19 (2000), pp. 366–374.

102. G.E. Miller, N.C. Banerjee, and C.M. Stowe, *Diffusion of certain weak organic acids and bases across the bovine mammary gland membrane after systemic administration*, J. Pharmacol. Exp. Ther. 157 (1967), pp. 245–253.

103. G.E. Miller, R.D. Peters, R.V. Engebretsen, and C.M. Stowe, *Passage of pentobarbital and phenobarbital into bovine caprine milk after systemic administration*, J. Dairy Sci. 50 (1967), pp. 769–772.

104. H.C. Atkinson and E.J. Begg, *Prediction of drug distribution into human milk from physicochemical characteristics*, Clin. Pharmacokinet. 18 (1990), pp. 151–167.

105. L.M. Hall, L.H. Hall, and L.B. Kier, *QSAR modeling of beta-lactam binding to human serum protein*, J. Comput. Aided Mol. Des. 10 (2003), pp. 103–118.

106. P.R. Gerber, *Topological pharmacophore description of chemical structures using MAB-force-field derived data and corresponding similarity measures*, in *Fundamentals of Molecular Similarity*, R. Carbó-Dorca, X. Gironés, and P.G. Mezey, eds., Kluwer Academic/Plenum Publishers, New York, 2001, pp. 67–82.

107. R.A. Jarvis, *Clustering using a similarity measure based on shared nearest neighbours*, IEEE Trans. Comput. C 22 (1973), pp. 1025–1034.

108. A.T. Hopfinger, S. Wang, J.S. Tobarski, B. Jin, M. Albuquerque, P.J. Madhav, and C. Duraiswami, *Construction of 3D-QSAR models using the 4D-QSAR analysis formalism*, J. Am. Chem. Soc. 119 (1997), pp. 10509–10524.

109. A.J. Hopfinger, A. Reaka, P. Venkatarangan, J.S. Duca, and S. Wang, *Construction of a virtual high throughput screen by 4D-QSAR analysis: Application to a combinatorial library of glucose inhibitors of glycogen phosphorylase b*, J. Chem. Inf. Comput. Sci. 39 (1999), pp. 1151–1160.

110. G.M. Maggiora and M.A. Johnson, *Concepts and Applications of Molecular Similarity*, Wiley, New York, 1990, pp. 99–117.

111. P.R. Gerber, Roche Internal Report B174895, 1995.

112. K. Mueller, H.J. Ammann, D.M. Doran, P.R. Gerber, K. Gubernator, and G. Schrepfer, *Complex heterocyclic structures—A challenge for computer-assisted molecular modeling*, Bull. Soc. Chim. Belg. 97 (1988), pp. 655–667.

113. P.R. Gerber and K. Müller, *MAB, a generally applicable molecular force field for structure modelling in medicinal chemistry*, J. Comput. Aided. Mol. Dis. 9 (1995), pp. 251–268.

114. P.R. Gerber, *Charge distribution from a simple molecular orbital type calculation and non-bonding interaction terms in the force field MAB*, J. Comput. Aided. Mol. Dis. 12 (1998), pp. 37–51.

115. G. Grassy, B. Calas, A. Yasri, R. Lahana, J. Woo, S. Iyer, M. Kaczorek, R. Floc'h, and R. Buelow, *Computer-assisted rational design of immunosuppressive compounds*, Nat. Biotechnol. 16 (1998), pp. 748–752.

116. D. Gorse, A. Rees, M. Kaczorek, and R. Lahana, *Molecular diversity and its analysis*, Drug Discov. Today 4 (1999), pp. 257–264.

117. S. Agatonovic-Kustrin, R. Beresford, and A.P.M. Yusof, *Theoretically derived molecular descriptors important in human intestinal absorption*, J. Pharm. Biomed. Anal. 25 (2001), pp. 227–237.

118. L.H. Hall and L.B. Kier, *The molecular connectivity chi indexes and kappa shape indexes in structure-property modeling*, in *Reviews of Computational Chemistry*, Volume 2, Chap. 9, D. Boyd and K. Lipkowitz, eds., VCH Publishers, Inc., New York, 1991, pp. 367–422.

119. L.H. Hall and L.B. Kier, *Design of molecules from quantitative structure-activity relationship models. 3. Role of higher order path counts: Path three*, J. Chem. Inf. Comput. Sci. 33 (1993), pp. 598–603.

120. L.B. Kier and L.H. Hall, *The generation of molecular structures from a graph-based equation*, Quant. Struct.-Act. Relat. 12 (1993), pp. 383–388.

121. A. Verloop, W. Hoogenstraaten, and J. Tysker, *Chapter 4, Development and Application of New Steric Substituent Parameters in Drug Design*, Vol. 7., E.J. Ariens, ed., Academic Press, New York, 1976.

122. E. Estrada, *Generalization of topological indices*, Chem. Phys. Lett. 336 (2001), pp. 248–252.

123. V.K. Gombar and D.V.S. Jain, *Quantification of molecular shape and its correlation with physico-chemical properties*, Indian J. Chem. 26A (1987), pp. 554–555.

124. A. Kovatcheva, G. Buchbauer, A. Golbraikh, and P. Wolschann, *QSAR modeling of α-campholenic derivatives with sandalwood odor*, J. Chem. Inf. Comput. Sci. 43 (2003), pp. 259–266.

125. S. Trohalaki and R. Pachter, *The utility of electron affinity as a QSAR descriptor in predictive toxicology*, Abstr. Pap. Am. Chem. S. 217 (1999), p. U674.

126. S.L. Rodgers, A.M. Davis, and H. van de Waterbeem, *Time-series QSAR analysis of human plasma protein binding data*, QSAR Comb. Sci. 26 (2007), pp. 511–521.

127. R.D. Cramer, D.E. Patterson, and J.D. Bunce, *Comparative molecular field analysis (CoMFA). 1. Effect of shape on binding of steroids to carrier proteins*, J. Am. Chem. Soc. 110 (1988), pp. 5959–5967.

128. M. Safdaria and H. Golmohammadi, *Prediction of n-octanol–water partition coefficient for polychlorinated biphenyls from theoretical molecular descriptors*, Eur. J. Chem. 1 (2010), pp. 266–275.

129. C. Zhao, H. Zhang, X. Zhang, R. Zhang, F. Luan, M. Liu, Z. Hu, and B. Fan, *Prediction of milk/plasma drug concentration (M/P) ratio using support vector machine (SVM) method*, Pharm. Res. 23 (2006), pp. 41–48.

130. N.S. Zefirov, M.A. Kirpichenok, F.F. Izmailov, and M.I. Trofimov, *Scheme for the calculation of the electronegativities of atoms in a molecule in the framework of Sanderson's Principle*, Dokl. Akad. Nauk SSSR. 296 (1987), pp. 883–887.

12 Approaches for Pediatric Developmental Drug-Induced Liver Injury*

William F. Salminen and Xi Yang

CONTENTS

---

* This chapter is not an official guidance or policy statement of the US Food and Drug Administration (FDA). No official support or endorsement by the US FDA is intended or should be inferred.

ABSTRACT

Children are not simply small adults and it follows that children may exhibit differential sensitivity to drug-induced adverse events. This also applies to drug-induced liver injury (DILI). As an embryo develops, leading to the birth of a child, and eventually maturation into an adult, the human body goes through many different developmental phases. Various factors may make the developing human more or less susceptible to DILI when compared to adults. Drug-induced hepatotoxicity in children is commonly presented as acute hepatitis, although almost any major clinical pathological pattern of liver disease can occur. In most cases, children spontaneously recover after the offending medication is discontinued. The differential DILI responses observed between children and adults can be partly explained by the developmental changes of absorption, distribution, metabolism, and excretion. This article reviews the major developmental phases of the maturing liver with an emphasis on phases that may pose unique sensitivities to DILI. A review of the pediatric DILI literature is then presented so that pediatricians can better understand the potential hepatotoxicity risks associated with some drugs, such as antituberculous and antiepileptic medications.

KEYWORDS

Pediatric, liver, hepatotoxicity, development, drug-induced liver injury

12.1 INTRODUCTION

Children are not simply small adults and it follows that children may exhibit differential sensitivity to drug-induced adverse events. This also applies to drug-induced liver injury (DILI). As an embryo develops, leading to the birth of a child, and eventually maturation into an adult, the human body goes through many different developmental phases. Various factors may make the developing human more or less susceptible to DILI when compared to adults. This chapter reviews the major developmental phases of the maturing liver with an emphasis on phases that may pose unique sensitivities to DILI. A review of the pediatric DILI literature is then presented.

DILI is one of the leading causes of drug attrition during development and is one of the leading causes of postmarketing drug withdrawal [1–4]. Although every drug goes through extensive preclinical (animal) and clinical testing, idiosyncratic DILI has proven to be very difficult to predict. Current testing paradigms often weed out drugs that have a clear threshold of toxicity and elicit clear and consistent clinical signs (e.g., acetaminophen). The major problem is with drugs that cause rare but serious hepatotoxicity. Often these drugs showed no to minimal signs of hepatotoxicity in a wide range of species and at doses much higher than the therapeutic dose. It isn't until these drugs reach the market that clear signs of clinical hepatotoxicity are observed. Therefore, better models are needed to screen out these types of drugs or at least identify their risk so that they can be used safely.

Pediatric safety of drugs has been recognized as an underserved area of research and drug development. It was only recently that the US Food and Drug Administration (FDA) was required by law to specifically consider the pediatric safety of drugs under the Pediatric Research Equity Act (Public Law 108-155) of 2003 [5]. Before this, drug registrants were not required to assess the safety of drugs in the pediatric population, which has resulted in a paucity of data concerning DILI in children. Some regulatory bodies, such as the US Environmental Protection Agency (EPA) under the Food Quality Protection Act of 1996 (FQPA), assume that children are significantly more sensitive than adults to chemical exposures unless there are clear data to the contrary [6]. In risk assessments for products such as pesticides, the EPA uses an additional 10-fold margin of safety to protect infants and children from potential pre- and postnatal toxicity. The 10-fold FQPA safety factor can be modified only if reliable data demonstrate that the resulting level of exposure would be safe for infants and children. The EPA's approach is a mixture of science (children may or may not be more susceptible depending on the chemical and type of exposure) and public health policy (assuming by default that children are always more sensitive than adults). This chapter focuses on the scientific assumptions and data concerning children and DILI and does not address public health policy decisions.

12.2 THE DEVELOPING HUMAN LIVER

The adult human liver serves many essential functions including, but not limited to, carbohydrate and lipid metabolism, glycogen storage, urea synthesis, plasma protein secretion, and detoxification and elimination of endogenous and exogenous compounds. For drugs, metabolism and elimination occur primarily through the Phase I (oxidative metabolism), Phase II (conjugation reactions such as glucuronidation and sulfation), and Phase III (active transporters such as the multidrug resistance [MDR] proteins) metabolic pathways. As the liver is formed and matures, it undergoes dramatic changes in structure and function that affect the rate and metabolic pathways used in the processing and disposition of drugs. These developing factors may lead to differential sensitivity to DILI at certain developmental stages.

12.2.1 In Utero

Fertilization of the egg by a sperm typically occurs in the midportion of the oviduct with subsequent migration to and implantation in the endometrium of the uterine wall. The outer layer of cells surrounding the implanted blastocyst erodes the endometrium and establishes the placental circulation. The inner cells of the blastocyst (embryoblast) divide and form the three essential germ layers (endoderm, ectoderm, and mesoderm) that develop into various organs. The developing conceptus undergoes three major development periods consisting of the preorganogenic period (first 2.5 weeks after fertilization), organogenesis (weeks 3 to 8 after fertilization), and the final fetal development period [7].

The liver develops from the endoderm, which also gives rise to the gastrointestinal tract, lungs, thyroid, and pancreas [8,9]. During foregut closure, a subset of cells is dedicated to a hepatic fate under the influence of various inductive signals and genetic regulatory factors that are highly conserved among vertebrates [10]. These dedicated cells are called hepatoblasts and eventually form the liver bud, which develops as an outgrowth from the ventral wall of the foregut [11]. Hepatoblasts express serum protein genes specific to hepatocytes such as albumin, transthyretin, and alpha-fetoprotein. These cells are bipotential and differentiate into hepatocytes (alpha-fetoprotein positive/albumin positive) or cholangiocytes (cytokeratin-19 positive) [12]. A large number of regulatory factors such as cytokines and growth factors maintain proliferation of the hepatoblasts and subsequent liver organogenesis throughout development [13,14]. As the liver develops its complex architecture, parenchymal and nonparenchymal cells extensively differentiate and form the extracellular matrix, biliary tract, sinusoidal capillaries and hepatic vasculature, and polarized epithelial cells [11].

During the early stages of liver bud formation and subsequent organogenesis, drug exposure can occur via transfer from the maternal circulation across the developing placenta into the embryo or directly from the uterine fluid surrounding the embryo. If the drug damages the liver at this early stage of development, spontaneous abortion or fetal death are likely to occur given the pivotal role the liver plays in organism development and the sensitivity of the embryo to malformation during organogenesis [10,15]. The developing human embryo and liver have been shown to metabolize various drugs via cytochrome P450 (CYP) enzymes, the heme-containing enzymes that play important roles in drug metabolism; however, the metabolic competence of the embryo and liver at this early stage of development is minimal compared to adult liver [16]. Phase II enzymes, which often contribute significantly to drug and reactive metabolite elimination, exhibit complex developmental expression patterns with increasing maturity [17,18]. Therefore, in general, the fetal liver has a limited capacity to metabolize drugs. If the drug itself is cytotoxic without bioactivation or a reactive metabolite accumulates due to poor elimination, the drug is more likely to damage the liver at this early stage. These drugs are also likely to damage other parts of the developing embryo, in addition to the developing liver, leading to spontaneous abortion, fetal death, or birth defects.

The placenta plays a key role in modulating drug exposure to the developing embryo and fetus. The placenta is formed from the outer layers of the implanted blastocyst and provides the developing embryo with nourishment via the maternal blood supply. The developing placenta consists of proliferating cells (trophoblasts) that infiltrate the maternal vascular channels permitting the exchange of nutrients, waste products, and gases. In the fully developed placenta, the main unit is the villous tree, which contains the fetal capillaries and associated endothelium that is in contact with the maternal blood supply allowing for efficient exchange.

Although the blood supplies of the mother and the developing fetus are separated by the placental–blood interface, many types of molecules can pass through the placenta, which is the method of nutrient and waste transfer between the mother and the fetus. In general, the placenta is readily permeable to molecules with molecular weights under 500 Da [19,20]. It was previously believed that the placenta served as a simple passive filter and that larger drugs and macromolecules could not physically pass through the placenta and reach the fetus. Recent research has shown that the placenta has many active drug transporters similar to the gastrointestinal tract and liver such as the MDR family of proteins [20,21]. These transporters play a key role in regulating drug exposure to the fetus. For example, 2.4-, 7-, or 16-fold higher transplacental transfer of digoxin, saquinavir, or paclitaxel to the fetus was measured in mdr1(a/b) knockout mice compared to wild type, clearly showing that this transporter plays a key role in decreasing fetal exposure to this drug [22]. Other factors that play a role in embryo and fetal exposure are pH gradients between maternal and conceptal compartments (ion trapping), differences in maternal and conceptal fluid protein binding of drugs, and differences in CYPs, dehydrogenases, and other metabolic enzymes found in the placenta [15].

In contrast to many lipid-based biological membranes, the placenta is highly permeable to hydrophilic compounds, which is necessary for transferring essential nutrients to the fetus [23]. The process also works in reverse where waste products are transferred to the mother for elimination. Therefore, models of drug penetration across biological barriers, such as the skin, may not accurately model drug penetration across the placenta. For most small-molecule drugs, the key factors that influence the rate of transfer and fetal exposure by passive diffusion across the placental membrane are molecular weight, pK_a, lipid solubility, and protein binding [24,25]. However, given the unique passive and active transport properties of the placenta, it is difficult to predict accurately drug exposure of the fetus and, unfortunately, this is an underserved area of research resulting in a paucity of empirical data to build accurate models. Therefore, no assumptions can be made about fetal exposure to a drug or metabolites. In addition, since the transport properties of the placenta change over the course of gestation, drug exposure is likely to vary over the course of pregnancy even for a single drug [26].

In utero exposure to maternally administered drugs can reach clinically significant levels. One of the first examples of this was the thalidomide tragedy in the 1960s where serious limb defects were induced by maternal consumption of thalidomide with subsequent passage to the developing fetus [27]. As another example, in 2004, a letter was sent to health care professionals and the label for Effexor, a serotonin and norepinephrine reuptake inhibitor (SNRI), was updated to include a warning about in utero exposure [28]. Some neonates born to mothers taking Effexor late in the third trimester experienced respiratory distress, cyanosis, apnea, seizures, hypoglycemia, hypertonia, tremor, and irritability, among other signs. These features are consistent with either a direct toxic effect of the SNRI or possibly a drug discontinuation syndrome, indicating clinically significant in utero exposure to the SNRI. This example shows that clinically significant levels of maternally administered drugs can reach the developing fetus and possibly lead to adverse clinical effects.

During the second and third trimesters, the structure of the liver is well defined; however, its function is still immature. Based on animal studies, the liver in utero has poor drug metabolizing capability and expresses many fetal proteins that recede after birth. In contrast, the human fetal liver has measurable CYP activity, although it is much less than the adult human liver [16,29]. The major CYP isozyme in fetal liver is CYP3A7, which has been shown to metabolize a wide array of drugs similar to CYP3A4, the major CYP isozyme in adult human liver. However, although both isoforms metabolize a wide spectrum of drugs with significant overlap, there are important differences such as a lower rate of testosterone 6β-hydroxylation by fetal

CYP3A7 compared to adult CYP3A4 [30]. After birth, CYP3A7 decreases to minimal levels and CYP3A4 increases dramatically. This changing spectrum of activity applies to other CYP isozymes and other metabolic enzymes that are involved in drug metabolism [31]. Most Phase II enzymes are expressed early in development with some isoforms exhibiting complex expression patterns [18]. For example, glutathione S-transferase P1 is high in the fetus and nondetectable in adults. In contrast, glutathione S-transferase M expression is low in the fetus and increases with age. Therefore, there are many drug metabolism factors that change over the course of gestation and after birth that greatly complicate being able to predict if a drug will cause DILI.

In addition to its changing metabolic pattern, the liver is more of a hematopoietic than a metabolic organ at the fetal stage [10,32]. This functional difference in utero is probably attributed to the fact that metabolism is not as important to the fetus since the mother supplies essential nutrients and eliminates wastes but cannot provide support for the developing hematopoietic system. After birth, the liver transitions from a hematopoietic support role to one of controlling metabolite and serum protein concentrations and detoxification and elimination. When the mother takes a drug, exposure of the fetus may or may not occur depending on the interplay of the factors mentioned above. Even if the drug, or a metabolite, crosses the placental–blood interface and reaches the fetal circulatory system, the drug may or may not have the potential to cause DILI depending on its mechanism of action. If a drug requires metabolic activation to a reactive metabolite that is the ultimate toxicant, it is less likely to cause any adverse effects in the fetal liver since the CYP metabolic system is poorly developed [33,34]. However, if the drug itself is hepatotoxic or its reactive metabolite is not efficiently eliminated by the delayed expression of Phase II enzymes, it could theoretically cause liver damage.

Assessing the potential of a compound to induce DILI in the developing fetus is very difficult and typically not studied in traditional toxicology testing paradigms. Therefore, the true potential of a compound to induce DILI during fetal development may be underestimated. Even if a compound does cause DILI in the developing fetus, if exposure time is restricted and the damage is not life threatening, it is likely that any damage that occurs can be repaired because of the great regenerative capacity of the liver. This would essentially mask any incidence of DILI in the developing fetus since it would be clinically normal at birth. Although the fetus may appear clinically normal, DILI and regeneration occurring in utero may theoretically make the developing fetal liver susceptible to future adverse effects. Also, the manifestation of in utero DILI after birth may not be diagnosed correctly since it might be attributed to other factors such as physiological jaundice.

In traditional toxicology testing batteries, acute through chronic toxicity studies are conducted in a range of species; however, these studies typically use young animals approximating the developmental stage of an adolescent. Therefore, they do not assess potential effects on the fetus. Reproductive and developmental studies are often a standard part of the testing battery and they are good at detecting gross defects of the developing fetus. However, routine histopathology of the developing liver and other endpoints of liver damage are not conducted and, therefore, any potential of a drug to cause DILI would be missed [35]. On the basis of this testing paradigm, it is difficult to conclude if the developing fetus is more resistant to DILI or if the low reported incidence is just attributed to a lack of data.

12.2.2 BIRTH

Birth represents an abrupt transition of the fetus being dependent on the mother for many life-supporting functions to being a self-supporting organism. Although maturation of many functions starts before parturition, they are often incomplete after birth. In particular, the metabolic capacity of the liver is still developing in the infant. A common problem seen in infants is jaundice owing to increased serum bilirubin levels [36,37]. Unconjugated bilirubin is produced by the degradation of heme, which is derived mostly from the breakdown of erythrocyte hemoglobin. The immature liver has a decreased capacity to conjugate the bilirubin with glucuronic acid to effect elimination. Depending on the rate of blood cell degradation, the liver may not be able to sufficiently conjugate the bilirubin, ultimately leading to its accumulation [37]. Fortunately, the liver typically matures quickly and, combined with phototherapy, infants often recover from the jaundice without major intervention.

The liver is a major site of hematopoiesis in the fetus. Shortly before and continuing after birth, hematopoiesis transitions from the liver to the other main hematopoietic organs such as the bone marrow. This transition coincides with the start of the maturation process of the liver into a highly metabolic organ that is responsible for many essential life-supporting functions. The liver is the main site of albumin and clotting factor synthesis, and cholesterol, amino acid, and glucose processing, among many others. The liver also becomes the primary site for the metabolism and elimination of many endogenous and exogenous compounds, including the majority of drugs [38,39]. This applies to drugs that are administered by various routes of exposure as long as they result in significant blood concentrations of the drug. Orally administered drugs pass through the liver before they reach other parts of the body, and therefore, they may be extensively metabolized before further distribution throughout the body (hepatic first-pass effect). Drugs administered by other routes of exposure and reaching the bloodstream are also typically metabolized by the liver since the liver has the highest metabolic capacity of any organ in the body. The liver is also a major site of elimination for many drugs via biliary elimination.

Phase I reactions (oxidative metabolism) are performed by a variety of different enzymatic systems; however, the largest number and variety of reactions are catalyzed by the CYP isoenzymes [38,40]. CYP enzymes are located in the liver endoplasmic reticulum (microsomes). They are a large family of highly conserved enzymes that catalyze the oxidation, reduction, or hydrolysis of drugs and introduce or uncover a functional group on the drug. The functional group increases hydrophilicity, aiding drug elimination from the body. More importantly, the functional group allows the metabolite to react in Phase II conjugation reactions. Besides playing a key role in drug metabolism, CYP enzymes catalyze a large variety of other reactions such as the metabolism of xenobiotics and the biosynthesis or catabolism of steroid hormones, bile acids, fat-soluble vitamins, fatty acids, and eicosanoids. A variety of other enzyme systems also catalyze Phase I reactions such as carboxylesterases, epoxide hydrolase, alcohol and aldehyde dehydrogenases, monoamine oxidase, and flavin-containing monooxygenase [41].

CYP enzymes undergo developmental changes in expression and activity levels after birth, which can lead to differential drug sensitivity. For example, CYP3A4 is present at high levels in adult human liver and is responsible for the metabolism of a wide variety of drugs. In contrast, in the neonatal liver, CYP3A7 is the predominant CYP. CYP3A4 expression increases during the first year of life, reaching approximately 50% of adult levels and continuing to increase with age [38]. Although many of the drugs metabolized by CYP3A4 are also metabolized by CYP3A7, there are some differences that could lead to altered metabolic profiles for a given drug and in turn differential sensitivity to DILI. This differential pattern of CYP expression and activity applies to many different CYPs that are involved with drug metabolism and elimination. In general, there are three patterns of CYP developmental expression: (1) CYPs that are expressed in the fetal liver and decline with age (e.g., CYP3A7), (2) CYPs that begin expression in the early neonatal period within hours after birth (e.g., CYP2D6 and CYP2E1), and (3) CYPs expressed later in neonatal development (e.g., CYP1A2 and CYP3A4) [42,43].

Although Phase I metabolism is typically considered a detoxifying reaction, many drugs are metabolized to reactive intermediates that can cause cellular injury. Acetaminophen metabolism by CYPs 1A2, 2E1, and 3A4 to the reactive metabolite *N*-acetyl-*p*-benzoquinoneimine (NAPQI) that binds cellular macromolecules, such as free sulfhydryl groups on proteins, is a classic example [44,45] (Figure 12.1). Therefore, depending on the role Phase I metabolism plays in activating or detoxifying a drug, the fetus or pediatric patient may be more or less susceptible to a given DILI depending on the interplay of the rate of metabolism and the toxicity of the drug or metabolite.

Most Phase II conjugation reactions involve adding a hydrophilic cofactor to the drug or metabolite to greatly expedite elimination of the drug via the kidneys or liver. Although Phase II reactions often attach cofactors to those functional groups added or exposed during Phase I reactions, Phase II reactions can occur directly on functional groups on the parent drug (e.g., on a free hydroxyl group). Phase II conjugation reactions include glucuronidation, sulfation, methylation, acetylation, and amino acid and glutathione conjugation. Most Phase II biotransforming enzymes are located in the cytosol with some exceptions such as UDP-glucuronosyltransferases located in the microsomes. Phase II reactions often proceed at a relatively fast rate, making the Phase I reactions the rate-limiting step. Table 12.1 summarizes the various cofactors that are used in the Phase II reactions and how they vary with age [39]. These changes are likely to affect the metabolic pathways a drug goes through as a child ages.

The metabolism of acetaminophen provides a great example of the interplay of the Phase I and II metabolic pathways. At therapeutic doses, acetaminophen is primarily glucuronidated and sulfated (Figure 12.1). At supratherapeutic doses, the Phase II cofactors are depleted and a larger portion of acetaminophen is metabolized by CYP enzymes to the reactive NAPQI. NAPQI is detoxified by glutathione; however, there is a finite supply and once that becomes depleted, NAPQI can bind and disrupt cellular macromolecules. In the young, glucuronidation is low, whereas

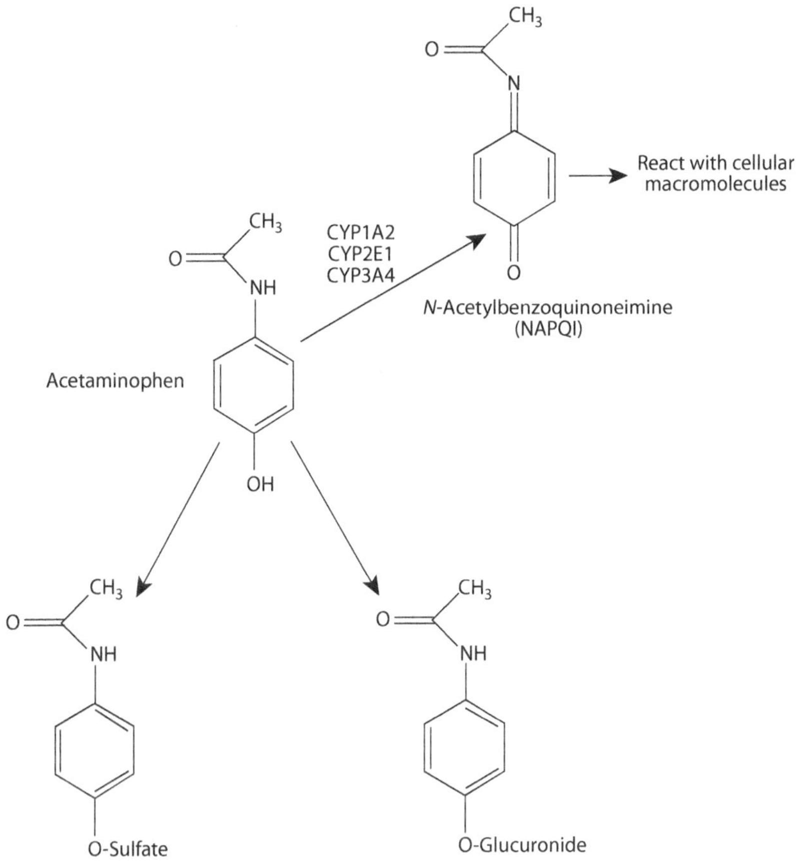

FIGURE 12.1 Hepatic metabolism of acetaminophen to the reactive metabolite NAPQI.

TABLE 12.1
Phase II Metabolism Cofactors and Changes with Age

Phase II Reaction	Cofactor	Change with Age
Glucuronidation	Uridine-5'-diphospho-α-D-glucuronic acid (UDP-GA)	Low at birth, increases with age
Sulfation	3'-Phosphoadenosine-5'-phosphosulfate (PAPS)	High at birth and remains relatively constant
Acetylation	Acetyl coenzyme A	Low at birth, slow or fast acetylator phenotype acquired by 4 years of age
Methylation	S-Adenosylmethionine (SAM)	Insufficient data
Glutathione conjugation	Glutathione	Insufficient data
Amino acid conjugation	Glycine, taurine, glutamine	Insufficient data

sulfation is similar to adult levels. Although acetaminophen is extensively glucuronidated in adults, the low activity in children results in the sulfation pathway being the predominant route of metabolism [31,46]. The complicated interplay of the Phase I and Phase II pathways is active for many drugs and the influence of age adds another layer of complexity to predicting the metabolism in pediatric versus adult patients. Two examples of neonatal sensitivity attributed to deficient metabolism are the gray baby syndrome observed in the 1950s as a result of reduced glucuronidation of the antibiotic chloramphenicol and gasping syndrome identified in the 1980s owing to benzyl alcohol used as an antibacterial agent in intravenous solutions [42,47]. Both of these produced very clear clinical signs of toxicity owing to poor metabolism of the compounds in children. It is likely that similar types of metabolic factors can lead to differential sensitivity to DILI in children.

Drug transporters are often referred to as Phase III metabolism although they do not alter the chemical structure of a drug or metabolite. Drug transporters are responsible for transporting (influx and efflux) drugs across various biological surfaces such as the gastrointestinal tract, blood–brain barrier, placenta, hepatocytes, and renal tubular cells. The MDR protein (also called P-glycoprotein) is the most famous of the drug transporters and it is a member of the ATP-binding cassette family. Many of the transporters prevent the accumulation of drugs and aid their elimination from the body. Inhibition of drug transporters by compounds such as grapefruit juice can result in serious drug interactions because of the accumulation of toxic levels of the drug. Unfortunately, very little data exist on the ontogeny of Phase III transporters; however, it is likely that there are development changes in the levels of these transporters that affect the exposure to drugs and in turn their potential to cause DILI [48,49].

The developing child undergoes various physiological changes that can influence the absorption, distribution, metabolism, and elimination of drugs, which could in turn lead to differential sensitivity to DILI. In general, the rate of absorption of most drugs is slower in neonates and infants and increases with age. There are also developmental differences in the activity of intestinal flora, intestinal drug-metabolizing enzymes, and Phase III transporters that can affect drug bioavailability. For example, immature intestinal CYP3A4 metabolism of midazolam reduces clearance in preterm infants [50]. Once absorbed, in general, the overall clearance of drugs appears to be reduced during the first year of life [46]. Metabolic clearance then increases and reaches a maximum between 2 and 10 years of age, declining thereafter with age. This higher clearance in young children often requires relatively higher weight-based doses and Table 12.2 highlights a few of these as adapted from Kearns et al. [48]. Clearance is the result of various factors such as liver metabolism, biliary excretion, and kidney elimination. The higher clearance in children is believed to be partly caused by the increased liver volume normalized to body weight. Some of the physiological factors and changes that influence drug absorption and clearance are highlighted in Table 12.3.

The previous sections dealt with developmental changes that affect the pharmacokinetics of a drug. Another key factor in determining child sensitivity to DILI is pharmacodynamics, which represents the interaction of a drug with the target, such as a receptor on a cell, and the subsequent response. Although it is likely that there are developmental differences in pharmacodynamics, very little data are available

TABLE 12.2

Examples of Age-Based Dosing Recommendations

Drug	Average Dose			
	Neonates	Infants	Children	Adults
Ceftazidime	50 mg/kg every 12 h	50 mg/kg every 8 h	50 mg/kg every 8 h	14–28 mg/kg every 8–12 h
Digoxin	4–8 µg/kg every 24 h	7.5–12.0 µg/kg every 24 h	3–8 µg/kg every 24 h	1.4–4.0 µg/kg every 24 h
Gentamicin	2.5 mg/kg every 12 h	2.5 mg/kg every 6–8 h	2.5 mg/kg every 8 h	1–2 mg/kg every 8 h
Phenobarbital	3–4 mg/kg every 24 h	2.5–3.0 mg/kg every 12 h	2–4 mg/kg every 12 h	0.5–1.0 mg/kg every 12 h
Ranitidine	0.75–1.0 mg/kg every 12 h	0.75–1.0 mg/kg every 12 h	1 mg/kg every 6–12 h	0.7 mg/kg every 6–8 h

Source: Adapted from G.L. Kearns et al., *N Engl J Med*, 349 (2003), pp. 1157–67.

Note: The information contained in this table is only for illustrating the influence of development on clearance rates of various drugs and is not for dosing recommendations.

about changes with age [51]. Combining age-specific pharmacokinetic/metabolism changes with potential pharmacodynamic changes results in a complex paradigm of drug sensitivity that changes over age and this is likely to vary for each type of drug.

Another important factor in the potential differential sensitivity of the developing human is the capacity of various defense mechanisms. It is well known that oxidized proteins accumulate with age, providing an indirect indication that defense mechanisms may be high in the young and decrease with age [52]. Glutathione is a sulfhydryl-containing compound that acts as an antioxidant by scavenging reactive metabolites. Glutathione can either directly bind reactive metabolites or be conjugated to reactive metabolites through an enzymatic process by glutathione S-transferase. Glutathione levels in human blood were reported to be inversely correlated with age, with the exception of infants <1 year of age, supporting a role of high levels of defense mechanisms in children [53]. Therefore, some defense mechanisms may be lower in certain ages. In addition, even if certain defense mechanisms are elevated in the young, they might also be more overwhelmed with normal physiological insults and less able to deal with exogenous insults from drugs. For example, Tsukahara reported that urinary biomarkers of oxidative stress were highest in young children and decreased with age to reach constant levels by early adolescence [54]. It was concluded that the markers of oxidative stress are probably attributed to the high metabolic rate of growing children compared to adults. Addition of a drug that causes oxidative stress on top of the high metabolic rate could overwhelm endogenous defense mechanisms and lead to DILI.

Exposures of children represent a unique factor in assessing differential sensitivity of the growing infant. Infants may be exposed to drugs that the mother consumes

TABLE 12.3

Physiological Changes Affecting Oral Drug Absorption and Clearance

Parameter	Neonates	Infants	Children	Potential Effects
Gastric pH	Neutral at birth, decrease to ~3 within 48 h, neutral for 10 days, then slow decline	Slow decline until reach adult values at ~2 years	Similar to adults	Differential absorption of drugs (ionization effects)
Gastric emptying	Delayed	Delayed, reaching adult values within 6–8 months	Similar to adults	Altered absorption affecting pharmacokinetics such as C_{max} and T_{max}
Intestinal transit time	Prolonged and irregular	Reduced in older infants owing to increased intestinal motility	Similar to adults	Altered absorption affecting pharmacokinetics such as bioavailability, C_{max}, and T_{max}
Total body water and fat	Water is high and fat is low	Water is high and fat is low	Water decreases and fat increases	Affects volume of distribution of drugs (increases for hydrophilic drugs)
Protein concentration and binding	Reduced	Reduced	Similar to adults	Affects unbound (active) fraction of drug
Hepatic blood flow	High	High, decreasing to adult level by 6 months	Similar to adults	Increased clearance of drugs
Hepatic architecture	Hepatocytes are arranged in plates ~3 cells thick	Hepatocytes thin to ~2 cells thick by 5 months	Adult pattern of single cell layer established at ~5 years	Limited entry and egress of drugs from the sinusoids
Renal excretion	Low	Low reaching adult levels by ~1 year	Similar to adults	Decreased clearance of drugs

via breast milk [43]. This is of increasing importance since breastfeeding women have been shown to be more likely to take multiple medications, including more prescription drugs, compared to pregnant women, probably because they feel that exposure via the milk is unlikely or minimal [55]. Fortinguerra et al. reported that of a variety of psychoactive drugs taken by breastfeeding mothers, 12 of them resulted in child blood drug levels that were ≥10% of the maternal dose, clearly showing the potential for significant drug transfer from mother to child via breast milk [56]. In a case report of a mother taking lamotrigine, a drug for epilepsy and bipolar disorder, a 16-day-old breastfeeding infant developed mild episodes of apnea that resulted in a severe cyanotic episode [57]. Blood levels of lamotrigine in the infant were in the upper therapeutic range for adults and the adverse effects were attributed to high lamotrigine exposure via the breast milk. This highlights the fact that children consuming breast milk could be exposed to sufficient concentrations of a drug (or metabolites) to cause pharmacological and toxicological effects, such as DILI. Even if the drug itself does not reach levels sufficient to cause clinical effects, it could interact with other drugs administered directly to the infant. Therefore, it is important for the breastfeeding mother to be aware of the potential of drug transfer to infants via breast milk.

For formula-fed infants, drug exposure via the mother's milk is not a factor; however, some components of formula might alter the sensitivity of the infant to concomitant drug exposures. Although it is a controversial subject, some practitioners advise against soy-based infant formulas since some of the isoflavones from soy have been shown to be estrogenic in various test systems [58,59]. Although it appears unlikely that the soy formula itself causes major adverse effects, it is possible that compounds in soy formula could have an additive or synergistic effect with concomitantly administered drugs leading to toxicity and possibly DILI. Another point to consider is that formula-fed infants, particularly during the first week after birth, are not getting essential factors from the mother such as maternal antibodies from the colostrum. These factors may make the formula-fed infant more sensitive to DILI and less able to adapt to any damage that may occur.

A typical toxicology study package for a new drug will include an assessment of the drug on the developing organism through weaning. These studies are often divided into developmental phases and focus on gross developmental defects and the reproductive capacity of the animals. However, the specific effect of the drug on the liver is rarely assessed. Therefore, if a drug causes liver toxicity in the developing animal, it would likely be missed since the current study designs are not specifically assessing DILI. As with the developing fetus, it is difficult to conclude if the developing organism is more resistant to DILI or if the low incidence is just attributed to a lack of data.

12.2.3 WEANING THROUGH ADOLESCENCE

After weaning, the developing human continues to mature. At this stage, a child is not just a small adult as evidenced by the large number of developmental stages that a child goes through. One of the most striking changes occurs in adolescence where there is a clear differentiation of the sexes; however, although this is one of the most clear phenotypic changes, many other changes go on during all the stages of maturation that could influence the ability of a drug to cause DILI. Many of the factors that were described in Sections 12.2.1 and 12.2.2 also play a role at this stage of development (e.g., CYP and defense mechanism levels).

Sexually mature males and females differentially express a large number of hepatic genes leading to differential effects such as the metabolism of drugs [60]. In mice and rats, more than 4000 male- and female-biased genes have been identified [61,62]. The sex difference in hepatic gene expression is largely attributed to the difference in growth hormone (GH) exposure between the sexes. Males have a pulsatile pattern of GH release from the pituitary gland with a large release every couple of hours. Between releases, circulating GH is at a negligible level. In stark contrast, females have nearly continuous release of GH; however, the peak concentration of GH is much lower than the peak level in males [60,63]. The combination of the sex-specific patterns of GH release combined with the differences in GH peak concentration causes differential activation of the target transcription factors, which directly control the expression of a variety of liver genes, thus leading to sex-biased hepatic gene expression patterns. It is likely that the sex-specific hepatic gene expression patterns may play a role in age- and sex-specific DILI.

A factor that is often overlooked in drug toxicity is concomitant environmental exposures that may have an additive or synergistic effect with the drug. The growing child is particularly vulnerable to these concomitant exposures since they are starting to probe their environment without knowledge of the potential risks. A good example is hand-to-mouth activity of young children. A child might touch the floor to pick up a toy and the floor was just treated with a household pesticide. The pesticide is transferred to the child's hand and subsequently to the child's mouth. Assuming a sufficient quantity of pesticide, or other chemical, is ingested, it might interact with a concomitantly administered drug resulting in DILI. An even more problematic time is adolescence where young adults may experiment with alcohol and illicit drugs. The drugs themselves might cause liver injury (e.g., alcohol, ecstasy, and cocaine) or the illicit drug could interact with concomitantly administered drugs resulting in DILI [64,65].

12.2.4 SUMMARY

The human liver goes through many stages of development as a child matures into an adult. Depending on the various ontogenic, environmental, and genetic factors, a child may be more, equally, or less susceptible to a drug when compared to an adult. Child sensitivity to DILI is best assessed on a drug-by-drug basis without any preconceived notions about a child's general inherent tolerance.

12.3 PEDIATRIC DILI

Section 12.2 highlighted potential factors that may make a child more or less sensitive to DILI. In general, DILI is considered rare in children. However, this could partly be attributed to a failure to diagnose and report drug hepatotoxicity. Other factors that are likely to play a role are the fact that children often take relatively fewer of the drugs commonly associated with hepatotoxicity (e.g., cardiovascular, antihypertensive, and antidepressant), they typically have a lean body mass and fewer concomitant diseases, and they often do not have other common predisposing factors such as ethanol consumption and cigarette smoking.

Before reviewing specific drugs, a basic overview of the manifestation and classification of drug-induced hepatotoxicity is needed. The liver is anatomically and physiologically complex and liver injury can affect one or more of these factors. Therefore, drug-induced hepatotoxicity presents as diverse biochemical, histological, and clinical abnormalities and can mimic all forms of liver injury ranging from asymptomatic elevations of hepatic enzymes to fulminant hepatic failure [66]. With the advent of various "omics" technologies (i.e., genomics, proteomics, and metabolomics), the spectrum of hepatotoxic effects caused by different drugs has widened exponentially such that it is unlikely that any two drugs will produce the exact same spectrum of effects using these exquisitely sensitive techniques that involve thousands of different endpoints [67]. However, although the spectra are unlikely to match up exactly, there are likely to be sufficient similarities to group drugs into broad categories [68].

The microscopic liver architecture is divided into zones (Figure 12.2). Blood flow enters the liver in Zone 1, which encompasses the portal triad (portal vein, bile duct, and hepatic artery). Zone 1 has the highest oxygen tension and nutrient supply. Blood flows through the sinusoids (spaces between the hepatocytes) through Zone 2 (often referred to as the midzonal region) and out the central hepatic vein in Zone 3. As the blood traverses the zones toward the central hepatic vein, oxygen and nutrients are used up, resulting in lower levels in Zone 3. Hepatotoxicity may manifest itself in a zonal pattern or it may produce more diffuse damage. Damage may also be restricted to certain cell types such as parenchymal cells (hepatocytes), cholangiocytes (bile duct cells), endothelial cells, stellate cells, or Kupffer cells [69]. Often, hepatotoxicants that cause damage primarily through a reactive metabolite will produce damage in a zonal pattern since although drug concentrations are highest in Zone 1, most CYPs, necessary for bioactivation, are highest in Zone 3. For a hepatotoxicant such as acetaminophen, CYPs in Zone 1 are at negligible levels and cannot form the reactive metabolite. In contrast, in Zone 3, CYPs are at much higher levels and form the reactive metabolite NAPQI, restricting covalent binding to this zone [45]. The location of covalent binding correlates with subsequent hepatocellular damage that is also restricted to this zone.

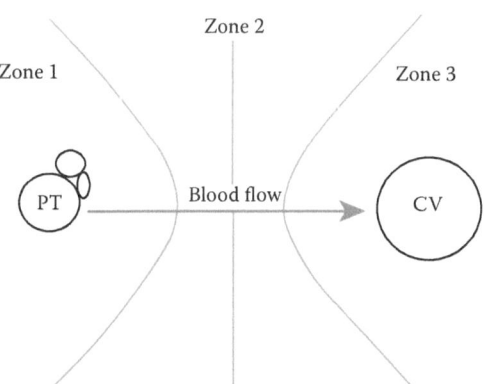

FIGURE 12.2 Microscopic architecture of the liver lobule. Blood flow enters the liver lobule through the portal triad (PT), which encompasses a portal vein, hepatic artery, and bile duct. Blood flows from Zone 1 through Zone 2 and out the central hepatic vein (CV) in Zone 3.

Clinically, DILI has been classified in terms of the clinical liver disease, which is hepatocellular, cholestatic, or mixed hepatocellular/cholestatic. Hepatocellular often involves direct damage of the hepatocytes such as the Zone 3 necrosis caused by acetaminophen. This type of damage is often associated with elevated serum transaminase levels (i.e., alanine and aspartate aminotransferases) owing to leakage from damaged hepatocytes. Cholestatic injury often involves damage to some part of the bile excretion apparatus resulting in impaired bile excretion. This type of injury is often associated with elevated serum levels of analytes such as gamma glutamyl peptidase, alkaline phosphatase, and bilirubin that denote bile duct injury. Mixed injury presents with a mixture of both types of effects. In addition to these three clinical types of injury, each one may be associated with a systemic syndrome. For example, in addition to the liver injury, some drugs may cause hypersensitivity reactions in other organ systems (e.g., Stevens-Johnson syndrome, renal dysfunction, myocarditis), suggesting an immunoallergic component to the injury [70,71]. Unfortunately, drugs rarely produce a single clear clinical picture, making the diagnosis of DILI difficult. For example, amoxicillin/clavulanic acid usually causes cholestatic injury but can also produce acute hepatocellular injury or a mixed type injury [66,72]. There are also other types of DILI such as microvesicular steatosis, nonalcoholic steatohepatitis, chronic hepatitis, cirrhosis, venoocclusive disease, and neoplasia; however, these are much less common.

Traditionally, hepatotoxicants have been separated into those that produce predictable hepatotoxicity in both humans and animal models (intrinsic hepatotoxicants) and those that produce unpredictable reactions (idiosyncratic hepatotoxicants) [71,72]. Acetaminophen is a classic example of an intrinsic hepatotoxicant. Most of the intrinsic hepatotoxic drugs are screened out during the early developmental phases since their hepatotoxicity can be assessed in typical preclinical screening models or the incidence of hepatotoxicity is high enough that it is readily detected in clinical trials. The other group of hepatotoxicants has been typically classified as "idiosyncratic" since they produce unpredictable incidences of hepatotoxicity that often cannot be reproduced in animal models. Most drugs in clinical use that have been associated with hepatotoxicity fall into this class such as isoniazid, leflunomide, dantrolene, erythromycin, phenytoin, propylthiouracil, flutamide, and diclofenac. Sometimes the clinical incidence of hepatotoxicity becomes too high in relationship to its benefit and the drug is removed from the market such as the case with lumiracoxib, bromfenac, and troglitazone [67,72]. These "idiosyncratic" hepatotoxicants present a major concern with drug development since current preclinical screening models cannot successfully predict the hepatotoxicity and people become unnecessarily exposed to these hazardous drugs. Although these drugs are tested in a phased approach in human clinical trials with increasing power at each phase, given the low incidence of most idiosyncratic hepatotoxic reactions (e.g., 1 in 10,000 patients), the probability of detecting clear signs of hepatotoxicity in a typical clinical trial involving 1000 people is low. When signs of hepatotoxicity are observed in clinical trials, this raises great concern since the probability of observing hepatotoxicity once the drug is marketed to a much larger population increases dramatically [73].

Drugs causing DILI are further categorized on the basis of both the timing of their effects and whether or not the immune system is likely to play a major role in the injury. Acute reactions are a quick flare-up of signs of liver injury and can be life threatening. Although they are termed acute, they may not occur until weeks or months after being on the drug. Chronic reactions are protracted and typically entail milder signs of liver injury while taking the drug over a prolonged period. Chronic reactions often recede after drug withdrawal, whereas serious acute reactions can lead to acute liver failure even after drug withdrawal. If DILI is also associated with clinical signs of a hypersensitivity reaction (e.g., fever, rash, and eosinophilia) and especially a rapid positive rechallenge, the overall picture will be considered to involve an immune-mediated hypersensitivity reaction [3,71,72].

Despite the broad classification of drugs as hepatocellular, cholestatic, or mixed with modifiers of acute, chronic, intrinsic, idiosyncratic, or immune-mediated hypersensitivity, they all are likely to share the common trait of having a biochemical basis to the injury. For a compound like acetaminophen, the biochemical basis has been worked out in detail and has explained the clinical picture of injury [45]. Even for compounds that appear to have an unknown mechanism of action with a very confusing clinical picture (e.g., mixed injury with systemic hypersensitivity reaction), it is likely that there is a biochemical basis behind the injury. The adaptation seen to the DILI caused by some drugs provides some clues about the molecular processes that might play a role in DILI. Some idiosyncratic hepatotoxicants have been shown to cause transient, asymptomatic increases in serum alanine aminotransferase (ALT) in a significant percentage of patients. The majority of the patients adapt to the drug (adapters) as indicated by a return of serum ALT to baseline levels despite continued treatment. A small percentage of these patients will continue to experience increased ALT levels (sensitive) evolving into other clinical signs of liver injury and possibly leading to liver failure. Although the clinical outcome is much different for the three groups of patients (nonresponders, adapters, and sensitive), the basic biochemical events are likely to be similar. The difference separating the three groups of people could be attributed to a multitude of reasons and include both genetic and environmental factors [74]. A theoretical "Drug X" causing immune-mediated hypersensitivity in a minority of patients highlights the interplay of the various possible factors. Drug X is metabolized to a reactive metabolite that binds cellular macromolecules. In nonresponders, either the reactive metabolite is produced at very low levels or they have higher levels of defense mechanisms (e.g., glutathione) to detoxify the metabolite. In adapters, the reactive metabolite is formed in a sufficient quantity to cause hepatocellular injury; however, with continued exposure, the quantity of reactive metabolite is reduced (e.g., downregulation of the CYP), defense mechanisms are increased (e.g., the level of glutathione or glutathione S-transferase is upregulated), or an unknown environmental factor increases the patient's tolerance. The damage in adapters never reaches a level sufficient to illicit an immune response. In sensitive patients, the reactive metabolite causes hepatocellular injury and continues to produce injury with continued drug exposure. Damaged cells or cellular components are presented to the immune system (e.g., via the human leukocyte antigen) and a hypersensitivity reaction ensues. This example highlights the many factors that could play a role in DILI and the difficulty in determining the critical factors that cause DILI in some people but not others.

12.4 SPECIFIC DRUGS ASSOCIATED WITH PEDIATRIC DILI

The following sections provide an overview of drugs that have been associated with DILI in children. A comprehensive review of every drug that has been associated with pediatric DILI is beyond the scope of this chapter. Instead, the more common drugs associated with pediatric DILI are briefly reviewed. When available, information on the differential sensitivity between children and adults, and the potential contributing factors, is discussed.

12.4.1 ANALGESICS AND ANESTHETICS

12.4.1.1 Acetaminophen

Acetaminophen is an over-the-counter analgesic and antipyretic that is widely used by both children and adults. Acetaminophen is often found in combination with other active drug ingredients in mixed formulations such as cough and cold remedies and high-strength prescription pain relievers (e.g., hydrocodone, oxycodone, and codeine with acetaminophen). Although acetaminophen is used safely by millions of people each year, it remains a leading cause of DILI in the United States and is by far the leading cause of acute liver failure in adults [75–77]. In children, acetaminophen is the second most frequent cause of acute liver failure only behind indeterminate causes [78,79]. Because of the high incidence of hepatotoxicity, a recent FDA Advisory Committee recommended lowering the daily therapeutic dose of acetaminophen [80]. The Committee concluded that, in addition to people unknowingly consuming multiple products containing acetaminophen, some individuals may be especially sensitive to liver injury from acetaminophen and that more research is needed to understand whether age, ethnicity, genetics, nutrition, or other factors might have a role in making some individuals more sensitive.

Acetaminophen has high bioavailability, exhibits low protein binding, and has an elimination half-life of approximately 2 h [81]. The mechanism of efficacy is unknown, although various targets have been proposed, such as a novel cyclooxygenase isoenzyme [82]. As mentioned in Section 12.2.2, at therapeutic doses, acetaminophen is metabolized by the Phase II metabolic pathways of glucuronidation and sulfation with the glucuronidation pathway accounting for ~60% of the metabolism after a therapeutic dose in adults (Figure 12.1) [81]. At higher doses, the Phase II cofactors are depleted and a larger portion of acetaminophen is metabolized by CYP enzymes to the reactive metabolite NAPQI. NAPQI is detoxified by glutathione; however, once glutathione levels become depleted, NAPQI can bind and disrupt cellular macromolecules resulting in cellular injury. In the young, the sulfation pathway is the predominant route of metabolism until approximately 12 years of age since glucuronidation activity is low compared to adults [31,46].

In general, children appear to be more resistant to acetaminophen-induced hepatotoxicity compared to adults [70,83]. This is backed up by a query of the USFDA Adverse Event Reporting System (AERS) database where children (ages 0–16 years) have predicted reporting incidences of hepatotoxicity attributed to acetaminophen that were similar to or less than those of adults (≥17 years). Although children appear to be less sensitive, acetaminophen hepatotoxicity still remains a major concern in both children and adults since the number of adverse event reports of hepatotoxicity is high compared to other drugs. The reason for children's general decreased sensitivity is likely attributed to a combination of factors such as differential metabolic pathways, higher capacity for producing glutathione, and fewer risk factors, such as alcohol consumption.

Hepatotoxicity may occur after ingestion of a single dose of >150 mg/kg [84]. When plasma concentrations are >300 µg/ml 4 h after ingestion or 45 µg/ml 15 h after ingestion, severe liver injury occurs in 90% of patients. Lower doses used repeatedly at regular intervals can also result in serious hepatotoxicity [82]. Conditions that result in CYP induction or glutathione depletion, such as concomitant medications, heavy alcohol use, recreational drugs, or fasting or malnutrition, can increase the toxicity of acetaminophen.

Acetaminophen overdose is a medical emergency. It is important to note that criteria used for assessing acetaminophen overdose in adults may not be entirely applicable for children, such as the King's College Hospital criteria [83]. In adults, the clinical course of an overdose starts with gastric distress, such as anorexia, nausea, vomiting, and abdominal pain, over the first couple of days. Unfortunately, these early signs do not foretell the underlying seriousness of the injury and this may delay intervention. Serum transaminase levels become elevated within 12 to 36 h after ingestion with peak levels within 3 to 4 days. Clinical signs of hepatotoxicity (right subcostal pain, tender hepatomegaly, coagulopathy, and jaundice) occur within 2 to 4 days. Poor prognosis is often associated with worsening coagulopathy, onset of hepatic encephalopathy, or renal failure. Biopsy reveals centrilobular necrosis with no to minimal damage to the periportal regions. If the injury is not fatal, the centrilobular necrosis is reversible with remodeling of the injured areas over the course of several weeks or months [81].

Early intervention is critical to treatment of acetaminophen overdose. When administered within 12 h of ingestion, N-acetylcysteine (NAC) is very effective in preventing hepatotoxicity [81,84]. Its effectiveness diminishes with time, most likely because of the irreversibility of the cellular injury caused by NAPQI, such as covalent binding and disruption of cellular macromolecules. The exact mechanisms of NAC detoxification are not entirely known; however, it is likely that NAC can directly conjugate with and neutralize NAPQI and also serve as a precursor for glutathione synthesis. Activated charcoal is useful to prevent absorption of acetaminophen if administered shortly after ingestion. Supportive care is administered as required.

12.4.1.2 Halothane

Halothane is an inhaled anesthetic that produces mild liver injury in 1 in 5 patients [85] and a more severe immune-mediated adverse drug reaction, or "HAL hepatitis," in 1 in 6,000–30,000 patients exposed to the drug [86]. Halothane is metabolized by CYP 2E1 to form trifluoroacetyl (TFA) chloride, which binds covalently to proteins and lipids, forming TFA adducts. Studies in which halothane hepatotoxicity in guinea pigs was ameliorated by SKF-525A, a broad-based P450 inhibitor, and exacerbated by 4-methylpyrazole, a CYP2E1 inducer, illustrated the requirement for metabolism in the development of hepatotoxicity [87,88]. Furthermore, susceptibility to HAL hepatotoxicity in a guinea pig model correlated with the formation of liver TFA–protein adducts [89]. Based on a review of the USFDA AERS database, children do not appear to have a unique sensitivity to halothane hepatotoxicity when compared to adults.

12.4.2 ANTI-INFECTIVES

12.4.2.1 Antituberculosis Medications

12.4.2.1.1 Isoniazid

Since 1952, isoniazid therapy has been used for treating tuberculosis (TB) and it has effectively prevented the disease progression in children with latent TB infection [90]. However, isoniazid carries a known risk for hepatotoxicity, most likely owing to reactive metabolite formation. Serum transaminase elevations occur in approximately 7%–14% of children receiving isoniazid monotherapy [91]. More severe isoniazid-induced hepatitis has been considered to be more common in adults than in children with a frequency of 0.1% in children versus 1.3% in adults [92]. However, the incidence of isoniazid-associated liver failure in children is estimated as 3.2 per 100,000 patients, approaching the 4.2–14 per 100,000 rate reported in adults [90]. Clinical symptoms consist of jaundice, nausea, fatigue, loss of appetite, abdominal pain, and vomiting. The histological findings resemble acute viral hepatitis, showing hepatocyte necrosis, hepatocellular degeneration more commonly in Zone 3, and inflammatory infiltrates. Current guidelines recommend immediate withdrawal of isoniazid and frequent measurement of serum transaminase levels at the first symptoms suggestive of hepatitis [93].

It has been speculated that isoniazid-associated hepatic injury is attributed to the isoniazid metabolites hydrazine or acetylhydrazine. In animal models, preexposure to phenobarbital increased the oxidative elimination of isoniazid active metabolites and resulted in a lower plasma hydrazine concentration. Therefore, medications that increase CYP enzyme activity may enhance isoniazid hepatotoxicity. Rifampicin is commonly used in combination with isoniazid for the treatment of TB and it rarely causes hepatotoxicity in children when used alone. In patients who received isoniazid and rifampicin, the incidence of hepatic dysfunction is 5% to 8% more frequent and earlier than with either medication alone. There are a few liver injury cases reported attributed to an interaction of isoniazid and carbamazepine [94,95]. Isoniazid itself has been reported to induce CYP2E1 expression, and one study reported that CYP2E1 polymorphisms correlated with increased hepatotoxicity in pediatric patients [96]. Plasma hydrazine concentrations are also dependent on whether the patient is a fast, intermediate, or slow acetylator, determined by the NAT2 genotype, which differs between ethnic groups and ages. Slow acetylators, who have a higher plasma concentration of hydrazine, have been associated with greater risk of isoniazid-induced hepatotoxicity [97–99]. Although children are slow acetylators until approximately 4 years of age, phenotypic studies in children failed to demonstrate the significant role of the slow acetylator phenotype in the development of hepatotoxicity [96].

Other risk factors associated with isoniazid hepatotoxicity are severe or meningeal TB, acetaminophen, viral hepatitis, ethanol, and pregnancy.

12.4.2.1.2 Rifampicin

At therapeutic doses, rifampicin is well tolerated for the treatment of TB. There are few data on rifampicin-associated hepatotoxicity in children when using rifampicin alone. However, children seem as vulnerable as adults to the combination of isoniazid and rifampicin treatment. O'Brien et al. [100] performed a retrospective review of hepatotoxicity rates in anti-TB therapy and reported a 3.3% incidence of drug-induced hepatotoxicity in 430 children receiving isoniazid and rifampicin. The histological appearance is characteristically patchy, and there is less periportal inflammation than is seen with isoniazid alone. There are two mechanisms that probably contribute to the synergistic hepatotoxicity of isoniazid and rifampicin: (1) interference with the clearance and excretion of bilirubin and bile acids contribute to the jaundice and alkaline phosphatase elevation and (2) acute hepatocellular injury attributed to immune-mediated allergic drug hepatotoxicity.

12.4.2.1.3 Pyrazinamide

Pyrazinamide is used more commonly in combination with other agents for active TB disease than as a monotherapy. There are reports of pyrazinamide-induced hepatotoxicity in adults that correlate with dosage and duration of treatment [99]. Some evidence suggests that age younger than 5 years was a factor that contributed to the development of severe hepatotoxicity in Japanese pediatric patients receiving pyrazinamide in combination with isoniazid and rifampicin [101]. Because pyrazinamide in combination with rifampicin has a high incidence of hepatotoxicity (7.7% severe hepatotoxicity), it is no longer recommended as latent TB treatment [102].

12.4.2.2 Antibiotics and Antifungals

Antibiotics are among the most widely prescribed drugs in the world. However, antibiotic-associated hepatitis is relatively rare considering they are frequently used. Most antibiotic-induced hepatotoxicity appears to be idiosyncratic and therefore unpredictable.

12.4.2.2.1 Amoxicillin/Clavulanic Acid

Amoxicillin/clavulanic acid (Augmentin) is one of the most frequently prescribed antibiotics and widely used for the treatment of respiratory infections and sinusitis/otitis. The incidence of hepatic disorder is estimated as 1.7 per 10,000 prescriptions [103] with a higher frequency in males [104]. The onset of symptoms starts after a mean period of 3 weeks and presents as nausea, vomiting, persistent jaundice, pruritus, abdominal pain, and signs of hypersensitivity such as fever and rash. Alterations of liver enzymes are often observed and histological changes consist of bile duct damage, spotty hepatocellular necrosis, and portal mononuclear inflammation. It was reported that rechallenge with amoxicillin/clavulanic acid led to the recurrence of hepatitis symptoms but not when amoxicillin was given alone. Further implicating clavulanic acid in eliciting hepatotoxic symptoms is that amoxicillin alone causes liver injury at a lower incidence of 0.3 per 10,000 prescriptions than the combination of amoxicillin and clavulanic acid.

12.4.2.2.2 Tetracyclines

Hepatotoxicity to tetracycline may be dose related, especially in women and patients with renal disease. The incidence of hepatotoxicity is estimated as 1.04 per 10,000 prescriptions [105]. With high doses of tetracycline, clinical evidence of hepatotoxicity appears 4 to 6 days into therapy and is characterized by nausea, vomiting, abdominal pain, and mild jaundice, with serum aminotransferase levels as high as 10 times the upper limit of normal and significant elevations of serum amylase. Histologically, the presence of microvesicular steatosis and hepatocellular necrosis is characteristic of tetracycline-induced liver injury. Tetracyclines inhibit the mitochondrial oxidation of fatty acids, resulting in an increased concentration of precursor free fatty acids in the liver. It is believed that both these free fatty acids and their oxidation metabolites are mitochondrial toxins, resulting in hepatotoxicity.

Minocycline has been associated with two types of hepatotoxicity: one is hypersensitivity hepatitis that occurs within days to weeks of the start of treatment and the other is a chronic autoimmune hepatitis that has a more delayed onset, generally months after the medication is started.

12.4.2.2.3 Erythromycin

Erythromycin-induced liver disease has been reported in adults and in children. Typically, symptoms begin 1 to 3 weeks after therapy has started and consist of right upper quadrant abdominal pain, anorexia, nausea, jaundice, and fever. Serum transaminase elevation less than 10 times the upper limit of normal and mild alkaline phosphatase and bilirubin elevations are usually seen [106,107]. Half of all patients experiencing hepatotoxicity will have peripheral eosinophilia. The pattern of hepatic injury is mostly cholestatic or mixed hepatocellular–cholestatic with mild hepatocyte necrosis and portal infiltration [108]. In pediatric patients, splenomegaly sometimes appears with hepatomegaly. The patterns of injury tend to be confined to Zone 3, which suggests the involvement of toxic metabolites owing to the high level of metabolizing enzymes in this region of the lobule such as the CYP3A subfamily that has been shown to metabolize erythromycin. The mechanism of injury has also been suggested as an immunoallergic reaction based on the recurrence of rash and fever with reexposure to the drug. In general, symptoms are reversible after discontinuing therapy; however, fatal liver disease can occur.

12.4.2.2.4 Sulfonamides

Multiple sulfonamide-containing medications (e.g., sulfanilamide, trimethoprim–sulfamethoxazole, and pyrimethamine–sulfadoxine) are routinely used for pediatric infections and have been reported to cause idiosyncratic hepatotoxicity. Acute liver failure is rare in children and most patients recover within 3 months after the discontinuation of the drug; however, instances of liver failure leading to the need for liver transplantation or death have been reported [109]. Symptoms of hepatotoxicity occur after a latency period of days to a month after therapy has begun. Most reported cases describe a centrilobular cholestasis with portal infiltration, but mixed hepatic–cholestatic and hepatocellular necrosis have also been observed. The mechanism of sulfonamide-induced hepatotoxicity has been suggested to be a hypersensitivity reaction, supported by symptoms of fever, rash, and peripheral eosinophilia.

12.4.2.2.5 Ketoconazole

Ketoconazole is associated with a high risk of hepatotoxicity. The incidence of keto-conazole-induced hepatotoxicity is estimated at 13.4 per 10,000 prescriptions [110]. Serum transaminase elevations occur in 2% to 17% of patients, with onset typically occurring after weeks of therapy. Hepatocellular injury is the dominant pattern; however, cholestatic or mixed patterns are also observed. Acute liver failure in the pediatric and adult populations has been observed with the use of ketoconazole. Older women appear to be the most sensitive to ketoconazole DILI [70].

12.4.3 ANTIEPILEPTICS

A variety of antiepileptic drugs have been associated with hepatotoxicity. Sections 12.4.3.1 through 12.4.3.5 provide a brief overview of the more problematic drugs.

12.4.3.1 Phenytoin

Phenytoin-induced hepatotoxicity occurs in children and 10%–38% of those cases will progress with hepatic necrosis and liver failure [111]. The interval between the initiation of phenytoin therapy and the onset of clinical abnormalities ranges from 1 to 6 weeks in the vast majority of patients [112]. The most common symptoms are fever, rash, and lymphadenopathy with jaundice and hepatosplenomegaly being found as well. Biochemical features are variable but generally include abnormal serum bilirubin, transaminases, and alkaline phosphatase levels, as well as eosinophilia and leukocytosis. The morphological and pathological abnormalities are nonspecific, including, but not limited to, primary hepatocellular degeneration and necrosis. Hepatic injury with phenytoin is most likely secondary to a hypersensitivity reaction rather than a direct hepatotoxic effect.

It is important to note that metabolites of aromatic antiepileptics (phenytoin, phenobarbital, and carbamazepine) are capable of binding to cellular proteins or forming antigens that could trigger an immunological response. Individuals with a defect of microsomal epoxide hydrolase may be unable to detoxify reactive metabolites and thus develop hepatotoxicity. Persons who develop hepatotoxicity from phenytoin are also likely to have hypersensitivity cross-reactions to other aromatic antiepileptics [113].

12.4.3.2 Carbamazepine/Oxcarbazepine

A transient and asymptomatic elevation of liver enzymes has been associated with carbamazepine, which occurs in 25%–61% of patients [112]. The onset of symptoms usually occurs within 3–4 weeks after the initiation of therapy and is independent of serum carbamazepine levels. In adults, two forms of hepatic injury have been reported: the predominant type is granulomatous hepatitis presenting with fever and right upper quadrant pain; the other is presented with small portal bile duct paucity and hepatocellular necrosis attributed to chronic toxicity from the drug. In children, the usual clinical picture is hepatitis, sometimes associated with a drug hypersensitivity syndrome similar to that of phenytoin. Serious adverse reactions may proceed even after early intervention and discontinuation of the drug, and liver failure may occur. Because carbamazepine causes autoinduction of hepatic enzymes, a higher dose may be required to ensure adequate serum drug concentrations. Oxcarbazepine is a derivative of carbamazepine; however, it does not induce hepatic enzymes. Drug–drug interactions may play a role in susceptibility to DILI. For example, phenytoin, felbamate, lamotrigine, and valproate have been shown to increase the serum concentration of the active metabolite of carbamazepine, carbamazepine-10,11 epoxide, thus causing potential toxicity [114,115].

12.4.3.3 Lamotrigine

Lamotrigine has been increasingly used in the pediatric population because it is an effective treatment for both focal and generalized epilepsies. Hepatotoxicity is uncommon, but hepatic failure and multiorgan failure have been described in adult and pediatric patients taking lamotrigine [116]. The dose of the drug needs to be reduced in patients with liver disease, who may have an altered metabolic clearance.

12.4.3.4 Felbamate

Felbamate has been associated with aplastic anemia and liver failure and side effects are much more common in polytherapy than in monotherapy. The incidence of liver failure is estimated to be 16.4 per 10,000 patients [117,118]. The usual clinical picture includes serious rash, aplastic anemia, and hepatic failure. Addition of felbamate will result in significant induction of CYP450 system and produce an increase in phenytoin, carbamazepine oxide, and valproate levels.

12.4.3.5 Sodium Valproate

Valproic acid (VPA) exposure is generally well tolerated but can cause hepatotoxicity. VPA-induced hepatotoxicity manifests as a variety of different adverse effects, including transient elevations of liver enzymes, hepatic dysfunction, hyperammonemia, and a more rare progressive liver failure that resembles Reye's syndrome. In 1996, L-carnitine was recommended by the Pediatric Neurology Advisory Committee as an antidote treatment for acute VPA toxicity in children; however, the outcome is still controversial [119]. The mechanism of VPA-induced hepatotoxicity is not fully understood; however, many theories have been postulated: first, acute VPA ingestion causes carnitine deficiency in the liver; second, VPA reduces coenzyme A thioester and inhibits mitochondrial oxidation of long-chain fatty acids; and third, the active metabolites (e.g., 4-ene-VPA) have been demonstrated to deplete glutathione and inhibit β-oxidation enzymes in the mitochondria.

Many studies have suggested that VPA-induced hepatotoxicity is much more frequent in children compared with adults and the risk of hepatotoxicity is estimated up to 1/600 for children less than 2 years of age and decreases thereafter [120,121]. In vitro testing has shown 4-ene-VPA to be toxic to human HepG2 cells [122]. The activity of CYP2C9, which is responsible for the production of 4-ene-VPA, is highest in young children and then declines with age. Concurrent with the changing activity of CYP2C9, the level of 4-ene-VPA decreases with age, suggesting that CYP2C9 and 4-ene-VPA are responsible for the differential sensitivity of young children to VPA.

12.4.4 ANTINEOPLASTICS

Many anticancer drugs can cause hepatotoxicity. A wide array of pathophysiological effects can occur including, but not limited to, cholestasis, hepatocellular necrosis, ductal injury, and steatosis [123]. Asymptomatic hepatocellular injury with elevated aminotransferases is common. Some drugs associated with this type of injury are nitrosoureas, 6-mercaptopurine, cytosine arabinoside, cis-platinum, and dacarbazine. Venoocclusive disease is another type of hepatotoxicity encountered with antineoplastic agents. Clinical signs include enlarged tender liver, ascites, and jaundice. Some drugs associated with this type of injury are thioguanine, dactinomycin, cytosine arabinoside, busulfan, and carmustine. Since venoocclusive disease most frequently develops after allogeneic bone marrow transplantation, it is possible that the liver toxicity is attributed to graft-versus-host disease and not a direct effect of the drugs [70]. Antineoplastic agents can also lead to more chronic effects such as fibrosis and cirrhosis [123].

12.4.5 HYPERTHYROIDISM

Propylthiouracil is an antithyroid medication widely used to treat hyperthyroidism owing to Graves' disease in both children and adults. Propylthiouracil interferes with the incorporation of iodine into thyroglobulin, thereby inhibiting thyroid hormone production [124]. Although hepatitis is rare, it is potentially dangerous and is the third most common cause of drug-induced acute liver failure in children (<18 years of age) [125]. Abnormal but asymptomatic liver function tests are sometimes found, particularly at higher dosages. When hepatitis occurs, symptoms typically begin within 2 months of starting treatment and consist of moderately elevated serum transaminases and nonspecific signs of hepatitis such as anorexia, nausea, vomiting, and jaundice [70].

Children appear to be more sensitive to propylthiouracil-induced hepatotoxicity than adults; however, the exact mechanism for the differential sensitivity is unknown. In a systematic review of adverse drug event reports submitted to the USFDA AERS, propylthiouracil had a higher adjusted reporting ratio for severe liver injury in children (<17 years of age) compared to adults [126]. In addition, propylthiouracil had a higher adjusted reporting ratio when compared to methimazole, another antithyroid medication, further supporting that children exhibit a unique sensitivity to propylthiouracil.

12.4.6 ALTERNATIVE MEDICATIONS

A large percentage of the US population consumes dietary supplements and the number of consumers continues to increase. Unlike FDA-regulated drugs, many dietary supplements are grandfathered under the regulatory law and thus do not require safety or efficacy information before marketing. Therefore, there are many products being consumed by the US population that have unknown safety. Compounding this is the fact that many people consider these "natural" dietary supplements to be safe. It is clear that at least some of the dietary supplements are not safe and have the potential to cause hepatotoxicity. Herbalife is a common weight loss product that has been associated with clinical cases of hepatotoxicity [127–130]. Hydroxycut manufactured several weight loss products that were recalled in 2009 because of concerns of hepatotoxicity [131]. Dietary supplements also have the potential to cause drug interactions by altering the activity of drug metabolism and transporter systems [84]. Therefore, it is important to be aware of the dietary supplements a patient is consuming since they alone may be the cause of the hepatotoxicity or affect the safety of a concomitant medication.

12.4.7   DRUGS OF ABUSE

Adolescence is a time when children may start experimenting with illicit drugs. Marijuana is a widely used drug that contains hundreds of different compounds, with some of them likely to be hepatotoxic. Liver enzyme abnormalities are commonly seen but serious hepatotoxicity is rare. Cocaine can cause ischemic necrosis of the liver and also centrilobular necrosis [70,84]. The amphetamine-derived compound ecstasy (3,4-methylenedioxymethamphetamine) is the most problematic. Ecstasy has been clearly associated with hepatotoxicity with clinical presentation ranging from asymptomatic elevations of serum liver enzymes to acute liver failure [132,133]. As with herbal medications, it is important to be aware of any illicit drugs a patient may be taking that might be the cause of the observed liver injury.

12.5   CONCLUSIONS

Drug-induced hepatotoxicity in children is commonly presented as acute hepatitis, although almost any major clinical pathological pattern of liver disease can occur. In most cases, children spontaneously recover after the offending medication is discontinued. The differential DILI responses observed between children and adults can be partly explained by the developmental changes of absorption, distribution, metabolism, and excretion. The dynamic balance between generation of a toxic metabolite and detoxification processes has a profound impact on the ultimate clinical outcome in children. Because of the complex pattern of biotransformation enzyme development, it is difficult to predict DILI potential in children. In addition, children's susceptibility to drug-induced hepatotoxicity is also influenced by genetic and environmental risk factors. Pediatricians should be aware of the potential hepatotoxicity risks associated with some drugs, such as antituberculous and antiepileptic medications. Mechanistic studies of hepatotoxic drugs are needed to better predict DILI potential and identify risk factors so that these drugs can be used safely in children.

REFERENCES

1. P.B. Watkins, P.J. Seligman, J.S. Pears, M.I. Avigan, and J.R. Senior, *Using controlled clinical trials to learn more about acute drug-induced liver injury*, Hepatology 48 (2008), pp. 1680–1689.
2. A. Suzuki, R.J. Andrade, E. Bjornsson, M.I. Lucena, W.M. Lee, N.A. Yuen, C.M. Hunt, and J.W. Freston, *Drugs associated with hepatotoxicity and their reporting frequency of liver adverse events in VigiBase: Unified list based on international collaborative work*, Drug Saf. 33 (2010), pp. 503–522.

3. N. Kaplowitz, *Drug-induced liver disorders: Implications for drug development and regulation*, Drug Saf. 24 (2001), pp. 483–490.

4. R.J. Temple and M.H. Himmel, *Safety of newly approved drugs: Implications for prescribing*, JAMA 287 (2002), pp. 2273–2275.

5. USFDA, *Draft Guidance for Industry: How to Comply with the Pediatric Research Equity Act*, US Food and Drug Administration, Washington, DC, 2005, pp. 1–18.

6. USEPA, *Determination of the Appropriate FQPA Safety Factor(s) in Tolerance Assessment*, US Environmental Protection Agency, Washington, DC, 2002, pp. 1–53.

7. A. Ornoy, *Valproic acid in pregnancy: How much are we endangering the embryo and fetus?* Reprod. Toxicol. 28 (2009), pp. 1–10.

8. F.P. Lemaigre, *Mechanisms of liver development: Concepts for understanding liver disorders and design of novel therapies*, Gastroenterology 137 (2009), pp. 62–79.

9. K.D. Tremblay and K.S. Zaret, *Distinct populations of endoderm cells converge to generate the embryonic liver bud and ventral foregut tissues*, Dev. Biol. 280 (2005), pp. 87–99.

10. K.S. Zaret, *Regulatory phases of early liver development: Paradigms of organogenesis*, Nat. Rev. Genet. 3 (2002), pp. 499–512.

11. N. Tanimizu and A. Miyajima, *Molecular mechanism of liver development and regeneration*, Int. Rev. Cytol. 259 (2007), pp. 1–48.

12. K.S. Zaret, *Genetic programming of liver and pancreas progenitors: Lessons for stem-cell differentiation*, Nat. Rev. Genet. 9 (2008), pp. 329–340.

13. R. Zhao and S.A. Duncan, *Embryonic development of the liver*, Hepatology 41 (2005), pp. 956–967.

14. J.W. Kung, I.S. Currie, S.J. Forbes, and J.A. Ross, *Liver development, regeneration, and carcinogenesis*, J. Biomed. Biotechnol. 2010 (2010), p. 984248.

15. E.W. Carney, A.R. Scialli, R.E. Watson, and J.M. DeSesso, *Mechanisms regulating toxicant disposition to the embryo during early pregnancy: An interspecies comparison*, Birth Defects Res. C Embryo Today 72 (2004), pp. 345–360.

16. J. Hakkola, O. Pelkonen, M. Pasanen, and H. Raunio, *Xenobiotic-metabolizing cytochrome P450 enzymes in the human feto-placental unit: Role in intrauterine toxicity*, Crit. Rev. Toxicol. 28 (1998), pp. 35–72.

17. J. Alcorn and P.J. McNamara, *Pharmacokinetics in the newborn*, Adv. Drug Deliv. Rev. 55 (2003), pp. 667–686.

18. D.G. McCarver and R.N. Hines, *The ontogeny of human drug-metabolizing enzymes: Phase II conjugation enzymes and regulatory mechanisms*, J. Pharmacol. Exp. Ther. 300 (2002), pp. 361–366.

19. J.A. Thomas, *Toxic responses of the reproductive system*, in *Casarett and Doull's Toxioclogy The Basic Science of Poisons*, C.D. Klaassen, M.O. Amdur, and J. Doull, eds., McGraw-Hill, New York, 1996, pp. 547–581.

20. D. Evseenko, J.W. Paxton, and J.A. Keelan, *Active transport across the human placenta: Impact on drug efficacy and toxicity*, Expert Opin. Drug Metab. Toxicol. 2 (2006), pp. 51–69.

21. G.R. Lankas, L.D. Wise, M.E. Cartwright, T. Pippert, and D.R. Umbenhauer, *Placental P-glycoprotein deficiency enhances susceptibility to chemically induced birth defects in mice*, Reprod. Toxicol. 12 (1998), pp. 457–463.

22. J.W. Smit, M.T. Huisman, O. van Tellingen, H.R. Wiltshire, and A.H. Schinkel, *Absence or pharmacological blocking of placental P-glycoprotein profoundly increases fetal drug exposure*, J. Clin. Invest. 104 (1999), pp. 1441–1447.

23. C.P. Sibley, *Understanding placental nutrient transfer—Why bother? New biomarkers of fetal growth*, J. Physiol. 587 (2009), pp. 3431–3440.

24. P. Myllynen, M. Pasanen, and K. Vahakangas, *The fate and effects of xenobiotics in human placenta*, Expert Opin. Drug Metab. Toxicol. 3 (2007), pp. 331–346.

25. G.W. Mihaly and D.J. Morgan, *Placental drug transfer: Effects of gestational age and species*, Pharmacol. Ther. 23 (1983), pp. 253–266.

26. D.R. Mattison, E. Blann, and A. Malek, *Physiological alterations during pregnancy: Impact on toxicokinetics*, Fundam. Appl. Toxicol. 16 (1991), pp. 215–218.

27. W.G. McBride, *Thalidomide embryopathy*, Teratology 16 (1977), pp. 79–82.

28. USFDA, *Safety Alerts for Human Medicinal Products: Effexor and Effexor XR (venlafaxine HCl)*, 2010. Available at http://www.fda.gov/Safety/MedWatch/Safety Information/SafetyAlertsforHumanMedicalProducts/ucm154975.htm.

29. R.N. Hines, *Ontogeny of human hepatic cytochromes P450*, J. Biochem. Mol. Toxicol. 21 (2007), pp. 169–175.

30. T. Shimada, H. Yamazaki, M. Mimura, N. Wakamiya, Y.F. Ueng, F.P. Guengerich, and Y. Inui, *Characterization of microsomal cytochrome P450 enzymes involved in the oxidation of xenobiotic chemicals in human fetal liver and adult lungs*, Drug Metab. Dispos. 24 (1996), pp. 515–522.

31. P.J. Gow, H. Ghabrial, R.A. Smallwood, D.J. Morgan, and M.S. Ching, *Neonatal hepatic drug elimination*, Pharmacol. Toxicol. 88 (2001), pp. 3–15.

32. A. Cumano and I. Godin, *Ontogeny of the hematopoietic system*, Annu. Rev. Immunol. 25 (2007), pp. 745–785.

33. S.N. Hart, Y. Cui, C.D. Klaassen, and X.B. Zhong, *Three patterns of cytochrome P450 gene expression during liver maturation in mice*, Drug Metab. Dispos. 37 (2009), pp. 116–121.

34. J. Hakkola, E. Tanaka, and O. Pelkonen, *Developmental expression of cytochrome P450 enzymes in human liver*, Pharmacol. Toxicol. 82 (1998), pp. 209–217.

35. ICH, *ICH harmonised tripartite guideline: Detection of toxicity to reproduction for medicinal products & toxicity to male fertility S5(R2)*, International Conference on Harmonisation, 1993, pp. 1–21.

36. R.S. Cohen, R.J. Wong, and D.K. Stevenson, *Understanding neonatal jaundice: A perspective on causation*, Pediatr. Neonatol. 51 (2010), pp. 143–148.

37. F.J. Cuperus, A.M. Hafkamp, C.V. Hulzebos, and H.J. Verkade, *Pharmacological therapies for unconjugated hyperbilirubinemia*, Curr. Pharm. Des. 15 (2009), pp. 2927–2938.

38. M.J. Blake, L. Castro, J.S. Leeder, and G.L. Kearns, *Ontogeny of drug metabolizing enzymes in the neonate*, Semin. Fetal. Neonatal. Med. 10 (2005), pp. 123–138.

39. K. Allegaert, J.N. van den Anker, G. Naulaers, and J. de Hoon, *Determinants of drug metabolism in early neonatal life*, Curr. Clin. Pharmacol. 2 (2007), pp. 23–29.

40. E.G. Hrycay and S.M. Bandiera, *Expression, function and regulation of mouse cytochrome P450 enzymes: Comparison with human P450 enzymes*, Curr. Drug Metab. 10 (2009), pp. 1151–1183.

41. A. Parkison, *Biotransformation of xenobiotics*, in *Casarett and Doull's Toxicology the Basic Science of Poisons*, C.D. Klaassen, M.O. Amdur, and J. Doull, eds., McGraw-Hill, New York, 1996, pp. 113–186.

42. T.N. Johnson, *The development of drug metabolising enzymes and their influence on the susceptibility to adverse drug reactions in children*, Toxicology 192 (2003), pp. 37–48.

43. G.D. Anderson, *Using pharmacokinetics to predict the effects of pregnancy and maternal-infant transfer of drugs during lactation*, Expert Opin. Drug Metab. Toxicol. 2 (2006), pp. 947–960.

44. R.B. Birge, J.B. Bartolone, S.D. Cohen, E.A. Khairallah, and L.A. Smolin, *A comparison of proteins S-thiolated by glutathione to those arylated by acetaminophen*, Biochem. Pharmacol. 42 (1991), pp. S197–S207.

45. J.G. Bessems and N.P. Vermeulen, *Paracetamol (acetaminophen)-induced toxicity: Molecular and biochemical mechanisms, analogues and protective approaches*, Crit. Rev. Toxicol. 31 (2001), pp. 55–138.

46. I.H. Bartelink, C.M. Rademaker, A.F. Schobben, and J.N. van den Anker, *Guidelines on paediatric dosing on the basis of developmental physiology and pharmacokinetic considerations*, Clin. Pharmacokinet. 45 (2006), pp. 1077–1097.

47. J. Gershanik, B. Boecler, H. Ensley, S. McCloskey, and W. George, *The gasping syndrome and benzyl alcohol poisoning*, N. Engl. J. Med. 307 (1982), pp. 1384–1388.

48. G.L. Kearns, S.M. Abdel-Rahman, S.W. Alander, D.L. Blowey, J.S. Leeder, and R.E. Kauffman, *Developmental pharmacology—Drug disposition, action, and therapy in infants and children*, N. Engl. J. Med. 349 (2003), pp. 1157–1167.

49. V.M. Pineiro-Carrero and E.O. Pineiro, *Liver*, Pediatrics 113 (2004), pp. 1097–1106.

50. S.N. de Wildt, G.L. Kearns, W.C. Hop, D.J. Murry, S.M. Abdel-Rahman, and J.N. van den Anker, *Pharmacokinetics and metabolism of oral midazolam in preterm infants*, Br. J. Clin. Pharmacol. 53 (2002), pp. 390–392.

51. W. Klinger, *Developmental pharmacology and toxicology: Biotransformation of drugs and other xenobiotics during postnatal development*, Eur. J. Drug Metab. Pharmacokinet. 30 (2005), pp. 3–17.

52. E.R. Stadtman, *Protein oxidation and aging*, Free Radic. Res. 40 (2006), pp. 1250–1258.

53. M. Erden-Inal, E. Sunal, and G. Kanbak, *Age-related changes in the glutathione redox system*, Cell Biochem. Funct. 20 (2002), pp. 61–66.

54. H. Tsukahara, *Biomarkers for oxidative stress: Clinical application in pediatric medicine*, Curr. Med. Chem. 14 (2007), pp. 339–351.

55. E.E. Stultz, J.L. Stokes, M.L. Shaffer, I.M. Paul, and C.M. Berlin, *Extent of medication use in breastfeeding women*, Breastfeed Med. 2 (2007), pp. 145–151.

56. F. Fortinguerra, A. Clavenna, and M. Bonati, *Psychotropic drug use during breastfeeding: A review of the evidence*, Pediatrics 124 (2009), pp. e547–e556.

57. E. Nordmo, L. Aronsen, K. Wasland, L. Småbrekke, and S. Vorren, *Severe apnea in an infant exposed to lamotrigine in breast milk*, Ann. Pharm. 43 (2009), pp. 1893–1897.

58. T.M. Badger, J.M. Gilchrist, R.T. Pivik, A. Andres, K. Shankar, J.R. Chen, and M.J. Ronis, *The health implications of soy infant formula*, Am. J. Clin. Nutr. 89 (2009), pp. 1668S–1672S.

59. D.R. Doerge, *Bioavailability of soy isoflavones through placental/lactational transfer and soy food*, Toxicol. Appl. Pharmacol. 254 (2011), pp. 145–147.

60. D.J. Waxman and M.G. Holloway, *Sex differences in the expression of hepatic drug metabolizing enzymes*, Mol. Pharmacol. 76 (2009), pp. 215–228.

61. X. Yang, E.E. Schadt, S. Wang, H. Wang, A.P. Arnold, L. Ingram-Drake, T.A. Drake, and A.J. Lusis, *Tissue-specific expression and regulation of sexually dimorphic genes in mice*, Genome Res. 16 (2006), pp. 995–1004.

62. J.C. Kwekel, V.G. Desai, C.L. Moland, W.S. Branham, and J.C. Fuscoe, *Age and sex dependent changes in liver gene expression during the life cycle of the rat*, BMC Genomics 11 (2010), p. 675.

63. A. Mode and J.A. Gustafsson, *Sex and the liver—A journey through five decades*, Drug Metab. Rev. 38 (2006), pp. 197–207.

64. R.J. Devlin and J.A. Henry, *Clinical review: Major consequences of illicit drug consumption*, Crit. Care 12 (2008), p. 202.

65. L.R. Cantilena, S.A. Cherstniakova, G. Saviolakis, R. Kahn, A. Elkashef, L. Rose, and F. Vocci, *Prevalence of abnormal liver-associated enzymes in cocaine experienced adults versus healthy volunteers during phase 1 clinical trials*, Contemp. Clin. Trials 28 (2007), pp. 695–704.

66. G. Stirnimann, K. Kessebohm, and B. Lauterburg, *Liver injury caused by drugs: An update*, Swiss Med. Wkly. 140 (2010), p. w13080.

67. S. Russmann, A. Jetter, and G.A. Kullak-Ublick, *Pharmacogenetics of drug-induced liver injury*, Hepatology 52 (2010), pp. 748–761.

68. Q. Shi, H. Hong, J. Senior, and W. Tong, *Biomarkers for drug-induced liver injury*, Expert Rev. Gastroenterol. Hepatol. 4 (2010), pp. 225–234.
69. R. Ramachandran and S. Kakar, *Histological patterns in drug-induced liver disease*, J. Clin. Pathol. 62 (2009), pp. 481–492.
70. E.A. Roberts, *Drug-induced liver disease*, in *Liver Disease in Children*, F.J. Suchy, R.J. Sokol, and W.F. Balistreri, eds., Lippincott William & Wilkins, Philadelphia, PA, 2001, pp. 463–491.
71. P.B. Watkins and L.B. Seeff, *Drug-induced liver injury: Summary of a single topic clinical research conference*, Hepatology 43 (2006), pp. 618–631.
72. N. Kaplowitz, *Idiosyncratic drug hepatotoxicity*, Nat. Rev. Drug Discov. 4 (2005), pp. 489–499.
73. USFDA, *Guidance for Industry: Drug-Induced Liver Injury- Premarketing Clinical Evaluation*, USFDA, Washington, DC, 2009, pp. 1–25.
74. N. Chalasani, R.J. Fontana, H.L. Bonkovsky, P.B. Watkins, T. Davern, J. Serrano, H. Yang, J. Rochon, and Drug Induced Liver Injury Network (DILIN), *Causes, clinical features, and outcomes from a prospective study of drug-induced liver injury in the United States*, Gastroenterology 135 (2008), pp. 1924–1934, 1934.e1–e4.
75. A.M. Larson, J. Polson, R.J. Fontana, T.J. Davern, E. Lalani, L.S. Hynan, J.S. Reisch, F.V. Schiødt, G. Ostapowicz, A.O. Shakil, W.M. Lee, and Acute Liver Failure Study Group, *Acetaminophen-induced acute liver failure: Results of a United States multicenter, prospective study*, Hepatology 42 (2005), pp. 1364–1372.
76. R.J. Fontana and L.G. Quallich, *Acute liver failure*, Curr. Opin. Gastroenterol. 17 (2001), pp. 291–298.
77. W. Norris, A.H. Paredes, and J.H. Lewis, *Drug-induced liver injury in 2007*, Curr. Opin. Gastroenterol. 24 (2008), pp. 287–297.
78. R.H. Squires Jr., B.L. Shneider, J. Bucuvalas, E. Alonso, R.J. Sokol, M.R. Narkewicz, A. Dhawan, P. Rosenthal, N. Rodriguez-Baez, K.F. Murray, S. Horslen, M.G. Martin, M.J. Lopez, H. Soriano, B.M. McGuire, M.M. Jonas, N. Yazigi, R.W. Shepherd, K. Schwarz, S. Lobritto, D.W. Thomas, J.E. Lavine, S. Karpen, V. Ng, D. Kelly, N. Simonds, and L.S. Hynan, *Acute liver failure in children: The first 348 patients in the pediatric acute liver failure study group*, J. Pediatr. 148 (2006), pp. 652–658.
79. L.P. James, E.V. Capparelli, P.M. Simpson, L. Letzig, D. Roberts, J.A. Hinson, G.L. Kearns, J.L. Blumer, J.E. Sullivan, and Network of Pediatric Pharmacology Research Units, National Institutes of Child Health and Human Development, *Acetaminophen-associated hepatic injury: Evaluation of acetaminophen protein adducts in children and adolescents with acetaminophen overdose*, Clin. Pharmacol. Ther. 84 (2008), pp. 684–690.
80. US-FDA, *Joint Meeting of the Drug Safety and Risk Management Advisory Committee with the Anesthetic and Life Support Drugs Advisory Committee and the Nonprescription Drugs Advisory Committee (June 29–30, 2009). Meeting Background Information, Presentations, and Minutes/Voting Results*, 2009. Available at http://www.fda.gov/AdvisoryCommittees/Calendar/ucm143083.htm.
81. A. Burke, E. Smyth, and G.A. Fitzgerald, *Chapter 26. Analgesic, antipyretic and antiinflammatory agents: Pharmacotherapy of gout*, in *Goodman & Gilman's The Pharmacological Basis of Therapeutics*, 11th ed., L.L. Brunton, J.S. Lazo, and K.L. Parker, eds., McGraw-Hill Companies, Inc., New York, 2006.
82. E. Kozer, R. Greenberg, D.R. Zimmerman, and M. Berkovitch, *Repeated supratherapeutic doses of paracetamol in children—A literature review and suggested clinical approach*, Acta Paediatr. 95 (2006), pp. 1165–1171.
83. S.B. Mahadevan, P.J. McKiernan, P. Davies, and D.A. Kelly, *Paracetamol induced hepatotoxicity*, Arch. Dis. Child 91 (2006), pp. 598–603.
84. K.F. Murray, N. Hadzic, S. Wirth, M. Bassett, and D. Kelly, *Drug-related hepatotoxicity and acute liver failure*, J. Pediatr. Gastroenterol. Nutr. 47 (2008), pp. 395–405.

85. J. Trowell, R. Peto, and A.C. Smith, *Controlled trial of repeated halothane anaesthetics in patients with carcinoma of the uterine cervix treated with radium*, Lancet 305 (1975), pp. 821–824.

86. W.W. Mushin, M. Rosen, and E.V. Jones, *Post-halothane jaundice in relation to previous administration of halothane*, Br. Med. J. 3 (1971), pp. 18–22.

87. G.C. Farrell, L. Frost, M. Tapner, J. Field, M. Weltman, and J. Mahoney, *Halothane-induced liver injury in guinea-pigs: Importance of cytochrome P450 enzyme activity and hepatic blood flow*, J. Gastroenterol. Hepatol. 11 (1996), pp. 594–601.

88. C.A. Lunam, M.J. Cousins, and P.D. Hall, *Guinea-pig model of halothane-associated hepatotoxicity in the absence of enzyme induction and hypoxia*, J. Pharmacol. Exp. Ther. 232 (1985), pp. 802–809.

89. M. Bourdi, H.R. Amouzadeh, T.H. Rushmore, J.L. Martin, and L.R. Pohl, *Halothane-induced liver injury in outbred guinea pigs: Role of trifluoroacetylated protein adducts in animal susceptibility*, Chem. Res. Toxicol. 14 (2001), pp. 362–370.

90. S.S. Wu, C.S. Chao, J.H. Vargas, H.L. Sharp, M.G. Martín, S.V. McDiarmid, F.R. Sinatra, and M.E. Ament, *Isoniazid-related hepatic failure in children: A survey of liver transplantation centers*, Transplantation 84 (2007), pp. 173–179.

91. J.R. Mitchell, H.J. Zimmerman, K.G. Ishak, U.P. Thorgeirsson, J.A. Timbrell, W.R. Snodgrass, and S.D. Nelson, *Isoniazid liver injury: Clinical spectrum, pathology, and probable pathogenesis*, Ann. Intern. Med. 84 (1976), p. 181.

92. D.E. Kopanoff, D.J. Snider, and G. Caras, *Isoniazid-related hepatitis: A U.S. Public Health Service cooperative surveillance study*, Am. Rev. Respir. Dis. 117 (1978), pp. 991–1001.

93. New Jersey Medical School Global Tuberculosis Institute, *Management of Latent Tuberculosis Infection in Children and Adolescents: A Guide for the Primary Care Provider*, Newark, NJ, 2009, pp. 12–20.

94. F.E. Berkowitz, S.L. Henderson, N. Fajman, B. Schoen, and M. Naughton, *Acute liver failure caused by isoniazid in a child receiving carbamazepine*, Int. J. Tuberc. Lung Dis. 2 (1998), pp. 603–606.

95. J.M. Wright, E.F. Stokes, and V.P. Sweeney, *Isoniazid-induced carbamazepine toxicity and vice versa: A double drug interaction*, N. Engl. J. Med. 307 (1982), pp. 1325–1327.

96. B. Roy, S.K. Ghosh, D. Sutradhar, N. Sikdar, S. Mazumder, and S. Barman, *Predisposition of antituberculosis drug induced hepatotoxicity by cytochrome P450 2E1 genotype and haplotype in pediatric patients*, J. Gastroenterol. Hepatol. 21 (2006), pp. 784–786.

97. A.R. Frydenberg and S.M. Graham, *Toxicity of first-line drugs for treatment of tuberculosis in children: Review*, Trop. Med. Int. Health 14 (2009), pp. 1329–1337.

98 I.G. Possuelo, J.A. Castelan, T.C. de Brito, A.W. Ribeiro, P.I. Cafrune, P.D. Picon, A.R. Santos, R.L. Teixeira, T.S. Gregianini, M.H. Hutz, M.L. Rossetti, and A. Zaha, *Association of slow N-acetyltransferase 2 profile and anti-TB drug-induced hepatotoxicity in patients from Southern Brazil*, Eur. J. Clin. Pharmacol. 64 (2008), pp. 673–681.

99. A. Tostmann, M.J. Boeree, R.E. Aarnoutse, W.C. de Lange, A.J. van der Ven, and R. Dekhuijzen, *Antituberculosis drug-induced hepatotoxicity: Concise up-to-date review*, J. Gastroenterol. Hepatol. 23 (2008), pp. 192–202.

100. R.J. O'Brien, M.W. Long, F.S. Cross, M.A. Lyle, and D.E. Snider Jr., *Hepatotoxicity from isoniazid and rifampin among children treated for tuberculosis*, Pediatrics 72 (1983), pp. 491–499.

101. K. Ohkawa, M. Hashiguchi, K. Ohno, C. Kiuchi, S. Takahashi, S. Kondo, H. Echizen, and H. Ogata, *Risk factors for antituberculous chemotherapy-induced hepatotoxicity in Japanese pediatric patients*, Clin. Pharmacol. Ther. 72 (2002), pp. 220–226.

102. *Update: Adverse event data and revised American Thoracic Society/CDC recommendations against the use of rifampin and pyrazinamide for treatment of latent tuberculosis infection—United States, 2003*, MMWR Morb. Mortal. Wkly. Rep. 52 (2003), pp. 735–739.

103. L.A. Garcia Rodriguez, B.H. Stricker, and H.J. Zimmerman, *Risk of acute liver injury associated with the combination of amoxicillin and clavulanic acid*, Arch. Intern. Med. 156 (1996), pp. 1327–1332.

104. U. Gresser, *Amoxicillin-clavulanic acid therapy may be associated with severe side effects—Review of the literature*, Eur. J. Med. Res. 6 (2001), pp. 139–149.

105. H.E. Seaman, R.A. Lawrenson, T.J. Williams, K.D. MacRae, and R.D. Farmer, *The risk of liver damage associated with minocycline: A comparative study*, J. Clin. Pharmacol. 41 (2001), pp. 852–860.

106. J.L. Carson, B.L. Strom, A. Duff, A. Gupta, M. Shaw, F.E. Lundin, and K. Das, *Acute liver disease associated with erythromycins, sulfonamides, and tetracyclines*, Ann. Intern. Med. 119 (1993), pp. 576–583.

107. L.E. Derby, H. Jick, D.A. Henry, and A.D. Dean, *Erythromycin-associated cholestatic hepatitis*, Med. J. Aust. 158 (1993), pp. 600–602.

108. E.S. Zafrani, K.G. Ishak, and C. Rudzki, *Cholestatic and hepatocellular injury associated with erythromycin esters: Report of nine cases*, Dig. Dis. Sci. 24 (1979), pp. 385–396.

109. B. Simma, B. Meister, J. Deutsch, W. Sperl, F. Fend, D. Ofner, R. Margreiter, and W. Vogel, *Fulminant hepatic failure in a child as a potential adverse effect of trimethoprim-sulphamethoxazole*, Eur. J. Pediatr. 154 (1995), pp. 530–533.

110. R. García, A. Duque, J. Castellsague, S. Pérez-Gutthann, and B.H. Stricker, *A cohort study on the risk of acute liver injury among users of ketoconazole and other antifungal drugs*, Br. J. Clin. Pharmacol. 48 (1999), pp. 847–852.

111. F.E. Dreifuss and D.H. Langer, *Hepatic considerations in the use of antiepileptic drugs*, Epilepsia 28 (1987), pp. S23–S29.

112. S.N. Ahmed and Z.A. Siddiqi, *Antiepileptic drugs and liver disease*, Seizure 15 (2006), pp. 156–164.

113. N.H. Shear and S.P. Spielberg, *Anticonvulsant hypersensitivity syndrome. In vitro assessment of risk*, J. Clin. Invest. 82 (1988), pp. 1826–1832.

114. T. Warner, P.N. Patsalos, M. Prevett, A.A. Elyas, and J.S. Duncan, *Lamotrigine-induced carbamazepine toxicity: An interaction with carbamazepine-10,11-epoxide*, Epilepsy Res. 11 (1992), pp. 147–150.

115. J.M. Potter and A. Donnelly, *Carbamazepine-10,11-epoxide in therapeutic drug monitoring*, Ther. Drug Monit. 20 (1998), pp. 652–657.

116. K. Overstreet, C. Costanza, C. Behling, T. Hassanin, and E. Masliah, *Fatal progressive hepatic necrosis associated with lamotrigine treatment: A case report and literature review*, Dig. Dis. Sci. 47 (2002), pp. 1921–1925.

117. F. Onat and C. Ozkara, *Adverse effects of new antiepileptic drugs*, Drugs Today (Barc.) 40 (2004), pp. 325–342.

118. J.M. Pellock, *Antiepileptic Drugs*, 5th ed., Raven Press, New York, 2002, pp. 301–318.

119. S. Russell, *Carnitine as an antidote for acute valproate toxicity in children*, Curr. Opin. Pediatr. 19 (2007), pp. 206–210.

120. A.E. Bryant 3rd and F.E. Dreifuss, *Valproic acid hepatic fatalities. III. U.S. experience since 1986*, Neurology 46 (1996), pp. 465–469.

121. J. Perrott, N.G. Murphy, and P.J. Zed, *L-carnitine for acute valproic acid overdose: A systematic review of published cases*, Ann. Pharm. 44 (2010), pp. 1287–1293.

122. M.G. Neuman, N.H. Shear, P.M. Jacobson-Brown, G.G. Katz, H.K. Neilson, I.M. Malkiewicz, R.G. Cameron, and F. Abbott, *CYP2E1-mediated modulation of valproic acid-induced hepatocytotoxicity*, Clin. Biochem. 34 (2001), pp. 211–218.

123. S. Castellino, A. Muir, A. Shah, S. Shope, K. McMullen, K. Ruble, A. Barber, A. Davidoff, and M.M. Hudson, *Hepato-biliary late effects in survivors of childhood and adolescent cancer: A report from the Children's Oncology Group*, Pediatr. Blood Cancer 54 (2010), pp. 663–669.

124. A.P. Farwell and L.E. Braverman, *Chapter 56. Thyroid and antithyroid drugs*, in *Goodman & Gilman's The Pharmacological Basis of Therapeutics*, 11th ed., L.L. Brunton, J.S. Lazo, and K.L. Parker, eds., McGraw-Hill Companies, Inc., New York, 2006.

125. M.W. Russo, J.A. Galanko, R. Shrestha, M.W. Fried, and P. Watkins, *Liver transplantation for acute liver failure from drug induced liver injury in the United States*, Liver Transpl. 10 (2004), pp. 1018–1023.

126. S.A. Rivkees and A. Szarfman, *Dissimilar hepatotoxicity profiles of propylthiouracil and methimazole in children*, J. Clin. Endocrinol. Metab. 95 (2010), pp. 3260–3267.

127. G.C. Chen, V.S. Ramanathan, D. Law, P. Funchain, G.C. Chen, S. French, B. Shlopov, V. Eysselein, D. Chung, S. Reicher, and B.V. Pham, *Acute liver injury induced by weight-loss herbal supplements*, World J. Hepatol. 2 (2010), pp. 410–415.

128. S. Chao, M. Anders, M. Turbay, E. Olaiz, L. Mc Cormack, and R. Mastai, *[Toxic hepatitis by consumption Herbalife products a case report]*, Acta Gastroenterol. Latinoam. 38 (2008), pp. 274–277.

129. A.M. Schoepfer, A. Engel, K. Fattinger, U.A. Marbet, D. Criblez, J. Reichen, A. Zimmermann, and C.M. Oneta, *Herbal does not mean innocuous: Ten cases of severe hepatotoxicity associated with dietary supplements from Herbalife products*, J. Hepatol. 47 (2007), pp. 521–526.

130. T. Sharma, L. Wong, N. Tsai, and R.D. Wong, *Hydroxycut((R)) (herbal weight loss supplement) induced hepatotoxicity: A case report and review of literature*, Hawaii Med. J. 69 (2010), pp. 188–190.

131. T.L. Fong, K.C. Klontz, A. Canas-Coto, S.J. Casper, F.A. Durazo, T.J. Davern 2nd, P. Hayashi, W.M. Lee, and L.B. Seeff, *Hepatotoxicity due to hydroxycut: A case series*, Am. J. Gastroenterol. 105 (2010), pp. 1561–1566.

132. M. Carvalho, H. Pontes, F. Remião, M.L. Bastos, and F. Carvalho, *Mechanisms underlying the hepatotoxic effects of ecstasy*, Curr. Pharm. Biotechnol. 11 (2010), pp. 476–495.

133. N. Brncic, I. Kraus, I. Visković, B. Mijandrusić-Sincić, and V. Vlahović-Palcevski, *3,4-methylenedioxymethamphetamine (MDMA): An important cause of acute hepatitis*, Med. Sci. Monit. 12 (2006), pp. CS107–CS109.

13 Adverse Outcome Pathways for Developmental Toxicity

Damiano Portinari and Philip N. Judson

CONTENTS

ABSTRACT

Developmental toxicity includes alterations in growth, functional abnormalities, structural abnormalities, and death of the immature organism. The various biological processes that can be modified include cell proliferation, cell movement, cell differentiation, cell signaling, and other cellular functions. Given the complex processes involved in both normal development and toxicity, understanding biological pathways that are disrupted is essential. This section reviews pathways that are important for development and how toxicity can modify the role of the pathway leading to developmental toxicity.

KEYWORDS

Adverse outcome pathway, developmental toxicity, fetal lung, glucocorticoids, labor, parturition, premature birth, prolonged gestation, risk assessment, sex development, toxicity pathway

13.1 INTRODUCTION

13.1.1 DEVELOPMENTAL TOXICOLOGY

Developmental toxicology is a branch of toxicology that studies mechanisms, pathogenesis, and the pregnancy outcome after exposure to chemicals or anomalous physiological conditions that lead to abnormal development. Developmental toxicity outcomes include functional deficit, structural malformation, growth retardation, and death of the organism. Basic processes leading to such adverse outcomes are related to disruption of enzymes or receptors, cell division, cell proliferation, protein synthesis, and also chromosomal aberrations and mutations. Developmental toxicity is a complex and ill-defined endpoint. Developmental toxicity and life stage at which exposure occurs are correlated. Developmental stages are vulnerable to agents during gametogenesis and fertilization, gastrulation where the single-cell zygote develops into a trilayered gastrula (ectoderm, endoderm, and mesoderm), and organogenesis where vulnerability is at its peak [1,2]. Major developmental defects such as death can be seen during early embryogenesis including during gastrulation; structural malformations can be seen during organogenesis; and growth retardation and functional deficiency can be seen if exposure occurs during fetal genesis.

13.1.2 ADVERSE OUTCOME PATHWAYS AS A POTENTIAL TOOL FOR RISK ASSESSMENT OF DEVELOPMENTAL TOXICANTS

A chemical category for risk assessment is a group of chemicals in which a defined set of properties are similar or show a similar trend. Properties that are relevant for such categories are physicochemical, human health, and environmental toxicity or environmental fate [3]. The process of building categories for hazard assessment becomes difficult when it is necessary to consider a particular toxicity endpoint. It cannot be assumed that chemicals with similar chemical or physicochemical properties will share the same toxicological endpoint without considering other factors. A toxicologically meaningful category (TMC) is formed by considering common reactivity mechanisms, biological mechanisms, and mode of toxic action, and the complexity of TMCs increase from those based on chemical reactivity to those based on biological activity [1]. It can be reasonably easy to form a TMC for hazard endpoints when covalent interactions act as a molecular initiating event (MIE) that leads to an in vivo outcome but it is more difficult to form a TMC when noncovalent interactions take place. MIE can be defined as the first and critical event in a sequence of chemical, biochemical, cellular, physiological responses that lead to an adverse in vivo outcome. The group of linkages within the biological cascade connecting the MIE and the in vivo adverse outcome at different levels of organization (from cellular to the whole animal) constitute an adverse outcome pathway (AOP) [1,4] (Figure 13.1). In complex endpoints such as developmental toxicity, different MIEs can lead to a very similar in vivo outcome but through different cellular, biochemical, and other system effects. A well-characterized AOP describes an initial chemical interaction and a series of measurable biological events from a cellular to a whole organism level leading to a final in vivo outcome. For complex hazard endpoints, it is important to identify which in vitro assays are essential in order to separate each pathway from the others [1]. The future use of the AOP approach would be to assess chemicals for which little or no information is available. If only molecular and biochemical processes are considered, there is not enough information for a quantitative assessment of a chemical. However, from its similarity to other chemicals, it may be possible to associate it with an AOP for which the in vivo outcome is well defined.

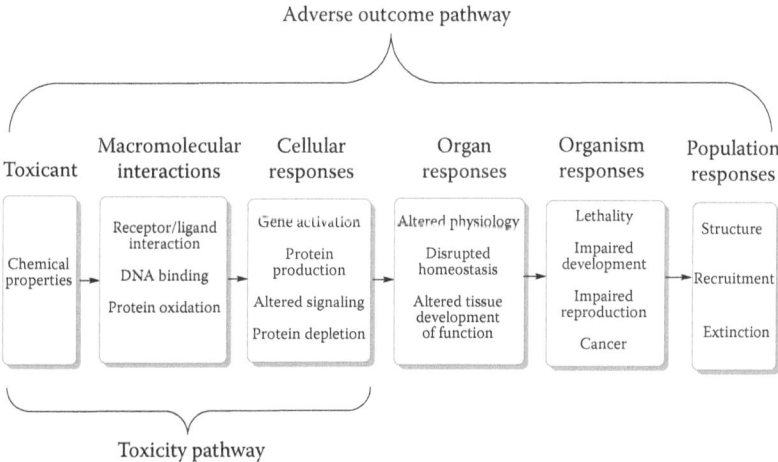

FIGURE 13.1 Diagram illustrating the key effects considered within the AOP conceptual framework. AOPs begin with an MIE, which is followed by a set of responses at different levels of organization from the cellular to whole organism and population levels. The first three steps within the AOP approach have been defined by the National Research Council as toxicity pathway.

With the AOP approach, it is possible to group chemicals that induce the same MIEs or that converge to the same key steps along a common pathway. Interestingly, the AOP approach could be used to help in solving problems related to interspecies extrapolation by aiding the identification of convergence or divergence points in pathways between species or clarifying which key processes are conserved between species. In this chapter, two case studies are presented, examining AOPs that interfere with the processes that occur toward the end of gestation (case study 1) by inhibition of 11β-hydroxysteroid dehydrogenase (11β-HSD) enzyme and that interfere with sexual development (case study 2), which have as MIE the inhibition of the 5α-reductase (5AR) and aromatase (CYP19A1) enzymes.

13.2   CASE STUDY 1

In this section, after a description of the role of glucocorticoids (GCs) in pregnancy and the interplay of the immune and endocrine systems during the late stages of gestation, AOPs for the inhibition of the two isoforms of the 11β-HSD enzyme, which, with other enzymes, regulate the metabolism of GC, are presented. It will be shown that for the inhibition of the type 1 isoform of this enzyme, the pathways that lead to an adverse outcome are different depending on which tissues this occurs in (e.g., decidual natural killer [NK] cells vs. amnion fibroblasts). These pathways are perhaps not conserved across species because most mammals other than humans do not produce placental corticotropin-releasing hormone (CRH). Thus, in order to test these pathways, in vitro experiments might be carried out for key steps using human cells.

13.2.1   Action of GCs during Late Pregnancy

The glucocorticoid receptor (GR) is present in two isoforms, GRα and GRβ. The function of GRβ is unclear because it does not bind GCs. GCs that diffuse through the plasma membrane passively bind to the intracellular receptor GRα that is present in the cytoplasm as an inactive complex [5]. Thus, GRα becomes an activated transcription factor, dissociating from the inactive complex and rapidly translocating to the nucleus where the ligand–receptor complex regulates transcriptional responses of inflammatory genes by binding to the DNA or by protein–protein interactions. Transcription is activated when GRα binds as a homodimer to glucocorticoid response elements (GREs) that are located on DNA. Through GRE-dependent activation, GC increases the synthesis of anti-inflammatory proteins such as interleukin (IL)-1 receptor antagonist, IL-10, mitogen-activated protein kinase, phosphatase-1, neural endopeptidase, lipocortin-1, serum leukoprotease inhibitor, and Clara cell protein 10. However, the most potent anti-inflammatory action of GC results from protein–protein interaction between GRα and other transcription factors such as NF-κB. The GRα and NF-κB interaction leads to repression of production of cytokines involved in inflammation such as TNF-α, GM-CSF, IL-1β, IL-3, IL-6, IL-8, and IL-11 [5]. The anti-inflammatory effect of GC is exerted also through decrease of expression/activity of phospholipase A2 with consequent decreased availability of arachidonic acid for cyclooxygenase 2 (COX-2), thus leading to inhibition of prostaglandin synthesis.

In human placenta at term, and fetal membranes, GC increases prostaglandin synthesis, while in the first-trimester human trophoblast, GC such as dexamethasone decreases prostaglandin synthesis. The dual effect (inhibitory/stimulatory) of GC on prostaglandin synthesis could depend on local changes on the response of placenta/fetal membranes to GC action, which may favor interaction, for instance, with GRE instead of with NF-κB [6]. Factors that perhaps play a pivotal role on the action of GC on placenta/fetal membranes are time of gestation, localization of GR (which is in the nucleus in amnion epithelial cells, amnion/chorion fibroblast cells, and chorion tropho-blast cells but in the cytoplasm in cells of the decidua), ability to activate the mineral corticoid receptor (MR) if it is present, and ability to bind to the progesterone receptor (PR). While cortisol or cortisone activates both GR and MR with comparable in vitro affinities, dexamethasone shows less affinity toward MR than toward GR [6].

13.2.2 Labor Is the Result of Interactions between the Immune and Endocrine Systems

Pregnancy can be viewed as a process divided into stages where pro-inflammatory and anti-inflammatory signals are modulated. During the implantation phase, an inflammatory microenvironment is required for tissue remodeling. In contrast, dur-ing the second trimester, there is equilibrium between pro- and anti-inflammatory agents such as chemokines and cytokines leading to uterine senescence. At the end of pregnancy, a partial immunosuppression and a significant increase of inflamma-tory mediators are needed for parturition [7].

It is thought that, after midgestation, the fetal adrenal gland is able to produce cortisol, which it is able to do also during the first trimester [8]. A "placental clock" that could define the pregnancy length is the CRH, which is made in the placenta only in primates [9,10]; it is produced in the syncytiotrophoblast cells and during the last 6 to 8 weeks of pregnancy shows a plasma concentration increase of 100-fold. Placental CRH produces most of the plasma cortisol through the hypothalamic–pituitary–adrenal (HPA) axis during pregnancy, stimulating maternal and fetal pituitaries to form adrenocorticotropic hormone (ACTH). A positive feedback loop regulates placental CRH and cortisol while a negative feedback is present between hypothalamic CRH and cortisol [11–13].

The increase of fetal cortisol production is also correlated to fetal lung matura-tion (Figure 13.2). Fetal lungs at the end of gestation are sufficiently mature to start to secrete phospholipids and surfactant proteins into the amniotic fluid leading to migration of fetal macrophages into the maternal uterus [14]. Fetal macrophages pro-ducing IL-1β activate NF-κB. Activation of NF-κB leads to upregulation of COX-2 enzyme transcription and also of pro-inflammatory cytokines (e.g., TNF-α, IL-1α, IL-6, and IL-8). Pro-inflammatory cytokines upregulated by NF-κB activate and promote contractility of the myometrial tissue. The NF-κB signaling pathway is crit-ical for cytokine-driven prostaglandin production [15,16]. At the end of gestation, the NF-κB signaling pathway becomes more active but is normally repressed by activa-tion of PR. However, NF-κB itself represses PR activity [15]. Progesterone is con-sidered an "anti-inflammatory steroid" and its plasma level in humans remains high

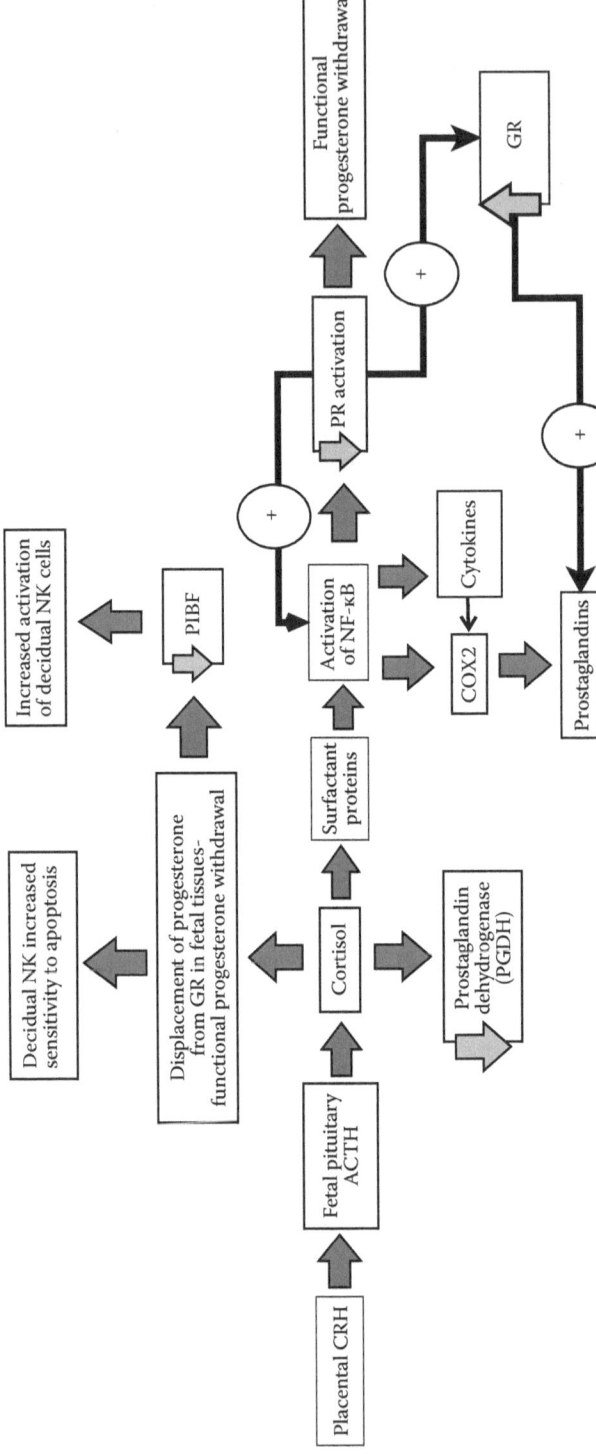

FIGURE 13.2 (See color insert.) Key effects occurring toward the end of gestation. The increase of placental CRH leads to an inflammatory process regulated by activation of the NF-κB factor and therefore production of prostaglandins and cytokines accompanied by a functional progesterone withdrawal. The increase of circulating cortisol, displacing progesterone from the GR, renders dNK cells more sensitive to apoptosis although more active because of a decrease of the PIBF factor. The increase of cortisol locally in dNK cells mediated by the 11β-HSD1 will induce their apoptosis in an autocrine fashion.

until the end of pregnancy in contrast to rat, mouse, and sheep where it is subject to a net decrease toward the end of gestation. Removal of the immunosuppressive activity of progesterone is achieved during pregnancy through local progesterone withdrawal such as in the myometrial cells—perhaps initiated by the activation of NF-κB, which upregulates inhibitory PR isoforms without affecting the total plasma concentration of progesterone [15,17]. Local progesterone withdrawal in a specific microenvironment such as the cervix takes place during the phases of cervical ripening and dilatation [18]. In the mouse, this phenomenon is caused by loss of progesterone synthesis in the ovary and an increase of progesterone metabolism within the cervix because of an increased expression of the 5α-reductase type 1 enzyme [18]. Progesterone withdrawal can also occur through an antiprogestinic effect of GCs that bind to the PR [19]. Progesterone could exert specific effects through binding to the GR. The increase of circulating GCs in late gestation may displace progesterone from the GR leading to inhibitory effects of the progesterone action. The increase of cortisol near term, displacing progesterone from the GR locally, would exert a pro-apoptotic effect because a higher number of GRs are available. This effect could perhaps be seen in decidual NK (dNK) cells where the balance between pro-apoptotic and anti-apoptotic factors would be toward the former, leading to decreased levels of dNK cells. Progesterone has been shown to possess an anti-apoptotic activity on fetal membranes [20,21] and after GC treatment in vitro and in vivo also on thymocytes [22]. One hypothesis is that these cells, in which PR and GR are expressed, are capable of binding a higher number of progesterone molecules that block the apoptosis pathway and cannot be all displaced by GC administration. Another hypothesis is that binding of progesterone to the PR would result in a decreased affinity of GR for cortisol where both PR and GR receptor are present. It has been found that the PR and, in particular, the PRA isoform repress the GR capabilities of inducing COX-2 expression. Therefore, toward term, a decreased activation of PR would lead to an upregulation of GR, which, upon binding of cortisol, would lead to increased prostaglandin production [23].

The increase of cortisol toward the end of pregnancy, locally displacing progesterone from the GR, could decrease the production of the progesterone-induced blocking factor (PIBF) [24]. PIBF has been shown to inhibit directly the phospholipase A_2 enzyme and thus to decrease arachidonic acid metabolism with a consequent reduction of IL-2 and IFN-γ production. IL-2 and IFN-γ and other pro-inflammatory cytokines activate NK cells [25,26]. At the end of gestation, therefore, dNK cells will be more activated through a local decrease of PIBF production but they will be more susceptible to apoptosis because of a decrease of progesterone.

13.2.3 The 11β-HSD Enzyme

In the early 1950s, it was found that the enzymatic conversion of cortisol in man was carried out by the 11β-HSD enzyme. During the 1970s and 1980s, active research demonstrated that the kidney was an important organ for the conversion of cortisol to cortisone while the liver mainly converts cortisone to cortisol. All these findings led to the discovery of two isoforms during the middle 1990s [27]. The 11β-HSD type 1 (11β-HSD1) isoform is bidirectional and NADP(H) dependent, capable of catalyzing dehydrogenase and 11-oxo-reductase reactions (Figure 13.3). 11β-HSD1 is located in the inner leaflet of the endoplasmic reticulum and is highly expressed in GC target tissues, specifically placenta (syncytiotrophoblast, chorion, and decidua), fetal testis, liver, gonad (oocyte and luteinized granulosa cells), bone, adrenal cortex, specific tissues of the eye, vascular smooth muscle, adipose tissue, and epidermis of the skin. The 11β-HSD type 2 (11β-HSD2) isoform is unidirectional and NAD dependent, converting cortisol to cortisone [27–32] (Figure 13.3). It is localized in the endoplasmic reticulum with the active site facing the cytosol and is highly expressed in placenta (syncytiotrophoblast, extravillous cytotrophoblast), kidney, gastrointestinal tract, several fetal organs (kidney, bone adrenal, gastrointestinal tract, peripheral nervous system, and muscle), gonad (nonluteinized granulosa cells), adrenal cortex, lung, eye (nonpigmented ciliary epithelium), mammary glands, vascular smooth muscle, salivary glands, and sweat glands. 11β-HSD2 isoform is colocalized with MR [33].

Deficiency of 11β-HSD1 manifests in women as hirsutism, acne, and adrenal androgen excess [34]. These effects could be attributed to an increased hypophyseal ACTH secretion caused by a state of systemic GC deprivation leading to increased androgen production. Deficiency of 11β-HSD2 has been related to the apparent mineral corticoid excess, which is an inherited form of hypertension, and it manifests with low birth weight, failure to thrive or short stature, and severe forms of hypertension and hypokalemia.

Studies found a positive correlation between fetal birth weight, the insurgence of intrauterine growth restriction (IUGR), and 11β-HSD2 activity, demonstrating the important role of this isoform in protecting the fetus from the excess of GC. It has been found that, toward the last weeks of gestation, there is a decrease of 11β-HSD2 and an increase of 11β-HSD1 activities. It has been shown that the decrease of 11β-HSD2 activity toward term is associated with a decrease of fetal growth rate and promotion of fetal organ maturation [35]. The 11β-HSD1 expression is stimulated by prostaglandins and GCs [6]. It is possible therefore that the 11β-HSD1 expression could increase through a positive feedback mechanism, which provides an increase of prostaglandin synthesis at term [36]. In contrast, upregulation of prostaglandins

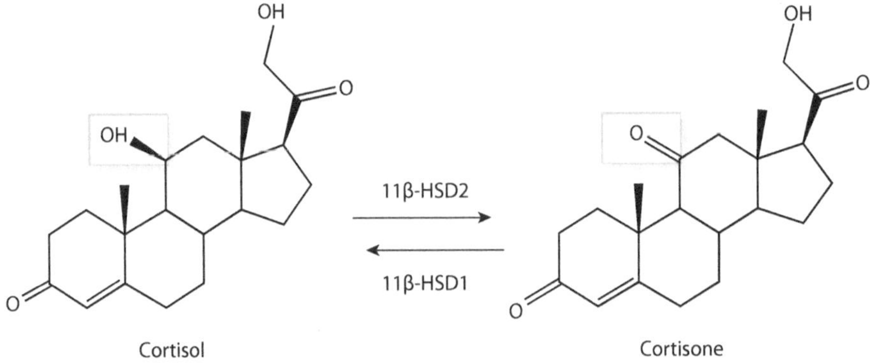

FIGURE 13.3   11β-HSD-catalyzed reactions. 11β-HSD1 catalyzes mainly the conversion of cortisone to cortisol (reductase activity) while 11β-HSD2 oxidizes cortisol to cortisone (oxidase activity).

and cytokines toward term decreases the oxidative capabilities of 11β-HSD2, the expression of which is normally upregulated during early-middle gestation [6,35].

The notion of the developmental origins of adult health and disease is called "developmental programming." It is accepted that variations within the feto-placental unit of the 11β-HSD2 enzyme have developmental programming effects as well as the detrimental effects of GCs on growth and maturation of the fetus. Deficiency or inhibition of 11β-HSD2 during gestation has been associated with alterations to pregnancy duration, birth weight, and other offspring outcomes. In contrast, there is no evidence of an adverse outcome when affecting the 11β-HSD1 activity. Most studies have been conducted on, for example, mice, which do not produce placental CRH, and this may be one reason for the lack of information about the effects of 11β-HSD1 deficiency on development. However, an increase in GC crossing the placenta caused by inhibition of the 11β-HSD2 enzyme does damage the fetus, and so an AOP for the withdrawal of GC has been studied.

13.2.4   dNK Cells and the 11β-HSD Enzyme

In humans, approximately 70% of lymphocytes that are in the decidua are NK cells (dNK) resulting from the maternal adaptation to pregnancy. The accumulation of dNK cells in the uterus originates from migration of NK cells from peripheral blood, which is facilitated by the expression of chemokine receptors [37]. NK cells are classified by their surface markers into two main phenotypically distinct subsets: $CD16^+CD56^{dim}$, which are in higher percentage in the peripheral blood (pNK), and $CD16^-CD56^{bright}$ located in the decidua. $CD16^-CD56^{bright}$ cells are considered less cytotoxic than $CD16^+CD56^{dim}$, which, in contrast, are poor cytokine producers [38]. Human dNK cells control invasion of the trophoblast and vascular remodeling by secreting angiogenesis-regulating molecules, chemokines, and cytokines [38,39]. It has been found that elevated pNK cell activity is observed in women with recurrent miscarriage and cases with recurrent miscarriage and infertility of unknown etiology have higher levels of activated pNK [24] and dNK cells [6]. GR and PR are expressed in pNK, whereas dNK do not possess PR [40]. It has been found that dNK cells express the 11β-HSD1 enzyme. Stress could decrease the levels of NK cells. Therefore, it has been speculated that these cells are more susceptible than other immune cells to stress, perhaps because of lack of 11β-HSD2. Moreover, some authors found that 11β-HSD2 mRNA was not expressed in dNK [41].

It has been suggested that 11β-HSD1 within intrauterine membranes may be related to the parturition process through a local generation of cortisol [42]. Apoptosis within decidual cells seems to be induced in an autocrine mode resulting from increased cortisol production from 11β-HSD1 [42]. Apoptosis has been found to be restricted to the nonstromal part of the decidua. It is possible to hypothesize that dNK cells toward the end of gestation may undergo apoptosis through an autocrine mechanism induced by cortisol production by 11β-HSD1, leading to a successful parturition. It could be that, toward term, their capabilities of secreting cytokines increase but their levels decrease. Perhaps this avoids their excessive activation, which would turn them from builders—as they are in early gestation when they show spiral artery remodeling properties—to cytotoxins with adverse outcomes such as premature birth.

13.2.5   INHIBITION OF THE 11β-HSD1 ENZYME AND DEVELOPMENTAL TOXICITY

The key effects of the inhibition of 11β-HSD1 have been divided in two distinct pathways on the basis of where the inhibition occurs (Figures 13.4 through 13.6). Inhibiting 11β-HSD1 in the placenta may lead to further activation of the maternal HPA, which is already compensating for a cortisol decrease by increasing ACTH production owing to the inhibition of 11β-HSD1 in the liver. Perhaps here the maternal hypothalamus, although further stimulated to produce CRH by a signal from the placenta, may reach a threshold beyond which additional stimulus does not lead to significant further production of CRH. In both AOPs, a decrease in cortisol production would

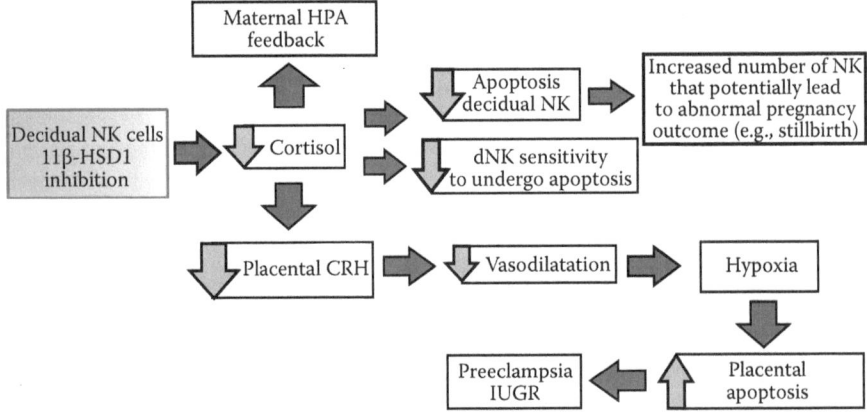

FIGURE 13.4   **(See color insert.)** Key effects related to the inhibition of 11β-HSD1 in dNK.

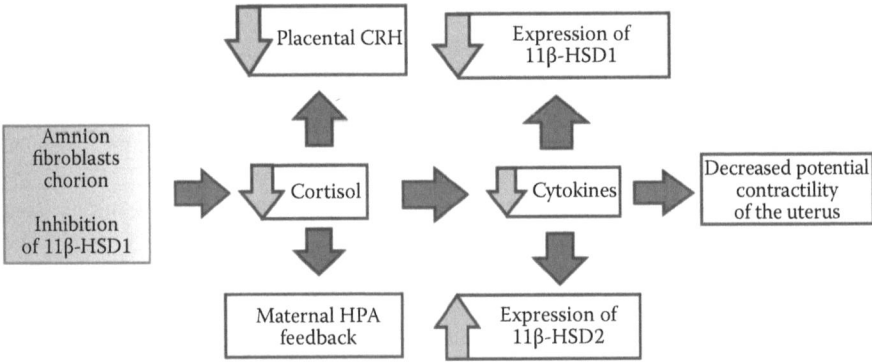

FIGURE 13.5   **(See color insert.)** Key effects related to the inhibition of 11β-HSD1 within the fetal membranes amnion and chorion.

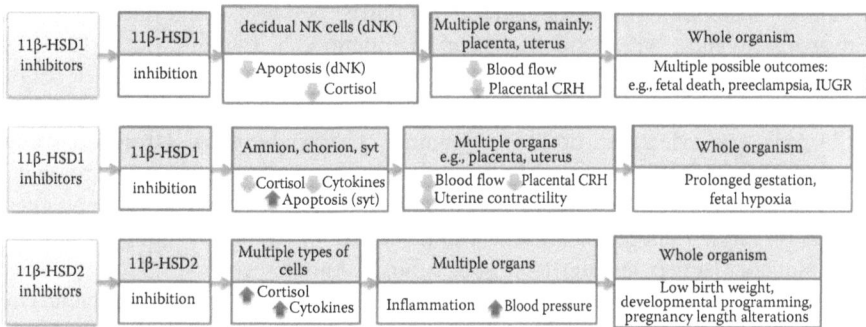

FIGURE 13.6 **(See color insert.)** Representation of three AOPs. MIEs triggered by chemical agents that inhibit the 11β-HSD enzyme isoforms lead to several outcomes. In the case of 11β-HSD1 enzyme inhibition, the final adverse outcomes are dependent on where the MIE is triggered (e.g., amnion vs. dNK cells). Abbreviations: dNK, decidual NK cells; syt, syncytiotrophoblast.

reduce placental CRH generation. Placental CRH acts as a vasodilator and regulates the feto-placental vascular tone in pregnant women but does not seem to play this role in most animals and nonpregnant women. Vasodilatation caused by CRH binding to its receptor is via a NO- and cGMP-mediated pathway [43]. A decreased secretion of placental CRH therefore leads to umbilical cord blood flow reduction, which eventually results in fetal hypoxia. This triggers placental apoptosis, resulting in an adverse outcome such as fetal growth restriction or preeclampsia.

If the inhibition of 11β-HSD1 occurs within dNK cells (Figures 13.4 and 13.6), the effect is a decrease in their apoptosis perhaps through an autocrine mechanism operated by cortisol. Decreasing cortisol in vivo in these cells would make them less susceptible to apoptosis because some GR receptors would be occupied by progesterone: less GR would be available to circulating GC that can trigger the apoptotic cascade. In a situation where local cortisol levels decrease within the decidua, PIBF concentration would not fall because of limited local progesterone withdrawal resulting in dNK cells not being strongly activated. The inhibition of 11β-HSD1 in dNK could lead to an increase in dNK levels, which potentially could become cytotoxic, giving stillbirth as an adverse outcome for instance.

If the inhibition of 11β-HSD1 takes place in membranes like the amnion and chorion (Figures 13.5 and 13.6), the inflammatory cascade that precedes parturition is interrupted, with a marked decrease in cytokine production leading to subsequent downregulation of 11β-HSD1 and upregulation of 11β-HSD2. A decline of the inflammatory process may lead to postterm birth with an abnormal labor owing to decreased contractility of the uterus. Chorioamnionitis is an inflammatory state of the placental membranes, caused in most cases by ascending infectious agents from the vaginal flora, that often leads to preterm delivery [44]. In women with chorioamnionitis, there is an increased production of pro-inflammatory cytokines with upregulation of 11β-HSD1 and downregulation of 11β-HSD2 [44]. These findings support the idea that inhibiting 11β-HSD1 in the amnion or chorion may lead to a prolonged gestation through decreased placental CRH (Figure 13.4).

In conclusion, the inhibition of 11β-HSD during gestation could result in several adverse outcomes, which are illustrated in Figure 13.5. Inhibiting the 11β-HSD2 isoform will expose the fetus to maternal GC, leading to an increased risk for IUGR, programming of postnatal hypertension, programming of postnatal activity of the HPA axis, and effects on brain development. The inhibition of 11β-HSD1 would instead cause a local decrease of GC concentration in specific microenvironments such as dNK cells or fetal membranes (amnion and chorion). In dNK cells, there would be an anti-apoptotic effect causing preterm birth and in the latter an anti-inflammatory action leading to postterm birth. Another general effect of a local decrease of GC concentration attributed to 11β-HSD1 inhibition within the uterine and placenta environment would be a decrease in placental CRH, which may, as a final outcome, cause preeclampsia or IUGR.

13.2.6 INHIBITION OF THE 11β-HSD ENZYME

The metabolite of glycyrrhizin is glycyrrhetinic acid (Figure 13.7), which is a nonselective inhibitor of 11β-HSD1 and 11β-HSD2. Hydroxamic acid derivatives of glycyrrhetinic acid have recently been synthesized as selective inhibitors of 11β-HSD2 [45].

Irreversible inhibition of 11β-HSD2 has been observed with dithiocarbamates (DTCs), which, every year, are released in large quantities into the environment as they are used as insecticides [46]. Exposure to these compounds leads to toxic effects in kidney, liver, placenta, and testis. DTCs cause kidney damage and low birth weight in rats' offspring. The irreversible modification of cysteine residues in 11β-HSD2 by covalent attachment of a carbamoyl group has been suggested as the MIE. 11-oxygenated derivatives of pregnenolone and progesterone are potent inhibitors of 11β-HSD1 and 11β-HSD2 while 7-keto steroids such as 7-keto-DHEA, 7-keto-pregnenolone, and 5α-androstane-3β-ol-7,17-dione are potent inhibitors only of 11β-HSD1 with IC_{50} within nanomolar range. 11β-HSD1 has a role in the metabolism of xenobiotics, catalyzing the reductive metabolism of metyrapone and its derivatives, ρ-nitroacetophenone and ρ-nitrobenzaldehyde. Metyrapone, which is a drug used to block the biosynthesis of adrenal GC by inhibiting 11β-hydroxylase, has

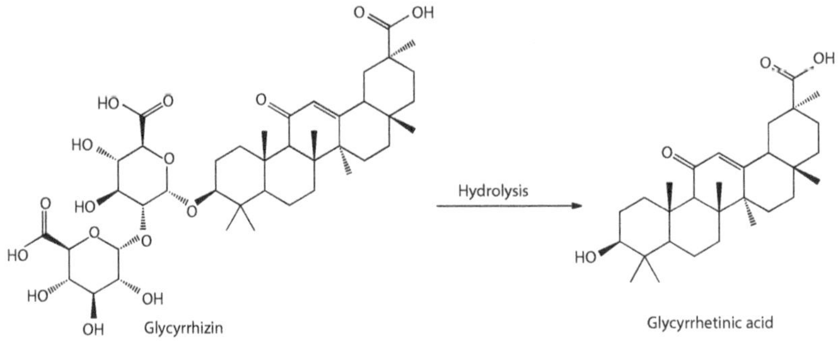

FIGURE 13.7 Hydrolysis of glycyrrhizin to glycyrrhetinic acid; glycyrrhizin is a natural product derived from licorice root.

been found also to inhibit 11β-HSD1 [46]. Inhibiting both enzymes therefore would decrease the total GC circulating pool, which would reinforce metyrapone action.

Simultaneous inhibition of cortisol-metabolizing enzymes such as 11β-HSD2 and 5α- and 5β-reductases may lead to a significant increase of circulating cortisol inducing fetal programming and other adverse effects, which may represent an amplification of those seen with 11β-HSD2 inhibition alone. An important consideration for the future might be the use of high-throughput in vitro testing strategies and computational methods to improve understanding of

- The adverse outcome (and, if possible, its severity) that results from a specific MIE
- How many MIEs can be triggered by the same chemical
- Whether mixtures of chemicals that trigger different MIEs can act synergistically or antagonistically if their pathways converge on the same key steps

This case study concerns the AOPs related to the inhibition of the 11β-HSD enzyme, which can be found in two isoforms. If the 11β-HSD1 enzyme isoform is inhibited locally in dNK cells that are present in the decidua, the capability of those cells to undergo apoptosis is significantly reduced, leading to increase in their levels. The higher levels of dNK cells perhaps lead to more activity with an increased risk for fetal death, preeclampsia, or IUGR. A local decrease of cortisol production may induce fetal hypoxia through a decrease of placental CRH, which may contribute to the fetal death as adverse outcome. If the 11β-HSD1 enzyme inhibition occurs in the amnion or chorion, a local decrease of cortisol may lead to prolonged gestation owing to a decrease in cytokine production and fetal hypoxia owing to a decrease in placental CRH. If the 11β-HSD2 enzyme is inhibited, the adverse outcomes may be similar to those arising from inhibition of the type 1 isoform, such as pregnancy length alterations, but there may also be outcomes peculiar to the inhibition of this isoform, such as the developmental programming of hypertension in the offspring.

13.3   CASE STUDY 2

In this case study, an introduction on human sex development and its disorders caused by different enzyme deficiencies is given, followed by the development of AOPs. These AOPs are related to the inhibition of the 5AR1 and 5AR2 isoforms of the 5α-reductase enzyme and the inhibition of the aromatase enzyme. The AOPs for the inhibition of 5AR2 and aromatase should be conserved across species, but there is some uncertainty regarding the inhibition of 5AR1.

13.3.1   PHYSIOLOGY OF SEX DEVELOPMENT

During the first and second week of gestation, the only difference between embryos of the two sexes is the karyotype. During the third week, the gonads differentiate and consequently produce hormones that induce anatomical and physiological differences (Figure 13.8). At the sixth to seventh week, the Müllerian (paramesonephric)

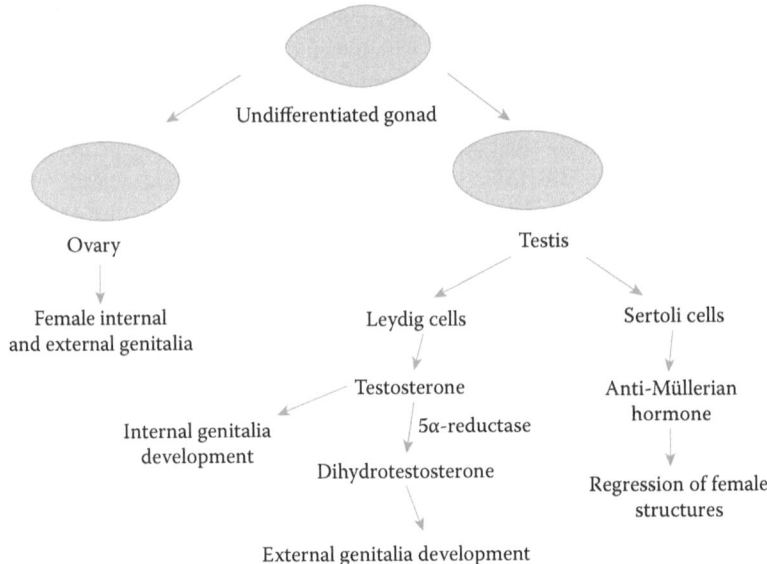

FIGURE 13.8   Hormonal control of human genitalia development.

duct develops close to the Wolffian (mesonephric) duct. Development of the testis leads to testosterone (T) secretion, which induces epididymis, vas deferens, and prostate differentiation and concomitant production of the anti-Müllerian hormone, which results in regression of the Müllerian duct [2,8,47]. If the testis does not develop, the Wolffian duct growth and differentiation are inhibited. Instead, the Müllerian duct differentiates and proliferates, leading to development of the fallopian tube, uterus, and the upper third of the vagina. The external genitalia, in contrast to the internal, develop from a common midline genital tubercle, which forms at around the eighth week of gestation. The genital tubercle develops urethral folds and labioscrotal swelling along its side. The genital tubercle continues to elongate to form the penis under the action of dihydrotestosterone (DHT) in the male while instead it forms the clitoris in the female. The urethral folds form the corpus spongiosum, which encloses the urethra in the male while they form the labia minora in the female. The labioscrotal swellings grow toward each other and fuse, forming the scrotum in the male while they remain separate and form the labia majora in the female. Female external genitalia development begins by the 11th week and is complete by the 20th week [2,8,47]. By the 14th week, the male external genitalia are complete except for testis descent and continuing phallic growth. At around the 32nd week, testis descent occurs.

13.3.2   Disorders of Sex Development

Congenital conditions where sex (chromosomal, gonadal, and anatomical) development is atypical are called disorders of sex development (DSDs) [48]. Often these conditions are observed at birth because of the ambiguous genitalia but sometimes the diagnosis is deferred to late childhood or even adulthood. In this section, the following common enzyme deficiencies [49,50] (Figure 13.9) lead to defects that produce DSD phenotypes and could be used to develop AOPs in the future.

- Steroidogenic acute regulatory protein (StAR) transports cholesterol into testicular and ovarian cells for steroid synthesis. Its deficiency causes the most severe form of congenital adrenal hyperplasia (CAH) producing phenotypic females independently of the gonadal sex, with slightly virilized external genitalia, internal male organs underdeveloped, and enlarged adrenal cortex.
- Cholesterol side-chain cleavage enzyme (P450scc or CYP11A1) converts cholesterol to pregnenolone and its deficiency in severe cases leads to complete underandrogenization and early-onset adrenal failure, while in less severe cases, it leads to mild masculinization and late-onset adrenal failure.
- 17α-Hydroxylase/17,20 lyase (P450c17 or CYP17A1) converts pregnenolone to 17α-hydroxypregnenolone and the latter to dehydroepiandrosterone. Deficiency of this enzyme in the majority of male patients is a phenotype female-like or slightly virilized external genitalia with blind vaginal pouch, cryptorchidism, and high blood pressure with hypokalemia. Deficiency of isolated 17,20 lyase activity can occur and patients show a phenotype that includes ambiguous genitalia, micropenis, perineal hypospadias, and cryptorchidism, with gynecomastia possibly occurring at puberty.
- 17β-Hydroxysteroid dehydrogenase type 3 catalyzes the reaction of androstenedione to T and is prevalently expressed in the testis. Patients affected

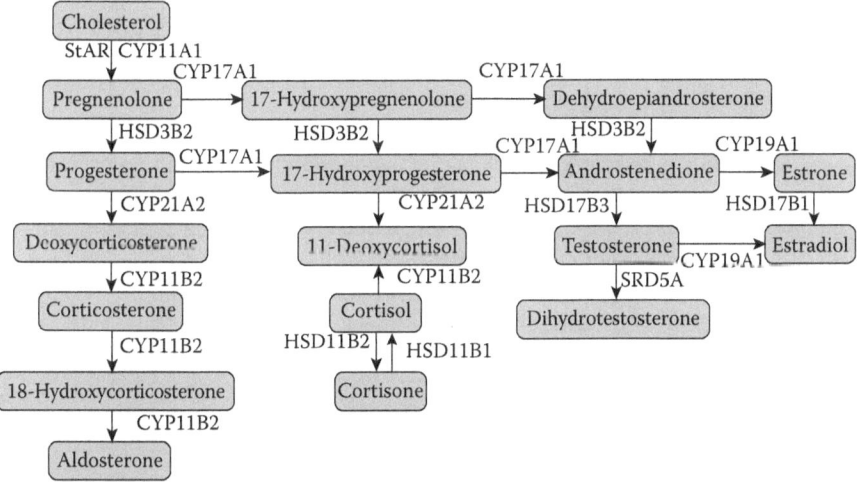

FIGURE 13.9   Human steroid biosynthesis regulated by steroidogenic acute regulatory protein (StAR), cholesterol side-chain cleavage enzyme (P450scc or CYP11A1), 17α-hydroxylase/17,20 lyase (CYP17A1), 3β-hydroxysteroid dehydrogenase/$\Delta^{5/4}$-isomerase type 2 (HSD3B2), aromatase (CYP19A1), 21-hydroxylase (CYP21A2), aldosterone synthase (CYP11B2), 11β-hydroxylase (CYP11B1), 5α-reductase (SRD5A, abbreviated in the text as 5AR), and 11β-hydroxysteroid dehydrogenase (HSD11B or 11β-HSD).

by deficiency of this enzyme show female-like or ambiguous genitalia, a blind vaginal pouch, and as a major feature the intra-abdominal or inguinal location of testes, epididymis, vasa deferentia, seminal vesicles, and ejaculatory ducts.

- P450 oxidoreductase (POR) is the required electron transfer protein for all microsomal (type II) P450 enzymes such as CYP17A1, CYP21A2, CYP19A1, CYP51A1, and so on. Thus, POR deficiency (PORD) can exhibit multiple partial defects in steroidogenesis and as a complex disease from a diagnostic and management perspective. PORD in males causes undermasculinization whereas, in females, it causes virilization. In females, it leads to a phenotype like the one associated with aromatase (CY19A1) deficiency alone. Often patients with this disorder present skeletal malformation in a pattern designated the Antley–Bixler syndrome.
- 5α-Reductase type 2 (SRD5A2, abbreviated to 5AR2 in Section 13.3.3) converts T to DHT. Its deficiency will be discussed further in Sections 13.3.3 to 13.3.5 as the basis for forming an AOP.
- 3β-Hydroxysteroid dehydrogenase/$\Delta^{5/4}$-isomerase type 2 is responsible for the production of delta-4 steroid, which is impaired in adrenals and gonads in cases where this enzyme is deficient. Males and females with this deficiency are born with ambiguous genitalia. Males are undermasculinized because of lack of T while females are virilized because of excess of dehydroepiandrosterone (DHEA), a substrate of 3β-hydroxysteroid dehydrogenase type 1, which converts it to androstenedione.
- 21-Hydroxylase (CYP21A2) catalyzes the next-to-last step in the synthesis of cortisol and aldosterone. Its deficiency produces the most common form of CAH. When the activity of CYP21A2 is reduced, androgen excess virilizes the external genitalia of 46,XX fetuses, causing DSD with different degrees of clitoromegaly and labioscrotal fusion.
- 11β-Hydroxylase (CYP11B1) catalyzes the conversion of deoxycortisol to cortisol. Its deficiency leads to the accumulation of steroid precursors, which are channeled into androgen pathways that induce virilization in 46,XX fetuses. 46,XY fetuses are not affected, but during postnatal life, virilization can occur in both sexes.
- Aromatase (CYP19A1) catalyzes the conversion of T to estradiol. Its deficiency will be discussed further below as the basis for forming an AOP.

13.3.3 5α-Reductase Enzyme

The 5α-reductase enzyme (5AR) can be found in two isoforms: types 1 and 2. The 5AR isoforms are NADPH dependent and catalyze the reduction of a double bond in the 4,5 position of androgens such as the reduction of T to DHT [51] (Figure 13.10). In general, 5AR1 is present in the liver, skin, and the initial segment (epithelial cells) of the epididymis with concentrations three- to sevenfold higher than those in other segments [52]. However, 5AR1 is found also in female mouse gonadal tissues where it reduces androstenedione to its 5α-reduced form within the decidual cells of the placenta and reduces progesterone to its 5α-reduced form within the epithelial cells

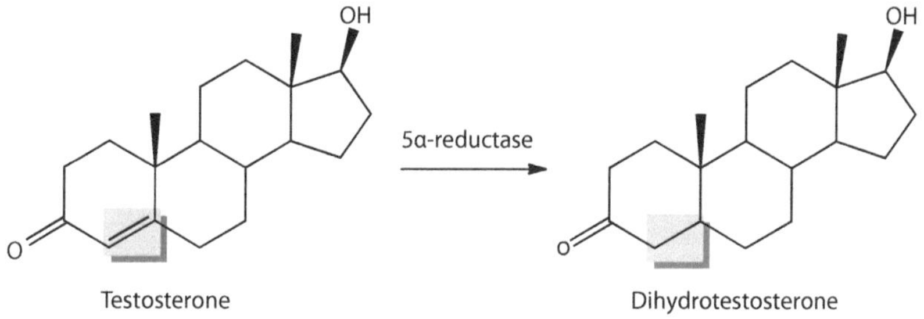

FIGURE 13.10 Reduction of T to DHT by the 5α-reductase enzyme. Highlighted is the double bond that is reduced by the enzyme.

of the cervix [53]. The 5AR2 isoform is prevalently distributed in the seminal vesicles, in stromal cells and the basal epithelium of the prostate (absent from acinar ductal epithelial cells of the prostate), and at higher concentrations than 5AR1 in the epididymis (stromal cells) [54]. In the epididymis, 5AR1 shows a dramatic increase during postnatal development while 5AR2 does not exhibit any particular changes. Torres and Ortega showed that both 5AR isoforms are also expressed in the prefrontal cortex of adult male rats [55]. The authors suggested that 5AR2 expression in rat's brain is positively regulated by DHT whereas 5AR1 is negatively regulated by testicular factors including partly T and DHT. Thus, 5AR1 may be involved in sexual differentiation of the brain favoring and maintaining female brain structures. 5AR1 activity in rat's liver is higher in females than in males. The 5AR isoforms are membrane bound and constitute less than 0.001% of the total protein in human prostate or skin tissue and their purification at present represents a difficult task [56].

13.3.4 5α-Reductase Deficiency/Inhibition

Inherited deficiency of the 5AR type 2 enzyme (5AR2) caused by mutations of the SRDA2 gene leads to karyotype-dependent (XY) genital birth defects. Genetic males presenting 5AR2 deficiency show disorders of sexual development (46 XY, DSD) which phenotypically manifest with ambiguous external genitalia, micropenis, prostate hypoplasia, testes with normal differentiation, and normal/reduced spermatogenesis usually located in the inguinal area. During puberty, the phenotype shows virilization, deepening of the voice, development of muscular mass, and enlargement of the penis without gynecomastia. Patients with this deficiency at/after puberty do not show baldness, acne, or prostate enlargement since these features depend on DHT, which is reduced in this disorder. Most patients are reared in the female social sex because of their female-like external genitalia at birth. Most patients not undergoing orchidectomy during childhood will have a male social sex change during puberty [50]. The inhibition of 5AR2 in mice leads to birth defects similar to the inherited deficiency, which are dose dependent, while inherited deficiency or probably inhibition (no experimental data) of 5AR1 leads to miscarriage or stillbirth [53,57,58] (Figure 13.11).

Well-known inhibitors of 5AR are finasteride (e.g., Proscar), which is more selective toward 5AR2, and dutasteride (e.g., Avodart), which inhibits both isoforms [51].

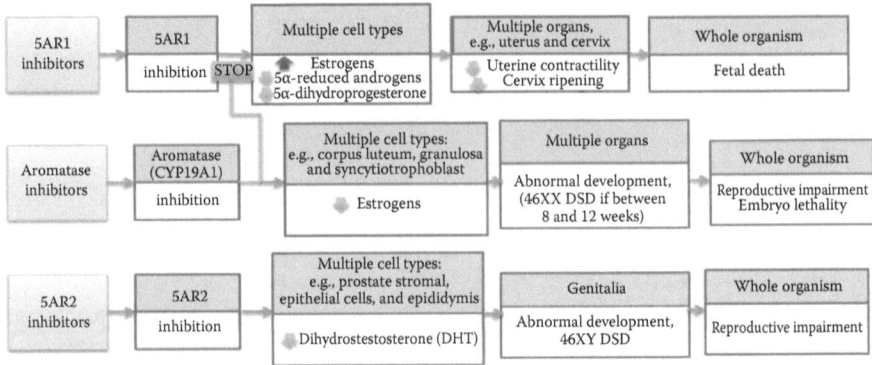

FIGURE 13.11   (See color insert.) Representation of the AOPs for the inhibition of the 5α-reductase enzyme isoforms and the aromatase enzyme. Aromatase enzyme inhibition prevents the adverse outcome related to the inhibition of the 5α-reductase enzyme type 1 (labeled as STOP) by diminishing the circulating estrogens. Estrogens are increased as a consequence of the increased concentration of T (T is a substrate of the aromatase enzyme and cannot be converted to DHT because of the 5α-reductase type 1 inhibition).

Both of these inhibitors are utilized for the treatment of benign prostatic hyperplasia. Finasteride and dutasteride are mechanism-based irreversible inhibitors of 5AR. These two compounds are 4-aza-3-oxo-steroids that, through their 1,2 double bond, form a covalent adduct with the pyridine-nucleotide cofactor [56]. In utero exposure to finasteride in rats at and above 0.1 mg/kg/day caused dose-dependent effects in the male offspring including decreased anogenital distance, nipple retention, ectopic testes, hypospadias, and a decrease in prostate gland weight [59]. A study on fetal development in Rhesus monkeys showed that when a high oral dose of finasteride (2 mg/kg) was administered, male fetuses presented external malformations of the genitalia [60].

13.3.5   RELATIONSHIP BETWEEN 5α-REDUCTASE INHIBITION AND DEVELOPMENTAL TOXICITY

Birth defects associated with inhibition of the 5AR2 enzyme, located within the stromal cells of the male reproductive tract, are related to decreased blood concentration of DHT (failure in converting T to DHT), which is a potent androgen and necessary to the development of male external genitalia [51].

Fetal death (in mice) associated with failure in converting T to DHT is related to a two- to threefold increase of estrogens. Fetal death occurs in rats lacking the 5AR1 enzyme at around the 11th day of gestation (midgestation). This phenomenon has been defined as a fecundity defect and can be prevented by reducing the formation of estrogens. Decreasing the conversion of T to estradiol by inhibition of the aromatase enzyme (CYP19) prevents the fecundity defect in mice lacking the 5AR1 enzyme. Thus, hypothetically, administration of a 5α-reductase inhibitor alone would cause fetal death, while administration of both 5AR1 and aromatase inhibitors might not cause fetal death [57] (Figure 13.11).

Fetal death (in mice) associated with failure in converting androgens such as androstenedione and T to their 5α-reduced form is related to a phenomenon defined as a parturition defect. Parturition defect consists of an aberrant parturition in which 70% of the pregnancies of rats deficient in 5AR1 fail to end at term and instead continue, thus leading to death in utero of the fetuses [58].

Fetal death (in mice) associated with failure in converting progesterone to its 5α-reduced form (dihydroprogesterone) is related to an aberrant parturition caused by defective cervical ripening. The incomplete progesterone catabolism in the cervix blocks ripening and prevents delivery [53].

13.3.6 AROMATASE ENZYME

The aromatase enzyme is part of the cytochrome P450 superfamily and therefore is a heme-binding protein. It is located within the microsomes of estrogen-producing cells. It has high substrate specificity and its activity is regulated by phosphorylation. A complex with a NADPH-cytochrome P450 reductase allows the aromatase to catalyze the conversion of androgens to estrogens (Figure 13.12). The placenta converts 16α-hydroxydehydroepiandrosterone to estriol, the ovary converts T to estradiol, and the adipose tissue aromatizes androstenedione to estrone. The aromatase in premenopausal women is found in granulosa cells, corpus luteum, the ovary, and syncytiotrophoblasts of the placenta. It has been detected also in several fetal tissues, testicular Leydig and Sertoli cells, epididymis, and germ cells. Other tissues where the aromatase is expressed are adipose mesenchymal tissue, skin fibroblasts, bone osteoclasts and osteoblasts, skeletal and smooth muscle, and vascular endothelium. Aromatase expression goes along with the estrogen receptor expression [61].

13.3.7 AROMATASE DEFICIENCY

Aromatase deficiency is rare in genetic females and causes 46XX, DSD. This condition manifests with ambiguous genitalia at birth with various degrees of virilization ranging from clitoral enlargement to complete labioscrotal fusion and a single meatus at the base of a phallus-like structure. At puberty, females affected by aromatase deficiency show amenorrhea, multiple large cysts of the ovaries, no breast development, and further enlargement of the clitoris along with virilization and acne.

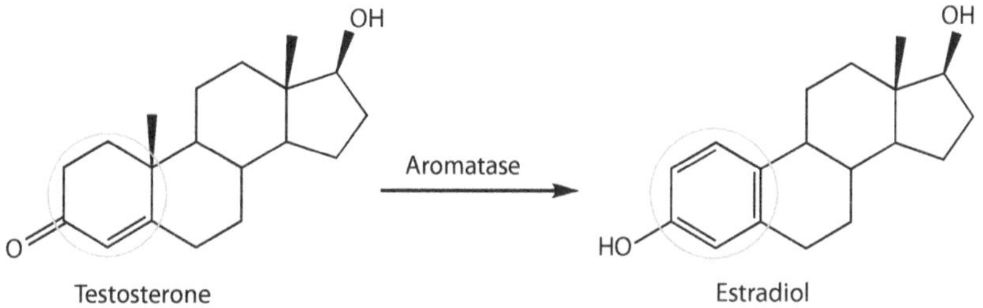

FIGURE 13.12 Action of the aromatase enzyme that converts T to estradiol as well as androstenedione to estrone and 16α-hydroxydehydroepiandrosterone to estriol.

In males, aromatase deficiency does not affect reproductive development to the same degree as in females. Common developmental symptoms in aromatase-deficient males are progressive infertility in adulthood, decreased sperm motility/count, and metabolic abnormalities. Aromatase knockout mice show a similar phenotype to the one observed in aromatase-deficient humans [49,61,62].

13.3.8   Aromatase Inhibitors

Aromatase inhibitors utilized to treat postmenopausal women presenting early breast cancer are letrozole (Femara), anastrozole (Arimidex), and exemestane (Aromasin). Letrozole inhibits peripheral aromatase at doses of 0.5–2.5 mg/day by 98%. Anastrozole and exemestane exhibit similar activity to letrozole.

13.3.9   Relationship between Aromatase Inhibition and Developmental Toxicity

Letrozole administered to rats (between 6 and 16 gestation days) was embryotoxic at doses that were not maternally toxic, showing that the compound could be considered a developmental toxicant. A significant dose-dependent increase in uterine lethality has been observed in rats exposed to doses lower than or equal to the recommended human therapeutic dose. It has been also found that letrozole increases the frequency of early and late resorptions and, at higher doses, increases the ratio of fetal mortality. Letrozole induced also minor structural anomalies of the vertebral bodies in fetuses. From these findings, it has been suggested that letrozole may not affect organogenesis or that, if it does, it induces malformations incompatible with the continuation of pregnancy [63]. Estradiol cyclopentylpropionate (ECP), when administered concomitantly with letrozole, gave protection against embryolethality (postimplantation pregnancy loss) at low doses more than at high ones. Postimplantation pregnancy loss when lower doses of ECP were administered with letrozole was similar to the control group. Moreover, ECP coadministration did not prevent the skeletal anomalies caused by letrozole. In baboons, letrozole increased significantly the number of pregnancies ending in spontaneous abortion, but when it was coadministered with estradiol, the induced increase of miscarriage rate was suppressed [63]. Anastrozole, like letrozole, caused dose-dependent increase in pregnancy loss; however, it did not cause the congenital anomalies that resulted with the letrozole treatment [63]. When the aromatase inhibitor fadrozole was administered to rats in the second half of pregnancy (from day 14), it caused obstruction of the development of the uterine tissue framework, which is required to render the uterus capable of accommodating the rapidly growing fetus. This causes fetuses to be exposed to high intrauterine pressure leading to hematoma on the head [64].

The AOP developed for the inhibition of the aromatase enzyme (Figure 13.8) has, as its main key effect, a significant decrease of estrogen concentration, which leads to embryo lethality if the decreases occur during the first month of gestation and DSDs in female if it occurs between 8 and 12 weeks of gestation. At present, it is not clear whether malformations (i.e., skeletal malformations, hematoma of the head) associated with aromatase inhibitors are related to decreased concentration of estrogens, and hence an AOP for those specific endpoints is not presented here.

It is clear that an excess of estrogens (caused by 5AR1 inhibition, see Figure 13.8) or a deficit (aromatase inhibition) during early pregnancy can lead to embryo lethality. Clearly there may be other MIEs that would cause a similar outcome, such as perhaps antagonism of the estrogen receptor, in which case effects would be attributed to a decrease in the action of estrogens rather than a decrease in their concentration.

13.4 CONCLUSIONS

This chapter has shown how the AOP framework can be used to treat complex endpoints such as developmental toxicity. Pathways for this complex endpoint can be conserved across species; even where they are not, key steps at a defined gestational stage may be. Near term, the immune and the endocrine systems cooperate in order to lead the pregnancy to term, and disruption of that delicate regulation by inhibition of one of these steps may lead to an adverse outcome. The AOP approach can be utilized in order to understand how the inhibition of one pathway influences another as in the case of the AOPs for 5AR1 inhibition and aromatase inhibition. It has been shown also how the AOPs related to the inhibition of 11β-HSD1 can lead to different outcomes depending on where the inhibition occurs. Thus, the permeability of a chemical through the decidua would be a limiting factor for triggering the 11β-HSD1 enzyme within the amnion or chorion. A chemical with high permeability would trigger both pathways, leading to an outcome that might be determined by the location in which more 11β-HSD1 is expressed.

The examples in this chapter show how the AOP approach can help with the building of TMCs for chemicals. There is a multitude of enzymes that, if inhibited at a defined life stage, can induce DSDs, presenting the opportunity for many more AOPs to be developed.

REFERENCES

1. M. Cronin and J. Madden, *In Silico Toxicology: Principles and Applications*, Royal Society of Chemistry, London, 2010.
2. K.L. Moore, *The Developing Human: Clinically Oriented Embryology*, 8th ed., Saunders, New York, 2007.
3. H. Sanderson, S.E. Belanger, P.R. Fisk, C. Schäfers, G. Veenstra, A.M. Nielsen, Y. Kasai, A. Willing, S.D. Dyer, K. Stanton, and R. Sedlak, *An overview of hazard and risk assessment of the OECD high production volume chemical category—Long chain alcohols [C6-C22] (LCOH)*, Ecotoxicol. Environ. Saf. 72 (2009), pp. 973–979.
4. G.T. Ankley, R.S. Bennett, R.J. Erickson, D.J. Hoff, M.W. Hornung, R.D. Johnson, D.R. Mount, J.W. Nichols, C.L. Russom, P.K. Schmieder, J.A. Serrrano, J.E. Tietge, and D.L. Villeneuve, *Adverse outcome pathways: A conceptual framework to support ecotoxicology research and risk assessment*, Environ. Toxicol. Chem. 29 (2010), pp. 730–741.
5. K.A. Smoak and J.A. Cidlowski, *Mechanisms of glucocorticoid receptor signaling during inflammation*, Mech. Ageing Dev. 125 (2004), pp. 697–706.
6. A.E. Michael and A.T. Papageorghiou, *Potential significance of physiological and pharmacological glucocorticoids in early pregnancy*, Hum. Reprod. Update 14 (2008), pp. 497–517.

7. J. Kwak-Kim, J.C. Park, H.K. Ahn, J.W. Kim, and A. Gilman-Sachs, *Review article: Immunological modes of pregnancy loss*, Am. J. Reprod. Immunol. 63 (2010), pp. 611–623.

8. N.A. Hanley and W. Arlt, *The human fetal adrenal cortex and the window of sexual differentiation*, Trends Endocrinol. Metab. 17 (2006), pp. 391–397.

9. J.A. Majzoub and K.P. Karalis, *Placental corticotropin-releasing hormone: Function and regulation*, Am. J. Obstet. Gynecol. 180 (1999), pp. S242–S246.

10. G. Mastorakos and I. Ilias, *Maternal and fetal hypothalamic–pituitary–adrenal axes during pregnancy and postpartum*, Ann. N.Y. Acad. Sci. 997 (2003), pp. 136–149.

11. M. Fadalti, I. Pezzani, L. Cobellis, F. Springolo, M.M. Petrovec, G. Ambrosini, F.M. Reis, and F. Petraglia, *Placental corticotropin-releasing factor. An update*, Ann. N.Y. Acad. Sci. 900 (2000), pp. 89–94.

12. S.N. Kalantaridou, E. Zoumakis, A. Makrigiannakis, L.G. Lavasidis, T. Vrekoussis, and G.P. Chrousos, *Corticotropin-releasing hormone, stress and human reproduction: An update*, J. Reprod. Immunol. 85 (2010), pp. 33–39.

13. A. Makrigiannakis, E. Zoumakis, S. Kalantaridou, and G. Chrousos, *Endometrial and placental CRH as regulators of human embryo implantation*, J. Reprod. Immunol. 62 (2004), pp. 53–59.

14. I. Garcia-Verdugo, Z. Tanfin, and M. Breuiller-Fouche, *Surfactant protein A: An immunoregulatory molecule involved in female reproductive biology*, Int. J. Biochem. Cell Biol. 42 (2010), pp. 1779–1783.

15. E. Golightly, H.N. Jabbour, and J.E. Norman, Endocrine immune interactions in human parturition. *Mol. Cell Endocrinol.* 335 (2010), pp. 52–59.

16. M. Lappas and G.E. Rice, *The role and regulation of the nuclear factor kappa B signalling pathway in human labour*, Placenta 28 (2007), pp. 543–556.

17. T. Zakar and F. Hertelendy, *Progesterone withdrawal: Key to parturition*, Am. J. Obstet. Gynecol. 196 (2007), pp. 289–296.

18. B. Timmons, M. Akins, and M. Mahendroo, *Cervical remodeling during pregnancy and parturition*, Trends Endocrinol. Metab. 21 (2010), pp. 353–361.

19. K. Karalis, G. Goodwin, and J.A. Majzoub, *Cortisol blockade of progesterone: A possible molecular mechanism involved in the initiation of human labor*, Nat. Med. 2 (1996), pp. 556–560.

20. Y. Wang, E. Norwitz, V. Abrahams, G. Luo, and S. Tadesse, *15: Progesterone inhibits basal apoptosis in fetal membranes by altering expression of both pro- and anti-apoptotic proteins*, YMOB 204 (2010), p. S9.

21. Y. Wang, E. Norwitz, V. Abrahams, G. Luo, and S. Tadesse, *536: Progesterone inhibits TNFα-induced apoptosis in fetal membranes by altering expression of pro- and anti-apoptotic proteins: Implications for inflammation-mediated preterm birth*, Am. J. Obstet. Gynecol. 204 (2011), p. S214.

22. R.W. McMurray, J.G. Wilson, L. Bigler, L. Xiang, and A. Lagoo, *Progesterone inhibits glucocorticoid-induced murine thymocyte apoptosis*, Int. J. Immunopharmacol. 22 (2000), pp. 955–965.

23. C.M. Guo, X.O. Zhu, X.T. Ni, Z. Yang, L. Myatt, and K. Sun, *Expression of progesterone receptor A form and its role in the interaction of progesterone with cortisol on cyclooxygenase-2 expression in amnionic fibroblasts*, J. Clin. Endocrinol. Metab. 94 (2009), pp. 5085–5092.

24. J. Szekeres-Bartho, A. Barakonyi, G. Par, B. Polgar, T. Palkovics, and L. Szereday, *Progesterone as an immunomodulatory molecule*, Int. Immunopharmacol. 1 (2001), pp. 1037–1048.

25. R. Druckmann and M.-A. Druckmann, *Progesterone and the immunology of pregnancy*, J. Steroid Biochem. Mol. Biol. 97 (2005), pp. 389–396.

26. J. Szekeres-Bartho, M. Halasz, and T. Palkovics, *Progesterone in pregnancy; receptor-ligand interaction and signaling pathways*, J. Reprod. Immunol. 83 (2009), pp. 60–64.

27. C.S. Wyrwoll, M.C. Holmes, and J.R. Seckl, *11β-Hydroxysteroid dehydrogenases and the brain: From zero to hero, a decade of progress*, Front. Neuroendocrinol. 32 (2010), pp. 265–286.

28. K.E. Chapman, A.E. Coutinho, M. Gray, J.S. Gilmour, J.S. Savill, and J.R. Seckl, *The role and regulation of 11β-hydroxysteroid dehydrogenase type 1 in the inflammatory response*, Mol. Cell. Endocrinol. 301 (2009), pp. 123–131.

29. A. Odermatt and L.G. Nashev, *The glucocorticoid-activating enzyme 11β-hydroxysteroid dehydrogenase type 1 has broad substrate specificity: Physiological and toxicological considerations*, J. Steroid Biochem. Mol. Biol. 119 (2010), pp. 1–13.

30. A. Odermatt, A.G. Atanasov, Z. Balazs, R.A.S. Schweizer, L.G. Nashev, D. Schuster, and T. Langer, *Why is 11β-hydroxysteroid dehydrogenase type 1 facing the endoplasmic reticulum lumen? Physiological relevance of the membrane topology of 11β-HSD1*, Mol. Cell. Endocrinol. 248 (2006), pp. 15–23.

31. J.W. Tomlinson and P.M. Stewart, *Cortisol metabolism and the role of 11β-hydroxysteroid dehydrogenase*, Best Pract. Res. Clin. Endocrinol. Metab. 15 (2001), pp. 61–78.

32. J.W. Tomlinson, E.A. Walker, I.J. Bujalska, N. Draper, G.G. Lavery, M.S. Cooper, M. Hewison, and P.M. Stewart, *11β-hydroxysteroid dehydrogenase type 1: A tissue-specific regulator of glucocorticoid response*, Endocr. Rev. 25 (2004), pp. 831–866.

33. J.R. Seckl, *Glucocorticoid programming of the fetus; adult phenotypes and molecular mechanisms*, Mol. Cell. Endocrinol. 185 (2001), pp. 61–71.

34. M. Quinkler, W. Oelkers, and S. Diederich, *Clinical implications of glucocorticoid metabolism by 11β-hydroxysteroid dehydrogenases in target tissues*, Eur. J. Endocrinol. 144 (2001), pp. 87–97.

35. V.E. Murphy and V.L. Clifton, *Alterations in human placental 11β-hydroxysteroid dehydrogenase type 1 and 2 with gestational age and labour*, Placenta 24 (2003), pp. 739–744.

36. J.R.G. Challis, S.G. Matthews, W. Gibb, and S.J. Lye, *Endocrine and paracrine regulation of birth at term and preterm*, Endocr. Rev. 21 (2000), pp. 514–550.

37. N. Acar, I. Ustunel, and R. Demir, *Uterine natural killer (uNK) cells and their missions during pregnancy: A review*, Acta Histochem. 113 (2011), pp. 82–91.

38. J. Hanna, D. Goldman-Wohl, Y. Hamani, I. Avraham, C. Greenfield, S. Natanson-Yaron, D. Prus, L. Cohen-Daniel, T.I. Arnon, I. Manaster, R. Gazit, V. Yutkin, D. Benharroch, A. Porgador, E. Keshet, S. Yagel, and O. Mandelboim, *Decidual NK cells regulate key developmental processes at the human fetal-maternal interface*, Nat. Med. 12 (2006), pp. 1065–1074.

39. I. Manaster and O. Mandelboim, *The unique properties of human NK cells in the uterine mucosa*, Placenta 29 (2008), pp. S60–S66.

40. T.A. Henderson, P.T.K. Saunders, A. Moffett-King, N.P. Groome, and H.O.D. Critchley, *Steroid receptor expression in uterine natural killer cells*, J. Clin. Endocrinol. Metab. 88 (2003), pp. 440–449.

41. S.E. McDonald, T.A. Henderson, C.E. Gomez-Sanchez, H.O.D. Critchley, and J.I. Mason, *11β-hydroxysteroid dehydrogenases in human endometrium*, Mol. Cell. Endocrinol. 248 (2006), pp. 72–78.

42. J. Chan, E.H. Rabbitt, B.A. Innes, J.N. Bulmer, P.M. Stewart, M.D. Kilby, and M. Hewison, *Glucocorticoid-induced apoptosis in human decidua: A novel role for 11β-hydroxysteroid dehydrogenase in late gestation*, J. Endocrinol. 195 (2007), pp. 7–15.

43. M. Vatish, H.S. Randeva, and D.K. Grammatopoulos, *Hormonal regulation of placental nitric oxide and pathogenesis of pre-eclampsia*, Trends Mol. Med. 12 (2006), pp. 223–233.

44. H.A. Johnstone, A. Wigger, A.J. Douglas, I.D. Neumann, R. Landgraf, J.R. Seckl, and J.A. Russell, *Attenuation of hypothalamic–pituitary–adrenal axis stress responses in late pregnancy: Changes in feedforward and feedback mechanisms*, J. Neuroendocrinol. 12 (2000), pp. 811–822.

45. C. Stanetty, L. Czollner, I. Koller, P. Shah, R. Gaware, T.D. Cunha, A. Odermatt, U. Jordis, P. Kosma, and D. Claßen-Houben, *Synthesis of novel 3-amino and 29-hydroxamic acid derivatives of glycyrrhetinic acid as selective 11β-hydroxysteroid dehydrogenase 2 inhibitors*, Bioorg. Med. Chem. 18 (2010), pp. 7522–7541.

46. A. Odermatt, C. Gumy, A.G. Atanasov, and A.A. Dzyakanchuk, *Disruption of glucocorticoid action by environmental chemicals: Potential mechanisms and relevance*, J. Steroid Biochem. Mol. Biol. 102 (2006), pp. 222–231.

47. A. Biason-Lauber, *Control of sex development*, Best Pract. Res. Clin. Endocrinol. Metab. 24 (2010), pp. 163–186.

48. I.A. Hughes, *Disorders of sex development: A new definition and classification*, Best Pract. Res. Clin. Endocrinol. Metab. 22 (2008), pp. 119–134.

49. R.J. Auchus and A.Y. Chang, *46,XX DSD: The masculinised female*, Best Pract. Res. Clin. Endocrinol. Metab. 24 (2010), pp. 219–242.

50. B.B. Mendonca, E.M.F. Costa, A. Belgorosky, M.A. Rivarola, and S. Domenice, *46,XY DSD due to impaired androgen production*, Best Pract. Res. Clin. Endocrinol. Metab. 24 (2010), pp. 243–262.

51. L.S. Marks, *5alpha-reductase: History and clinical importance*, Rev. Urol. 6 (2004), S11–S21.

52. B. Robaire and N.A. Henderson, *Actions of 5α-reductase inhibitors on the epididymis*, Mol. Cell. Endocrinol. 250 (2006), pp. 190–195.

53. M.S. Mahendroo and D.W. Russell, *Male and female isoenzymes of steroid 5α-reductase*, Rev. Reprod. 4 (1999), pp. 179–183.

54. W.D. Steers, *5alpha-reductase activity in the prostate*, Urology 58 (2001), pp. 17–24.

55. J.M. Torres and E. Ortega, *Differential regulation of steroid 5α-reductase isozymes expression by androgens in the adult rat brain*, FASEB J. 17 (2003), pp. 1428–1433.

56. H.G. Bull, M. Garcia-Calvo, S. Andersson, W.F. Baginsky, H.K. Chan, D.E. Ellsworth, R.R. Miller, R.A. Stearns, R.K. Bakshi, G.H. Rasmusson, R.L. Tolman, R.W. Myers, J.W. Kozarich, and G.S. Harris, *Mechanism-based inhibition of human steroid 5α-reductase by finasteride: Enzyme-catalyzed formation of NADP–dihydrofinasteride, a potent bisubstrate analog inhibitor*, J. Am. Chem. Soc. 118 (1996), pp 2359–2365.

57. M.S. Mahendroo, K.M. Cala, D.P. Landrum, and D.W. Russell, *Fetal death in mice lacking 5α-reductase type 1 caused by estrogen excess*, Mol. Endocrinol. 11 (1997), pp. 917–927.

58. M.S. Mahendroo, A. Porter, D.W. Russell, and R.A. Word, *The parturition defect in steroid 5α-reductase type 1 knockout mice is due to impaired cervical ripening*, Mol. Endocrinol. 13 (1999), pp. 981–992.

59. C.J. Bowman, N.J. Barlow, K.J. Turner, D.G. Wallace, and P.M. Foster, *Effects of in utero exposure to finasteride on androgen-dependent reproductive development in the male rat*, Toxicol. Sci. 74 (2003), pp. 393–406.

60. S. Prahalada, A.F. Tarantal, G.S. Harris, K.P. Ellsworth, A.P. Clarke, G.L. Skiles, K.I. MacKenzie, L.F. Kruk, D.S. Ablin, M.A. Cukierski, C.P. Peter, M.J. vanZwieten, and A.G. Hendrickx, *Effects of finasteride, a type 2 5α-reductase inhibitor, on fetal development in the rhesus monkey (Macaca mulatta)*, Teratology 55 (1997), pp. 119–131.

61. W.C. Boon, J.D. Chow, and E.R. Simpson, *The multiple roles of estrogens and the enzyme aromatase*, Prog. Brain Res. 181 (2010), pp. 209–232.

62. L. Zirilli, V. Rochira, G. Diazzi, G. Caffagni, and C. Carani, *Human models of aromatase deficiency*, J. Steroid Biochem. Mol. Biol. 109 (2008), pp. 212–218.

63. G.M. Tiboni, F. Marotta, A.P. Castigliego, and C. Rossi, *Impact of estrogen replacement on letrozole-induced embryopathic effects*, Hum. Reprod. 24 (2009), pp. 2688–2692.

64. H. Tamada, Y. Shimizu, T. Inaba, N. Kawate, and T. Sawada, *The effects of the aromatase inhibitor fadrozole hydrochloride on fetuses and uteri in late pregnant rats*, J. Endocrinol. 180 (2004), pp. 337–345.

Index